Imaging Spectrometry – a Tool for Environmental Observations

EURO COURSES

A series devoted to the publication of courses and educational seminars organized by the Joint Research Centre Ispra, as part of its education and training program.
Published for the Commission of the European Communities, Directorate-General Telecommunications, Information Industries and Innovation, Scientific and Technical Communications Service.

The EUROCOURSES consist of the following subseries:

- Advanced Scientific Techniques
- Chemical and Environmental Science
- Energy Systems and Technology
- Environmental Impact Assessment
- Environmental Management
- Health Physics and Radiation Protection
- Computer and Information Science
- Mechanical and Materials Science
- Nuclear Science and Technology
- Reliability and Risk Analysis
- Remote Sensing
- Technological Innovation

REMOTE SENSING

Volume 4

The publisher will accept continuation orders for this series which may be cancelled at any time and which provide for automatic billing and shipping of each title in the series upon publication. Please write for details.

Imaging Spectrometry – a Tool for Environmental Observations

Edited by

Joachim Hill

and

Jacques Mégier

Commission of the European Communities,
Joint Research Centre,
Institute for Remote Sensing Applications,
Ispra, Italy

KLUWER ACADEMIC PUBLISHERS

DORDRECHT / BOSTON / LONDON

Based on the lectures given during the Eurocourse on
'Imaging Spectrometry – a Tool for Environmental Observations'
held at the Joint Research Centre, Ispra, Italy
November 23–27, 1992

A C.I.P. Catalogue record for this book is available from the Library of Congress.

ISBN 0-7923-2965-1

Publication arrangements by
Commission of the European Communities
Directorate-General Telecommunications, Information Industries and Innovation,
Scientific and Technical Communications Unit, Luxembourg

EUR 15679

Published by Kluwer Academic Publishers,
P.O. Box 17, 3300 AA Dordrecht, The Netherlands.

Kluwer Academic Publishers incorporates the publishing programmes of
D. Reidel, Martinus Nijhoff, Dr W. Junk and MTP Press.

Sold and distributed in the U.S.A. and Canada
by Kluwer Academic Publishers,
101 Philip Drive, Norwell, MA 02061, U.S.A.

In all other countries, sold and distributed
by Kluwer Academic Publishers Group,
P.O. Box 322, 3300 AH Dordrecht, The Netherlands.

Printed on acid-free paper

Printed in the Netherlands

CONTENTS

PREFACE

The technique of imaging spectrometry has now passed its infancy and entered into a new phase of application oriented research. Advanced sensor systems (such as Nasa/JPL's AVIRIS) have become available for international research programmes (MAC Europe 1991), new imaging spectrometers are under development in several European countries or have already passed their acceptance tests, and first high spectral resolution imaging systems are already operated by private industry. On European level, the EARSEC programme of the Joint Research Centre has provided considerable financial investments for the development of an imaging spectrometer which covers the reflective and important parts of the emissive spectrum (DAIS-7915), and the European Space Agency has initiated an important airborne remote sensing campaign (EMAC 1994/95) in which imaging spectrometry will constitute one of the most important components.

The increasing sensor capabilities also reflect the fact that imaging spectrometry has advanced in many application fields of earth remote sensing. Progress has been made in the development of data pre-processing methods, spectral signature modeling and semi-empirical approaches for retrieving surface parameters. It therefore appeared important to further disseminate information about new approaches in the application-oriented analysis of imaging spectrometry data.

This volume presents the lectures of the second EUROCOURSE on imaging spectrometry which was held in November 1992 at the Joint Research Centre (a first course on "Fundamentals and Prospective Applications" of imaging spectrometry had been organised in October 1989, the lectures being published as EUROCOURSES in Remote Sensing, vol. 2). It was our intention to complete the information on principles of imaging spectrometry but, at the same time, emphasis was also given to the presentation of further application-oriented case studies. While the first course included lectures on marine applications of imaging spectrometry, the selected topics this time concentrated on the assessment of terrestrial ecosystems (vegetation studies, agriculture, soil erosion, and inland water quality). The lectures were presented by invited experts from universities and research institutes in Europe and the United States.

It is believed that this type of seminar has an important function for the dissemination of information from research institutes to a wider audience, in particular in view of the increasing importance of airborne remote sensing with imaging spectrometers and the future launch of spaceborne instruments like MERIS and MODIS on the first Polar Platforms.

J. Hill and J. Mégier, Editors

IMAGING SPECTROMETRY - ITS PRESENT AND FUTURE RÔLE IN ENVIRONMENTAL RESEARCH

PAUL J. CURRAN
Department of Geography
University of Southhampton
Highfield
Southhampton SO9 5 NM
United Kingdom

ABSTRACT. A basic aim of remote sensing is to identify and characterise objects on the Earth's surface by means of radiation that has interacted with that surface. In the optical region of the spectrum this could best be achieved using an imaging spectrometer that records a finely-sampled and continuous spectrum of radiation over the entire 400 nm to 2400 nm wavelength range.

This chapter outlines the airborne imaging spectrometers of today and the space borne imaging spectrometers of tommorow, the techniques for processing data from imaging spectrometers and the rôles that imaging spectrometry is finding in geological, aquatic, ecological and atmospheric research.

1. Introduction

The remotely sensed radiation (R) received by a sensor is, after atmospheric correction and to a first approximation, a function (f) of the location (x), time (t), wavelength (λ) and viewing geometry (θ) of a given ground resolution element (Verstraete and Pinty, 1992),

$$R = f(x, t, \lambda, \theta). \qquad (1)$$

It follows, therefore, that remote sensing can only provide information on environmental phenomena that change x, t, λ or θ by an amount that will result in a detectable change in R. Until recently many of the subtle changes in R with λ were not detectable, as the spectrum was sampled using wavebands that were too broad. Imaging spectrometry enables R to be measured in many narrow wavebands, thus providing a means of estimating those physical variables (e.g., sediment in water) and chemical variables (e.g., chlorophyll in leaves) that result in subtle changes in R with λ (Goetz *et al.*, 1985).

This chapter will introduce the characteristics of the fourteen major imaging spectrometers, the techniques used to process imaging spectrometry data and the present and future rôles of imaging spectrometry in environmental research.

J. Hill and J. Mégier (eds.), Imaging Spectrometry - a Tool for Environmental Observations, 1–23.

2. Imaging spectrometers

Spectroscopy is a standard technique for chemical assay (Banwell, 1972). For example, to determine the presence and amount of iron in a sample of blood a hospital technician would not precipitate the iron and weigh it, that would be very time consuming and would require rather a lot of blood! Rather the technician would use a laboratory spectrometer to illuminate and then to measure the radiation spectrum of the sample. The level of radiation at the absorption wavelengths for iron could then be used to both identify iron and to estimate its amount.

In a similar way the spectrum of solar radiation reflected from a point on the Earth's surface could be measured using a field spectroradiometer and for a region using an airborne imaging spectrometer. Subsequently, the radiation at known wavelengths could be used to identify and then estimate the amount of a particular feature on the Earth's surface.

In the laboratory there is typically a constant, controllable and strong radiation source illuminating a homogeneous sample, located a few centimeters from a detector. In the field there is typically a variable, uncontrollable and relatively weak radiation source illuminating a heterogeneous sample, located a few metres from a detector. Fortunately, the measurement (or dwell) time in both the laboratory and the field can be adjusted to ensure that a large supply of photons reaches a detector and therefore the signal is large. In contrast, an imaging spectrometer has a variable, uncontrollable and relatively weak radiation source, illuminating typically a heterogeneous sample located several kilometres from a detector. In addition, the imaging spectrometer has a very short dwell time, reducing further the number of photons that could reach a detector from a point on the Earth's surface. Together, these factors suggest that an imaging spectrometer is destined to have a low signal-to-noise ratio (SNR) (Curran and Dungan, 1989; Curran *et al.*, 1991a). The SNR could be increased by (i) restricting the wavelengths sensed to the solar radiation peak in visible wavelengths, (ii) decreasing the spectral resolution, (iii) increasing the spatial resolution or (iv) increasing the dwell time. The first three options are possible but limit the utility of the instrument for environmental applications. The final option is the most acceptable from a users point of view but was not possible until developments in charge-coupled-devices (CCDs) during the late 1970s made available reliable one and two-dimensional detector arrays. The result has been a wide range of imaging spectrometers each designed around CCDs and optimised to a particular set of specifications (Slater, 1985; Curran and Dungan, 1990).

Imaging spectrometers recording in visible to near infrared wavelengths tend to use two-dimensional arrays, either singularly (e.g., 612 x 288 array in the CASI) or in blocks (e.g., five, 770 x 576 arrays in the MERIS) to ensure a column of specific waveband detectors for each ground resolution element in the scene (table 1). Imaging spectrometers recording in visible to middle infrared wavelengths tend to use several one-dimensional arrays (e.g., one, 1 x 34, plus three, 1 x 64 in the AVIRIS) or a singular two-dimensional array (e.g., 64 x 64 array in the MODIS) fronted by a traditional linescanner that passes the beam of radiation from a ground-resolution-element along the array (Table 1). This approach ensures that each ground resolution element is sensed simultaneously in as many bands as there are detectors in the array (Vane and Goetz, 1988). The imaging spectrometers were also designed around certain geometric criteria, the most important being altitude, swath width and spatial resolution (Gower, 1990). At one extreme is the CASI, a low altitude (2 km) airborne sensor with a narrow swath (1.2 km) and a very fine spatial resolution (2.5 m), while at the other extreme is the MODIS, a high altitude (705 km) space borne sensor with a broad swath (1500 km) and a coarse spatial resolution (1.0 km) (Table 1). The AVIRIS falls between the two with a high altitude (20 km), a medium swath width (11 km) and a fine spatial

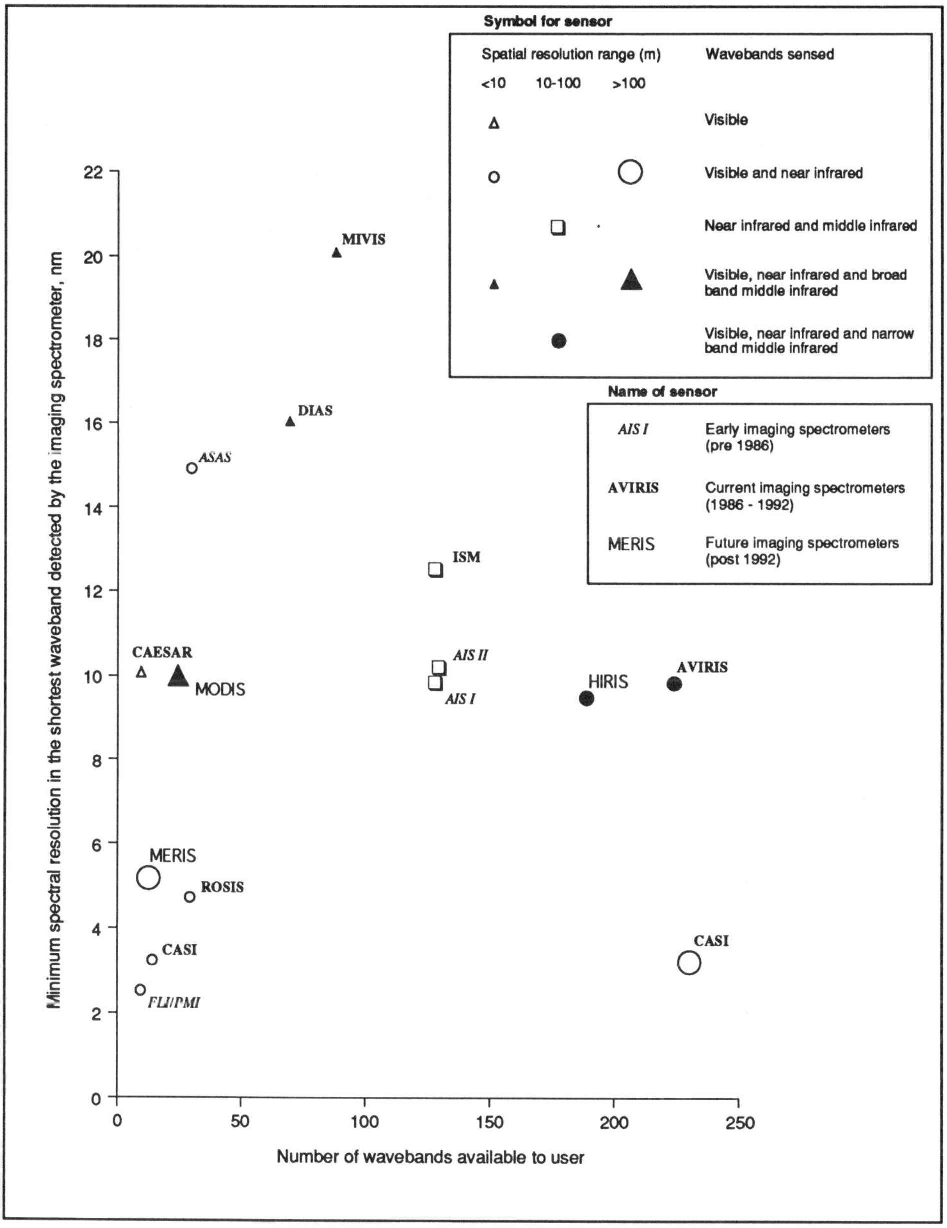

FIGURE 1. The relationship between the minimum spectral resolution and the number of wavebands available to the user. Note, (i) a typical airborne multispectral but broadband scanner would be located at the top left of the figure and the sensors designed for aquatic applications tend to cluster to the left of the figure, (ii) the spatial resolution quoted in the figure is that most commonly used by the sensor operators and (iii) the definition of acronyms and the sources of information are given in table 1.

resolution (20 m). Figure 1 is a simplified summary of this variability in which certain details have been omitted for clarity; for example, some sensors have more than one sensing mode and sensitivity to thermal infrared wavelengths.

The spaceborne imaging spectrometers pose many challenges for sensor designers because of their huge data rates and uncompromising accuracy requirements. At the time of writing (November 1992) the MERIS has a place on the first European Polar Platform (Envisat) and the MODIS has a place on the first US Polar Platform, both of which are due for launch in 1998. Unfortunately, the HIRIS has been deselected from any of the US Polar Platforms. The scientific arguments for HIRIS are compelling (compare figure 1 and table 2) and so a place on a different platform (e.g., Landsat 8) may be a possibility.

From a user's point of view the characteristics of the imaging spectrometer that are of most interest are those that differentiate it from the widely used broadband sensors (Kerekes and Landgrebe, 1991). The four key characteristics being the spectral coverage, the number of wavebands available to the user, the spectral resolution and the dynamic range of the signal (table 1).

2.1 Spectral Coverage

The strength and ease of measurement of a remotely sensed signal is related mainly to the wavelength of that signal. In visible and near infrared wavelengths the signal is strong but in middle infrared wavelengths the signal is weak. As this signal is only a few percent of irradiance over land and zero over water many imaging spectrometers were designed to record in only visible and near infrared wavelengths (figure 1). While this restricted spectral range is adequate for the study of aquatic and certain atmospheric and terrestrial phenomena it is a severe limitation for certain tasks (table 2), most notably the identification of lithology and the estimation of canopy chemistry (section 4).

2.2 Number of Wavebands Available to the User

The number of wavebands available to the user varies between 8 for the FLI/PMI to 224 for the AVIRIS (table 1). The NASA series of imaging spectrometers (AIS I, AIS II, AVIRIS and HIRIS) record all of the spectrum that is sensed (table 1). However, their data rates are large (e.g., 17 Mbits sec^{-1} for the AVIRIS) and there is considerable autocorrelation between wavebands. The majority of sensors sample the recorded spectrum and so the number of wavebands available to the user is not the same as the number of wavebands recorded. This sampling varies from approximately 25% for two-dimensional arrays with a linescanner (e.g., MODIS) to a tenth of that for two- dimensional arrays without a linescanner (e.g., MERIS). This can be useful if data in only a few wavelengths are required (as when sensing the fluorescence peak of water with the FLI/PMI) or if spectra can be interpolated using a spline, or similar, function (e.g., when sensing the 'red edge' of vegetation with the CASI).

2.3 Spectral Resolution

The spectral resolution is the bandwidth over which the radiation is recorded. As such it determines the accuracy with which features in the radiation spectra can be measured. To detect a change in radiation across the double absorption feature that is so characteristic of kaolinite requires a

TABLE 1. The fourteen major imaging spectrometers designed for environmental research. Some of the information is portrayed graphically in figure 1 (Slater, 1985; Goetz, 1987; Gower and Borstad, 1981; Hollinger, *et al.*, 1987; Huegel, 1987; Porter and Enmark, 1987; Goetz and Herring, 1989; Salomonson *et al.*, 1989; Daedalus Enterprises Inc., 1990; Gower, 1990; Irons *et al.*, 1991; Kunkel *et al.*, 1991; Rast, 1991; personal communication; ESA, 1992; GER, 1992; King *et al.*, 1992; Mustard *et al.*, 1992; Barale *et al.*, 1994).

Sensor name (A = airborne, S = space borne)	Acronym	Spectral coverage (nm)	Number of wavebands available to user	Minimum spectral resolution (nm)		
				visible wavelengths	near infrared wavelengths	middle infrared wavelengths
Fluorescence Line Imager/Programmable Multispectral Imager (A)	FLI/PMI	430 - 805	8	2.5	2.5	-
CCD Airborne Experimental Scanner for Applicators in Remote Sensing (A)	CAESAR	520 - 780	9	10.0	-	-
Medium Resolution Imaging Spectrometer (S)	MERIS	400 - 1050	15	5.0	5.0	-
Moderate Resolution Imaging Spectrometer (S)	MODIS	400 - 1040	19	10.0	1.2	5.0
Reflective Optics System Imaging Spectrometer (A)	ROSIS	430 - 880	28	4.7	4.7	-
Advanced Solidstate Array Spectroradiometer (A)	ASAS	465 - 870	30	15.0	15.0	-
Digital Airborne Imaging Spectrometer (A)	DIAS	400 - 2500	72	16.0	100.0	16.0
Multispectral Infrared and Visible Imaging Spectrometer (A)	MIVIS	400 - 2500	92	20.0	50.0	8.0
Imaging Spectroscopic Mapper (A)	ISM	760 - 3160	128	-	12.5	24.0
Airborne Imaging Spectrometer I (A)	AIS I	900 - 2400	128	-	9.6	10.6
Airborne Imaging Spectrometer II (A)	AIS II	800 - 2400	128	-	10.0	10.6
High Resolution Imaging Spectrometer (S)	HIRIS	400 - 2500	192	9.4	9.4	11.4
Airborne Visible/Infrared Imaging Spectrometer (A)	AVIRIS	410 - 2450	224	9.4	9.7	9.7
Compact Airborne Spectral Imager (A)	CASI	410 - 925	228/15	2.9	2.9	-

spectral resolution of around 10 to 15 nm (Hunt, 1980). In contrast, to estimate the cloud-top height by means of radiation measured across the O_2 absorption features in near-infrared wavelengths, or to estimate the size of a chlorophyll fluorescence peak in visible wavelengths requires a spectral resolution nearer to 3 nm. Spectral resolution is related positively to the SNR of the sensor. Therefore, imaging spectrometers are designed with a spectral resolution that is as large as is possible within the constraints imposed on the design criteria by the likely application. In general, sensors with a spectral resolution of 10 nm and larger were designed for primarily terrestrial applications while sensors with a spectral resolution of 10 nm and smaller were designed primarily for aquatic and atmospheric applications.

After *recording* at a given spectral resolution the spectrum is *resampled* to a new bandwith before the data are made available to the user. This ensures a constant bandwidth with wavelength and avoids the effect of spectral overlap between CCDs. The resampled bandwidth can be larger than the spectral resolution (e.g., 9.4 nm to 10 nm for the AVIRIS), the same as the spectral resolution (e.g., 5 nm for the MODIS), or smaller than the spectral resolution (e.g., 2.9 nm to 1.3 nm for the CASI).

2.4 Dynamic Range of the Signal

The variations in radiation with λ that are to be detected by an imaging spectrometer are usually at least two order of magnitude smaller than the potential radiation range of the environment (Goetz and Calvin, 1987). To measure these variations in radiation with λ requires signal digitisation of around 12 bits (4096 intensity levels) (Rast, 1991). System designers typically adjust the digitisation to reflect the SNR of the sensor and the required data rate. This can vary from 8 bit (e.g., ROSIS), to 16 bit (e.g., DIAS) with the majority using a 12 bit (e.g., CASI) digitisation (table 1).

Each potential application of imaging spectrometry has different requirements in terms of sensor performance. To make the compromises that were necessary during sensor design each sensor was designed for a specific group of applications. For example, the FLI/PMI, the CAESAR and the MERIS were designed around aquatic studies; the MIVIS, the DIAS, the MODIS, the ROSIS and the CASI were designed around aquatic and ecological studies and the AIS, the AVIRIS, the HIRIS, the ISM and the ASAS were designed around geological and ecological studies (table 1). This trend continues with several new imaging spectrometers being designed to secure the needs of specific fields. These instruments vary from the new airborne imaging spectrometer from Analytical Spectral Devices (ASD) that records in 204 contiguous channels throughout visible and near infrared wavelengths (ASD, 1992) to ESA's proposed spaceborne imaging spectrometer (Prism) with specifications that are very similar to those of NASA's HIRIS (ESA, 1991; Rast, 1991).

3. Processing imaging spectrometry data

Processing the large amount of data recorded by imaging spectrometers proved to be a problem for environmental researchers throughout the 1980s. The Jet Propulsion Laboratory's Workshops of the period (Vane and Goetz, 1985; 1986; Vane, 1987; 1988) reveals researchers viewing three waveband imagery on one system while trying to view, simplify, compress or combine spectra on another. Today there are several software packages on the market that avoid this inefficient divorce between images and spectra (Mehl, 1994). Some of these software packages are general purpose (e.g., GENESIS) but most are designed for a specific application (e.g., SPAM), or user group (e.g., T Spectra) (Donoghue and Robinson, 1990). A characteristic of current software is the creative fusion of basic data processing techniques in spectroscopy and image processing with recent developments in data visualisation. Using even the basic packages a user can now explore an image to extract both radiation and derivative radiation spectra for further analysis. The ability to calculate derivative radiation spectra (which are a measure of radiation change with λ) is of importance, as such spectra are relatively unaffected by atmospheric haze and can resolve the fine spectral detail that is obscured in the radiation spectra (Dixit and Ram, 1985).

Two AVIRIS images processed using PV WAVE software (Precision Visuals, 1992) (figures 2 and 3) give an insight into imaging spectrometry data and the flexibility of the available exploratory software. Processing techniques designed for specific applications and the software needed to perform such processing is beyond the scope of this paper, but they are reviewed in Mehl (1994).

4. Imaging spectrometry in environmental research

Environmental research has a wide range of aims and techniques but uses only three methodologies. These methodologies are the *inductive*, where phenomena are observed, generalisations made and conclusions drawn; the *deductive*, where hypotheses based on theory are tested by observation and the *technological,* where the solution to a 'real' problem is designed using inputs from all relevant fields (Curran, 1987). The first two methodologies are scientific, with the first used for description and exploration and the second used for understanding. The third methodology is common to engineering and planning with an emphasis placed on human need and management. Imaging spectrometry can be a part of all three methodologies. Initially much of the imaging spectroscopy research was inductive, as users learnt to work with and appreciate the strengths and limitations of the data (Vane and Goetz, 1985; 1986; Vane, 1987; 1988). Recent years have seen a move towards overtly deductive research (e.g., Wessman, 1994a) where data from imaging spectrometers are being used to help understand the environment and overtly technological research (e.g., Hill *et al.*, 1994) where data from imaging spectrometers are being used to help better manage our environment. A common thread to this research is the use of radiation spectra to determine the characteristics of material on the Earth's surface. These characteristics may be identity (e.g., rock type), amount (e.g., foliar chemistry) or a variable related to both of these (e.g., vegetation stress) (table 2).

Imaging spectrometry was used initially in geological, aquatic and later ecological and atmospheric research (Johnson and Melfi, 1989). Surprisingly, imaging spectrometers have been little used for the study of snow and ice (Goetz, 1987; 1991), bare soil (Escadafal, 1994), urban areas (Ridd *et al.*, 1992), or agricultural crops (Baret and Jacquemoud, 1994; Clevers, 1994).

4.1 GEOLOGICAL APPLICATIONS OF IMAGING SPECTROMETRY

A large number of minerals bearing Fe^{2+}, Fe^{3+}, and OH^- ions have unique radiation spectra as a result of electronic and vibrational transitions throughout the 400 nm to 2400 nm spectral region (Hunt and Salisbury, 1970; 1971; Hunt *et al.*, 1971). Where these minerals are found at the surface, it may be possible to identify them with imaging spectrometry (Hunt, 1980; Goetz and Herring, 1989; Rivard and Arvidson, 1992). A considerable amount of research has now been

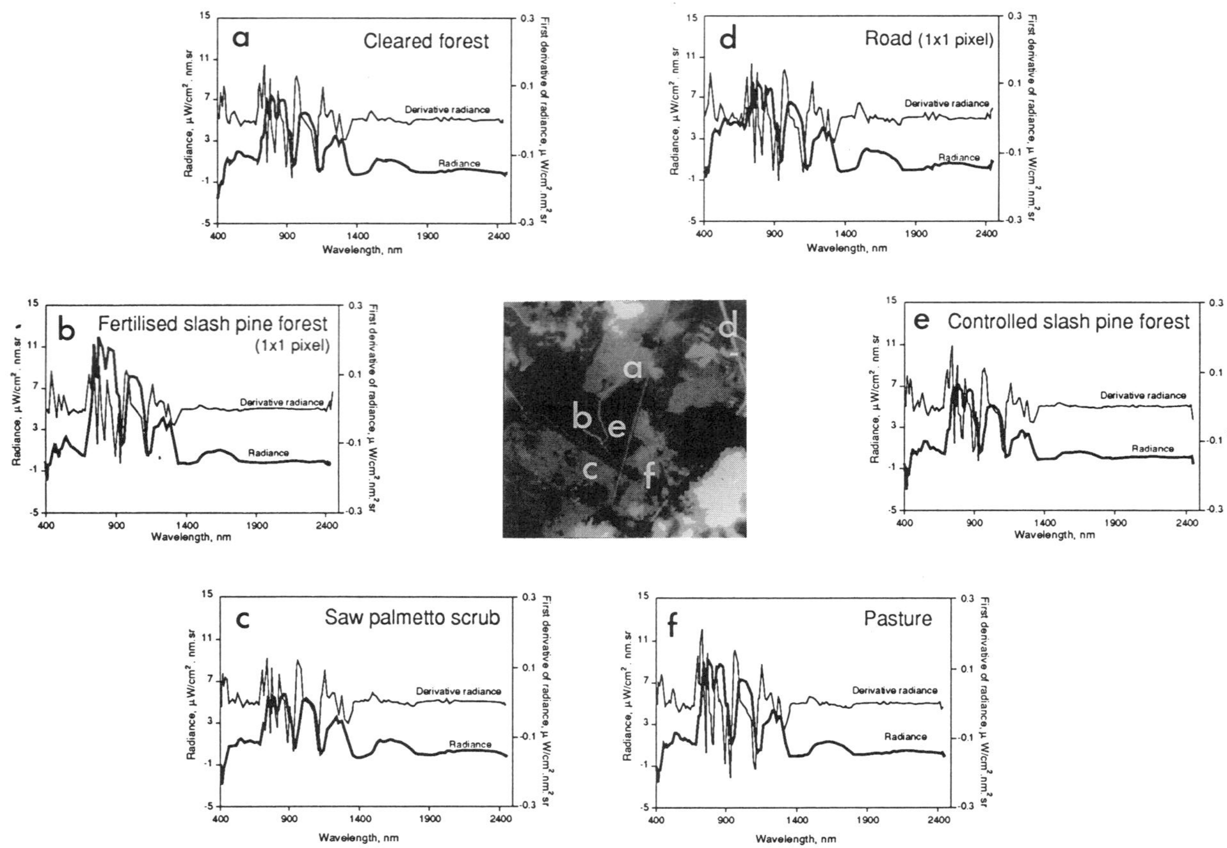

FIGURE 2. Radiance spectra and first derivative radiance spectra for 3 x 3 pixel areas (unless otherwise stated) of six land covers. The AVIRIS image was recorded on 6 September 1990 and covers approximately 25 km² to the north east of Gainesville, Florida, U.S.A. Image data courtesy of NASA/JPL.

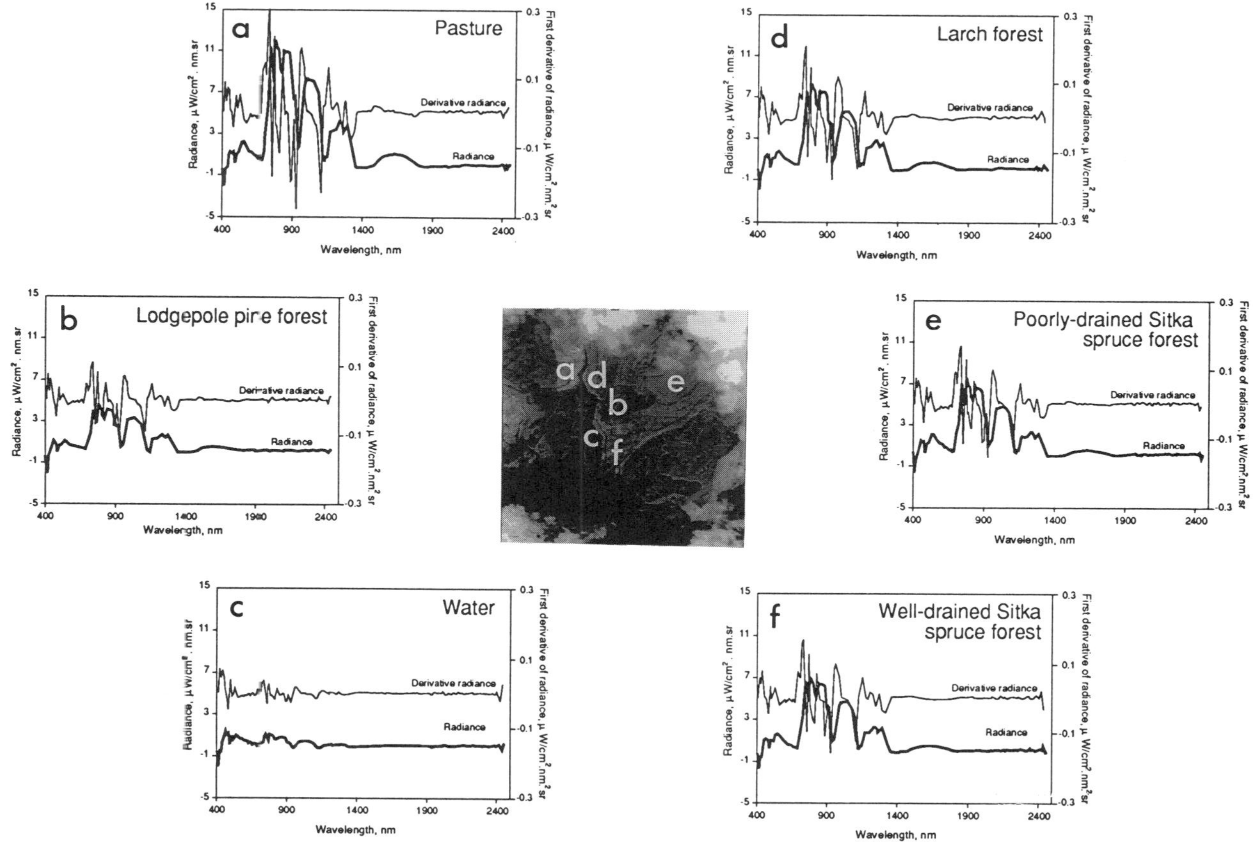

FIGURE 3. Radiance spectra and first derivative radiance spectra for 3 x 3 pixel areas of six land covers. The AVIRIS image was recorded on 26 August 1991 and covers approximately 25 km^2 to the north of Llyn Brianne, Wales, U.K. Image data courtesy of NASA/JPL.

undertaken on the use of imaging spectrometry for the identification of minerals (Mustard and Pieters, 1986; Kruse *et al.*, 1988; Kruse and Hauff, 1991), particularly those associated with hydrothermal alteration (Huntspiller and Taranik, 1986; Abrams, *et al.*, 1977; Lyon, 1987; Kruse, 1988). Experience with mineral identification has lead to the use of imaging spectrometry for geological mapping (Pieters and Mustard, 1988; Boardman and Goetz, 1991), using such data analysis techniques as spectral mixture modelling (Drake, 1990), expert systems (Kruse, 1990) and kriging (Meer, 1992). Imaging spectrometry has many potential geological applications (Farrand and Singer, 1991; Rast *et al.*, 1991), one of the more novel in recent years has been the estimation of emitted radiation as a means of deriving the thermal budget of volcanoes (Rothery and Oppenheimer, 1990; Oppenheimer *et al.*, 1992).

4.2 Aquatic Applications of Imaging Spectrometry

The fine spectral resolution of an imaging spectrometer makes possible the identification and estimation of chlorophyll, tannin and sediment concentrations within water (Buxton, 1988; Chen *et al.*, 1992). All three materials have unique characteristics in radiation spectra and in addition chlorophyll fluorescences slightly within a waveband centred at 685 nm (Gower, 1990). Imaging spectrometers have been used extensively in near-coastal environments (Boxall and Matthews, 1990; Wilson, 1990; Carder *et al.*, 1993) to estimate the concentration of sediment (Boxall and Reilly, 1989) and chlorophyll (Moore and Aiken, 1990). Estimates of chlorophyll concentration have also been used to monitor algal blooms (Pettersson, 1990) and to infer the distribution of phytoplankton (Gower and Borstad, 1990) for the study of aquatic productivity and the location of fish (Nakashima *et al.*, 1989). Research on inland waters has demonstrated the value of imaging spectrometry for the mapping of submerged vegetation (Mouchot *et al.*, 1988) and chlorophyll concentration (Melack and Gastil, 1990; 1992). More promisingly, it has been shown to have value as a tool for the integrated analysis of inland water quality (George, 1990; Dekker *et al.*, 1991; Dekker and Donze, 1994).

4.3 Ecological Applications of Imaging Spectrometry

There are strong relationships between the concentration of key elements/biochemicals (e.g., nitrogen, lignin etc.) and the level of radiation across spectral absorption features in dry ground leaves (Card *et al.*, 1988; Elvidge, 1990; McLellan *et al.*, 1991) and fresh whole leaves (Curran *et al.*, 1992), when measured in the laboratory. Researchers have been trying to derive similar relationships for canopies using imaging spectrometry (Curran, 1989). Initial work met with success (Peterson *et al.*, 1988; Wessman *et al.*, 1988a; 1988b) but there is still a long way to go before such relationships can be inverted reliably for the estimation of canopy chemical concentrations (Peterson, 1991; Janetos *et al.*, 1992; Wessman, 1992b). Information on canopy chemistry is of great importance for the study of nutrient cycling, productivity, vegetation stress and for input to ecosystem simulation models (Wessman, 1992a; Curran, 1994) and as a result a considerable amount of research is being undertaken in this field (Johnson and Peterson, 1991; Curran, 1992). The concentration of chlorophyll and water in vegetation canopies has been estimated using imaging spectrometry data. Chlorophyll concentration is related positively to a minor deepening and a major widening of the two chlorophyll absorption features (Rock *et al.*, 1988; Curran *et al.*, 1990; 1991b) whereas water concentration is related positively to a major deepening of the five major water absorption features (Riggs and Running, 1991; Green *et al.*,

TABLE 2. The regions of the spectrum within which environmental characteristics can be determined using imaging spectrometry (Curran 1989; Rast, 1991; Ardanuy et al., 1991; King, *et al.*, 1992).

Wavelength, nm	waveband	Environmental components				
		Rock and soil	Vegetation	Snow and ice	Water	Atmosphere
400, 600	visible	• *Iron bearing minerals*	• *Chlorophyll*	• *Snow grain size* • *Snow depth*	• *Organic matter* • *Suspended sediment* • *Chlorophyll*	• *Aerosol properties* • *Cloud thickness* • *Aerosol optical thickness*
800, 1000, 1200	near infrared	• *Iron bearing minerals* • *Soil water*	• *Protein* • *Starch* • *Oil* • *Water* • *Lignin* • *Cellulose*	• *Snow grain size* • *Snow water*		• *Water vapour* • *Cloud top height* • *Aerosol optical thickness* • *Precipitable water*
1400	middle infrared	water absorption band in atmosphere				
1600	middle infrared	• *Carbonates* • *Sulphates*	• *Lignin* • *Cellulose* • *Sugar / starch* • *Protein / nitrogen*	• *Water / ice differentiation* • *Cloud / snow differentiation*		
1800	middle infrared	water absorption band in atmosphere				
2000, 2200, 2400	middle infrared	• *Clays* • *Micas* • *Hydrates* • *Carbonates* • *Sulphates* • *Temperature (e.g., volcanic)*	• *Protein / nitrogen* • *Sugar / starch* • *Cellulose* • *Oil*			• *Cloud particle radius* • *Aerosol optical thickness*

1991). Imaging spectrometry has also been used for vegetation mapping (Miller and Hare, 1989; McDonald *et al.*, 1989), either directly or via spectral mixture modelling (Smith *et al.*, 1994a; 1994b). The aim of this mapping has been to either locate vegetation associations or to estimate the distribution of live and dead vegetation in space (Elvidge *et al.*, 1991; Fitzgerald and Ustin, 1992; Roberts *et al.*, 1993) and time (Ustin *et al.*, 1992).

4.4 ATMOSPHERIC APPLICATIONS OF IMAGING SPECTROMETRY

The radiation spectra recorded by an imaging spectrometer is modified by aerosols and gasses in the atmosphere. Some of the changes, for example those caused by aerosols, affect a large number of contiguous wavebands while some of the changes for example, those caused by 0_2 and water are specific to a few wavebands (Schanzer and Staenez, 1992; King *et al.*, 1992). Optical thickness is estimated readily using the first of these two effects (Berendes *et al.*, 1991; Feind *et al.*, 1992). The four most promising applications for imaging spectrometry in atmospheric research are the mapping of cloud cover in the water vapour wavebands (Gao and Goetz, 1991; 1992), estimating columnar water content, again in the water vapour wavebands (Conel *et al.* 1989; Gao and Goetz, 1989), estimating cloud top height via the relationship between this, atmospheric pressure and radiation in the O_2 absorption band (Rast, 1991) and finally, estimating aerosol concentration via its effect on image haze in different wavebands (Hickman and Duggin, 1992). Space borne imaging spectrometers offer the potential of estimating several further atmospheric variables (table 2) using the methods outlined by King *et al.* (1992).

5. Concluding comment

Imaging spectrometry has its *origins* in laboratory spectroscopy, field spectroscopy and the optical imaging sensors of the 1970s and its *future* in the space borne sensors onboard the forthcoming Polar Platforms (CEOS, 1991; ESA, 1991). The airborne imaging spectrometers of today are a fundamental technological advance in remote sensing. The challenge before us is to turn this technological advance into a useful tool for understanding and managing our environment.

It is hoped that this chapter will serve both as a primer to the field of imaging spectrometry and as a framework into which the following technological and environmental chapters can be placed.

6. Acknowledgements

I wish to acknowledge the support I have been given by the NASA, the ESA and the NERC (UK) to pursue my interest in imaging spectrometry; the image processing skills of Geoff Smith and the helpful comments on this chapter I have received from Dr Giles Foody, Geoff Smith and John Kupiec.

7. References

Abrams, M.J., Ashley, R.P., Rowan, L.C., Goetz, A.F.H. and A.B. Kahle (1977) 'Mapping of hydrothermal alteration in the Cuprite Mining District, Nevada, using aircraft scanner images for the spectral region 0.46 - 2.36 μm', *Geology* 5, 713-718.

Ardanuy, P.E., Han, D. and V.V. Salomonson (1991) 'The Moderate Resolution Imaging Spectrometer (MODIS) Science and data system requirements', *IEEE Transactions on Geoscience and Remote Sensing* 29, 75-88.

ASD (1992) The *ASD Imaging Spectrometer,* ISPRS XVII Congress New Product Announcement, Analytical Spectral Devices, Boulder, Co.

Banwell, C.N. (1972) *Fundamentals of Molecular Spectroscopy,* McGraw Hill, London.

Barale, V., Curran, P.J., Deschamps, P.Y., Fischer, J., Grassl, H., Malingreau, J.P., Morel, A. and Verstraete, M. (1994) *The Medium Resolution Imaging Spectrometer (MERIS),* European Space Agency, Paris (in press).

Baret, F. and S. Jacquemoud (1994) 'Modeling canopy spectral properties to retrieve biophysical and biochemical characteristics', *in* J. Hill and J. Mégier (eds.) *Imaging Spectrometry - a Tool for Environmental Observations,* Kluwer Academic Publishers, Dordrecht (this volume).

Berendes, T.A., Feind, R.E., Kuo, K.S. and Welch, R.M. (1991) 'Cloud base height and optical thickness retrievals using AVIRIS data', *in* R.O. Green (ed.) *Proceedings, Third AirborneVisible/Infrared Imaging Spectrometer (AVIRIS) Workshop,* National Aeronautics and Space Administration, Jet Propulsion Laboratory Publication, Pasadena, Ca., pp. 232-247.

Boardman, J.W. and Goetz, A.F.H. (1991) 'Sedimentary facies analysis using AVIRIS data: a geographysical inverse problem', *in* R.O. Green (ed.) *Proceedings, Third Airborne Visible/Infrared Imaging Spectrometer (AVIRIS) Workshop,* National Aeronautics and Space Administration, Jet Propulsion Laboratory Publication, Pasadena, Ca., pp. 4-13.

Boxall, S.R. and Matthews, A. (1990) 'Results of the CASI campaign in the West Solent', *Proceedings, NERC Workshop on Airborne Remote Sensing,* Natural Environment Research Council, Swindon, pp. 255-257.

Boxall, S.R. and Reilly, J.E. (1989) 'Results of the fluorescence line imager marine campaign in the West Solent', *Proceedings, NERC Workshop on Airborne Remote Sensing,* Natural Environment Research Council, Swindon, pp. 93-108.

Buxton, R.A.H. (1988) 'The FLI airborne imaging spectrometer: a highly versatile sensor for many applications', *Proceedings, ESA Workshop on Imaging Spectrometry,* ESA SP-1101, European Space Agency, Noordwijk, pp. 11-16.

Card, D.H., Peterson, D.L., Matson, P.A. and Aber, J.D. (1988) 'Prediction of leaf chemistry by the use of visible and near infrared reflectance spectroscopy', *Remote Sensing of Environment* 26, 123-147.

Carder, K.L., Reinersman, P., Chen, R. and Muller-Karger, F. (1993) 'AVIRIS calibration and application in coastal oceanic environments', *Remote Sensing of Environment,* 44, 205-216.

CEOS (1992) *The Relevance of Satellite Missions to the Study of the Global Environment,* Committee on Earth Observation Satellites, London.

Chen, Z., Curran, P.J. and Hansom, J.D. (1992) 'Derivative reflectance spectroscopy to estimate suspended sediment concentration', *Remote Sensing of Environment* 40, 67-77.

Clevers, J. (1994) Imaging spectrometry in agriculture - plant vitality and yield indicators, *in* J. Hill and J. Mégier (eds.) *Imaging Spectrometry - a Tool for Environmental Observations*, Kluwer Academic Publishers, Dordrecht (this volume).

Conel, J.E., Green, R.O., Carrere, V., Margolis, J.J., Vane, G., Breugge, C. and Alley, R. (1989) 'Spectroscopic measurements of atmospheric water vapour and schemes for determination of evaporation from land and water surfaces using AVIRIS', *Proceedings, IGARSS '89*, IEEE, New York, pp. 2658-2663.

Curran, P.J. (1987) 'Remote sensing methodologies and geography', *International Journal of Remote Sensing* 8, 1255-1275.

Curran, P.J. (1989) 'Remote sensing of foliar chemistry', *Remote Sensing of Environment*, 29, 271-178.

Curran, P.J. (1992) 'Estimating foliar chemical concentrations with the Airborne Visible/Infrared Imaging Spectrometer (AVIRIS)', *International Archives of Photogrammetry and Remote Sensing,* Commission VII, ISPRS, Washington DC, pp. 705-708.

Curran, P.J. (1994) 'Attempts to drive ecosystem simulation models at local to regional scales', *in* G.M. Foody and P.J. Curran (eds.) *Environmental Remote Sensing from Regional to Global Scales,* Wiley and Sons, Chichester, 149-166.

Curran, P.J. and Dungan, J.L. (1989) 'Estimation of signal-to-noise: a new procedure applied to AVIRIS data', *IEEE Transactions on Geoscience and Remote Sensing* 27, 620-628.

Curran, P.J. and Dungan, J.L. (1990) 'An image recorded by the Airborne Visible/Infrared Imaging Spectrometer (AVIRIS)', *International Journal of Remote Sensing* 11, 929-931.

Curran, P.J., Dungan, J.L. and Gholz, H.L. (1990) 'Exploring the relationship between reflectance red edge and chlorophyll content in slash pine', *Tree Physiology* 7, 33-48.

Curran, P.J., Dungan, J.L. and Smith, G.M. (1991a) 'Increasing the signal-to-noise ratio of AVIRIS imagery through repeated sampling', *in* R.O. Green (ed.) *Proceedings of the Airborne Geoscience Workshop*, National Aeronautics and Space Administration, Jet Propulsion Laboratory Publication, Pasadena, Ca., pp. 164-167.

Curran, P.J., Dungan, J.L., Macler, B.A. and Plummer, S.E. (1991b) 'The effect of a red leaf pigment on the relationship between red edge and chlorophyll concentration', *Remote Sensing of Environment* 35, 69-76.

Curran, P.J., Dungan, J.L., Macler, B.A., Plummer, S.E. and Peterson, D.L. (1992) 'Reflectance spectroscopy of fresh whole leaves for the estimation of chemical composition', *Remote Sensing of Environment* 39, 153-166.

Daedalus Enterprises Inc. (1990) *The Daedalus MIVIS: Multispectral Infrared and Visible Infrared Imaging Spectrometer*, Daedalus Enterprises Inc., Ann Arbor, Michigan.

Dekker, A.G. and Donze, M. (1994) 'Imaging spectrometry as a research tool for inland water resources analysis', *in* J. Hill and J. Mégier (eds.) *Imaging Spectrometry - a Tool for Environmental Observations,* Kluwer Academic Publishers, Dordrecht (this volume).

Dekker, A.G., Malthus, T.J. and Seyhan, E. (1991) 'Quantitative modelling of inland water quality for high-resolution MSS systems', *IEEE Transactions on Geoscience and Remote Sensing* 29, 89-95.

Dixit, L. and Ram, S. (1985) 'Quantitative analysis by derivative electronic spectroscopy', *Applied Spectroscopy Reviews* 21, 311-418.

Donoghue, D.N.M. and Robinson, D.R. (1990) 'A flexible data analysis system for high spectral resolution data for MS-DOS computers', *in* S.E. Plummer (ed.) *Applications and Developments in Imaging Spectrometry*, Remote Sensing Society, Nottingham, pp. 54-60.

Drake, N.A. (1990) 'Mapping rocks, soils and vegetation communities with the GERIS using linear mixture modelling and post-processing techniques', *in* S.E. Plummer (ed.) *Applications and Developments in Imaging Spectrometry*, Remote Sensing Society, Nottingham, pp. 61-69.

Elvidge, C.D. (1990) 'Visible and infrared reflectance characteristics of dry, plant materials', *International Journal of Remote Sensing* 11, 1775-1796.

Elvidge, C.D., Chen, Z., Portigal, F.P. and Groeneveld, D.P. (1991) 'Detection of trace quantitites of green vegetation in AVIRIS data', *in* R.O. Green (ed.) *Proceedings, Third Airborne Visible/Infrared Imaging Spectrometer (AVIRIS) Workshop,* National Aeronautics and Space Administration, Jet Propulsion Laboratory Publication, Pasadena, Ca., pp. 183-189.

ESA (1991) *Report of the Earth Observation User Consultation Meeting*, European Space Agency, Noordwijk.

ESA (1992) *MERIS : Medium Resolution Imaging Spectrometer,* European Space Agency, Noordwijk.

Escadafal, R. (1994) 'Soil spectral properties and their relationship with environmental parameters - examples from arid regions', *in* J. Hill and J. Mégier (eds.) *Imaging Spectrometry - a Tool for Environmental Observations,* Kluwer Academic Publishers, Dordrecht (this volume).

Farrand W.H. and Singer, R.B. (1991) 'Analysis of altered volcanic pyroclasts using AVIRIS data', *in* R.O. Green (ed.) *Proceedings, Third Airborne Visible/Infrared Imaging Spectrometer (AVIRIS) Workshop*, National Aeronautics and Space Administration, Jet Propulsion Laboratory Publication, Pasadena, Ca., pp. 248-257.

Feind, R.E., Christopher, S.A. and Welch, R.M. (1992) 'Spatial resolution and cloud optical thickness retrievals', *in* R.O. Green (ed.) *Third JPL Airborne Geoscience Workshop*, National Aeronautics and Space Administration, Jet Propulsion Laboratory Publication, Pasadena, Ca., 1, pp. 14-16.

Fitzgerald, M. and Ustin, S. (1992) 'Measuring dry plant residues in grasslands: a case study using AVIRIS', *in* R.O. Green (ed.) *Third Airborne Geoscience Workshop,* National Aeronautics and Space Administration, Jet Propulsion Laboratory Publication, Pasadena, Ca., 1, pp. 91-93.

Gao, B.C. and Goetz, A.F.H. (1989) 'Column atmospheric water vapour retrievals for airborne imaging spectrometer data', *Proceedings, IGARSS '90/12th Canadian Symposium on Remote Sensing,* IEEE, New York, 4, pp. 2664-2668.

Gao, B.C. and Goetz, A.F.H. (1991) 'Cloud area determination from AVIRIS data using water vapour channels near 1 μm'. *Journal of Geophysical Research* 96, 2857-2864.

Gao, B.C. and Goetz, A.F.H. (1992) 'Separation of cirrus cloud from clear surface from AVIRIS data using the 1.38 μm water vapour band', *in* R.O. Green (ed.) *Third Airborne Geoscience Workshop*, National Aeronautics and Space Administration, Jet Propulsion Laboratory Publication, Pasadena, Ca., 1, pp. 98-100.

George, D.G. (1990) 'Results of the 1989 imaging spectroscopy surveys of Windermere and Esthwaite Water', *Proceedings, NERC Workshop on Airborne Remote Sensing,* Natural Environment Research Council, Swindon, pp. 297-302.

GER (1992) *GER Corporation's Digital Airborne Imaging Spectrometer DAIS-7915,* Geophysical and Environmental Research Corporation, New York.

Goetz, A.F.H. ed. (1987) *HIRIS High Resolution Imaging Spectrometer: Science Opportunities for the 1990s,* National Aeronautics and Space Administration, Washington, DC.

Goetz, A.F.H. (1991) 'Imaging spectrometry for studying Earth, air, fire and water', *EARSeL Advances in Remote Sensing* 1, 3-15.

Goetz, A.F.H. and Calvin, W.M. (1987) 'Imaging spectrometry: spectral resolution and analytical identification of spectral features', *Proceedings, Society of Photo-Optical Instrumentation Engineers,* 834, SPIE, Bellingham, Wa., pp. 158-165.

Goetz, A.F.H. and Herring, M. (1989) 'The High Resolution Imaging Spectrometer (HIRIS) for EOS', *IEEE Transactions on Geoscience and Remote Sensing* 27, 137-144.

Goetz, A.F.H., Vane, G., Solomon, J.E. and Rock, B.N. (1985) 'Imaging spectrometry for Earth remote sensing', *Science* 228, 1147-1153.

Gower, J.F.R. (1990) 'New results in coastal remote sensing with imaging spectroscopy', *in* S.E. Plummer (ed.) *Applications and Developments in Imaging Spectrometry,* Remote Sensing Society, Nottingham, pp. 1-10.

Gower, J.F.R. and Borstad, G.A. (1981) 'Use of the *in vivo* fluorescence line imager at 865 nm for remote sensing of surface chlorophyll a', *in* J.F.R. Gower (ed.) *Oceanography from Space*, Plenum Press, New York, pp. 329-338.

Gower, J.F.R. and Borstad, G.A. (1990) 'Mapping of phytoplankton by solar-stimulated fluorescence using an imaging spectrometer', *International Journal of Remote Sensing* 11, 313-320.

Green, R.O, Conel, J.E., Margolis, J.S., Bruegge, C.J. and Hoover, G.C. (1991) 'An inversion algorithm for retrieval of atmospheric and leaf water absorption from AVIRIS radiance with compensation for atmospheric scattering', *in* R.O. Green (ed.) *Proceedings, Third Airborne Visible/Infrared Imaging Spectrometer (AVIRIS) Workshop,* National Aeronautics and Space Administration, Jet Propulsion Laboratory Publication, Pasadena, Ca., pp. 51-61.

Hickman, G.D. and Duggin, M.J. (1992) 'Hyperspectral modeling for extracting aerosols from aircraft/satellite data', *in* R.O. Green (ed.) *Third JPL Airborne Geoscience Workshop,* National Aeronautics and Space Administration, Jet Propulsion Laboratory Publication, Pasadena, Ca., 1, pp. 20-22.

Hill, J., Mehl, W., and Altherr, M. (1994) 'Land degradation and soil erosion mapping in a Mediterranean ecosystem', *in* J. Hill and J. Mégier (eds.) *Imaging Spectrometry - a Tool for Environmental Observations,* Kluwer Academic Publishers, Dordrecht (this volume).

Hollinger, A.B., Gray, L.H., Gower, J.F.R. and Edel, H.R. (1987) 'The fluorescence line imager: an imaging spectrometer for ocean and land remote sensing', *Proceedings, Society of Photo Optical Instrumentation Engineers*, 834, SPIE, Bellingham, pp. 2-11.

Huegel, F.G. (1987) 'Advanced Solidstate Array Spectroradiometer: sensor and calibration improvements', *Proceedings, Society of Photo-Optical Instrumentation Engineers,* 834, SPIE, Bellingham, Wa., pp. 12-15.

Hunt, G.R. (1980) 'Electromagnetic radiation: the communication link in remote sensing', *in* B. Siegal and A. Gillespie (eds), *Remote Sensing in Geology*, Wiley, New York, pp. 5-45.

Hunt, G.R. and Salisbury, J.W. (1970) 'Visible and near-infrared spectra of minerals and rocks - I : Silicate minerals', *Modern Geology* 1, 238-300.

Hunt, G.R. and Salisbury, J.W. (1971) 'Visible and near infrared spectra of minerals and rocks - II : Carbonates', *Modern Geology* 2, 23-30.

Hunt, G.R., Salisbury, J.W. and Lenhoff, C.J. (1971) 'Visible and near-infrared spectra of minerals and rocks - II : Oxides and hydroxides', *Modern Geology* 2, 195-205.

Huntspiller, A. and Taranik, J.V. (1986) 'Detection of hydrothermal alteration at Virginia City, Nevada, using Airborne Imaging Spectrometry AIS', *in* G. Vane and A.F.H. Goetz (eds) *Proceedings, Second Airborne Imaging Spectrometer Data Analysis Workshop* National Aeronautics and Space Administration, Jet Propulsion Laboratory Publication, Pasadena, Ca., pp. 102-108.

Irons, J.R., Ranson, K.J., Williams, D.L., Irish, R.R. and Huegel, F.G. (1991) 'An off-nadir pointing imaging spectroradiometer for terrestrial ecosystem studies', *IEEE Transactions on Geoscience and Remote Sensing* 29, 66-74.

Janetos, A.C., Aber, J.D. and Wickland, D.E. (1992) *Workshop Report: Measuring Canopy Chemistry with High Spectral Resolution Remote Sensing Data,* NASA White Paper, National Aeronautics and Space Administration, Headquarters, Washington, DC.

Johnson, W.B. and Melfi, S.H. (1989) *Airborne Geoscience : The Next Decade*, National Aeronautics and Space Administration, Washington, DC.

Johnson, L.F. and Peterson, D.L. (1991) 'AVIRIS observation of forest ecosystems along the Oregon transect', *in* R.O. Green (ed.) *Proceeding, Third Airborne Visible/Infrared Imaging Spectrometer (AVIRIS) Workshop*, National Aeronautics and Space Administration Jet Propulsion Laboratory Publication, Pasadena, Ca., pp. 190-199.

Kerekes, J.P. and Landgrebe, D.A. (1991) 'Parameter trade-offs for imaging spectrometer systems', *IEEE Transactions on Geoscience and Remote Sensing* 29, 57-65.

King, M.D., Kaufman, Y.J., Menzel, W.P. and Tanré D. (1992) 'Remote sensing of cloud, aerosol and water vapour properties from the Moderate Resolution Imaging Spectrometer (MODIS)', *IEEE Transactions on Geoscience and Remote Sensing* 30, 2-27.

Kruse, F.A. (1988) 'Use of Airborne Imaging Spectrometer data to map minerals associated with hydrothermally altered rocks in the Northern Grapevine Mountains, Nevada and California', *Remote Sensing of Environment* 24, 31-51.

Kruse, F.A. (1990) 'Thematic mapping with an expert system and imaging spectrometers', *Proceedings, International Workshop on Advances in Spatial Information Extraction and Analysis for Remote Sensing*, American Society for Photogrammetry and Remote Sensing, Bethesda, Ma., pp. 59-68.

Kruse, F.A. and Hauff, P.L. (1991) 'Identification of illite polytype zoning in disseminated gold deposits using reflectance spectroscopy and X-ray diffraction - potential for mapping with imaging spectrometers', *IEEE Transactions on Geoscience and Remote Sensing* 29, 101-104.

Kruse, F.A., Taranik, D.L. and Kierein-Young, K.S. (1988) 'Preliminary analysis of Airborne Visible/Infrared Imaging Spectrometer (AVIRIS) for mineralogic mapping at sites in Nevada and Colorado', *in* G. Vane (ed.) *Proceedings, Airborne Visible/Infrared Imaging Spectrometer (AVIRIS) Performance Evaluation Workshop*, National Aeronautics and Space Administration, Jet Propulsion Laboratory Publication, Pasadena, Ca., pp. 76-87.

Kunkel, B., Blechinger, F., Viehmann, D., Van der Piepen, H. and Doerffer, R. (1991) 'ROSIS imaging spectrometer and its potential for ocean parameter measurements (airborne and space-borne)', *International Journal of Remote Sensing* 12, 753-761.

Lyon, R.J.P. (1987) 'Evaluation of AIS-2 (1986) data over hydrothermally altered granitoid rocks of the Singatse Range (Yerington) Nevada and comparison with 1985 AIS-1 data', *in* G. Vane (ed.) *Proceedings, Third Airborne Imaging Spectrometer Data Analysis Workshop,* National Aeronautics and Space Administration, Jet Propulsion Laboratory Publication, Pasadena, Ca., pp. 107-119.

McDonald, A.J.W., Wadge, G. and Murphy, R.J. (1989) 'Imaging spectroscopy for geobotanical mapping: porphyry copper mineralisation covered by pasture in Dyfed', *Proceedings, NERC Workshop on Airborne Remote Sensing,* Natural Environment Research Council, Swindon, pp. 59-75.

McLellan, T.M., Martin, M.E., Aber, J.D., Melillo, J.M., Nadelhoffer, K.J. and Dewey, B. (1991) 'Comparison of wet chemistry and near infrared reflectance measurements of carbon-fraction chemistry and nitrogen concentration of forest foliage', *Canadian Journal of Forest Research* 21, 1689-1693.

Meer, Van der F.D. (1992) 'A comparison of conventional classification methods and a new indicator kriging based method using high-spectral resolution images (AVIRIS)', *International Archives of Photogrammetry and Remote Sensing,* Commission VII, Washington, D.C., 11, 72-79.

Mehl, W. (1994) 'Imaging spectrometry data analysis - processing requirements and available software tools', *in* J. Hill and J. Mégier (eds.) *Imaging Spectrometry - a Tool for Environmental Observations,* Kluwer Academic Publishers, Dordrecht (this volume).

Melack, J.M. and Gastil, M. (1990) 'Reflectance spectra from eutrophic Mono Lake, California measured with the Airborne Visible and Infrared Imaging Spectrometer (AVIRIS)', *Society for Photo-Optical Instrumentation Engineers*, SPIE 1298, Bellingham, Wa., pp. 202-212.

Melack, J.M. and Gastil, M. (1992) 'Seasonal and spatial variations in phytoplanktonic chlorophyll in eutroptric Mono Lake, California, measured with the Airborne Visible and Infrared Imaging Spectrometer (AVIRIS)', *in* R.O. Green (ed.) *Third JPL Airborne Geoscience Workshop,* National Aeronautics and Space Administration, Jet Propulsion Laboratory Publication, Pasadena, Ca., 1, pp. 53-55.

Miller, J.R. and Hare, E.W. (1989) 'Imaging spectrometry as a tool for botanical mapping', *Proceedings, Society of Photo-Optical Instrumentation Engineers*, 834, SPIE, Bellingham, Wa., pp. 108-113.

Moore, G. and Aiken, J. (1990) 'Aircraft multispectral remote sensing of water colour off Helgoland' *in* S.E. Plummer (ed.) *Applications and Developments in Imaging Spectrometry,* Remote Sensing Society, Nottingham, pp. 18-31.

Mouchot, M.C., Sharp, G. and Lambert, E. (1988) 'L'utilisation du 'Fluorescence Line Imager' (FLI) pour la cartographie thematique des vegetaux marins submerges', *Proceedings, 11th Canadian Symposium on Remote Sensing,* CRSS, Waterloo, pp. 699-708.

Mustard, J.F. and Pieters, C.M. (1986) 'Abundance and distribution of mineral components associated with Moses Rock (Kimberlite) diatreme', *in* G.Vane and A.F.H. Goetz (eds) *Proceedings, Second Airborne Imaging Spectrometer Data Analysis Workshop,* National Aeronautics and Space Administration, Jet Propulsion Laboratory Publication, Pasadena, Ca., pp. 81-85.

Mustard, J.F., Hurtrez, S., Pinet, P. and Scotin, C. (1992) 'First results from coordinated AVIRIS, TIMS and ISM (French) data for the Ronda (Spain) and Beri Bousera (Morocco) Peridotites', *in* R.O. Green (ed.) *Third JPL Airborne Geosciences Workshop,* National Aeronautics and Space Administration, Jet Propulsion Laboratory Publication, Pasadena, Ca., 1, pp. 26-28.

Nakashima, B.S., Borstad, G.A., Hill, D.A. and Kerr, R.C. (1989) 'Remote sensing of fish schools: early results from a ditigal imaging spectrometer', *Proceedings IGARSS '89/12th Canadian Symposium on Remote Sensing,* IEEE, New York, 4, pp. 2044-2047.

Oppenheimer, C., Pieri, D., Carrere, V., Abrams, M., Rothery, D. and Francis, P. (1992) 'Volcanic thermal features observed by AVIRIS', *in* R.O. Green (ed.) *Third JPL Airborne Geoscience Workshop,* National Aeronautics and Space Administration, Jet Propulsion Laboratory Publication, Pasadena, Ca., 1, pp. 41-43.

Peterson, D.L. (1991) *Report on the Workshop Remote Sensing of Plant Biochemical Content: Theoretical and Empirical Studies,* NASA White Paper, NASA Ames Research Center, Ca.

Peterson, D.L., Aber, J.D., Matson, P.A., Card, D.H., Swanberg, N., Wessman, C. and Spanner, M. (1988) 'Remote sensing of forest canopy and leaf biochemical contents', *Remote Sensing of Environment* 34, 85-108.

Pettersson, L.H. (1990) 'Norwegian remote sensing spectrometry for mapping and monitoring of algal blooms and pollution - NORSMAP '89', *in* S.E. Plummer (ed.) *Applications and Developments in Imaging Spectrometry*, Remote Sensing Society, Nottingham, pp. 11-17.

Pieters, C.M. and Mustard, J.F. (1988) 'Exploration of crustal/mantal material for the Earth and Moon using reflectance spectroscopy', *Remote Sensing of Environment* 24, 151-178.

Porter, W.M. and Enmark, H.T. (1987) 'A system overview of the Airborne Visible/Infrared Imaging Spectrometer (AVIRIS)', *Proceedings, Society of Photo-Optical Instrumentation Engineers*, 834, SPIE, Bellingham, Wa., pp.22-31.

Precision Visuals (1992) *PV WAVE,* Precision Visuals Inc., Boulder, Co.

Rast, M. (1991) *Imaging Spectroscopy and its Application in Spaceborne Systems*, ESA SP-1144, European Space Agency, Noordwijk.

Rast, M., Hook, S.J., Elvidge, C.D. and Alley, R.E. (1991) 'An evaluation of techniques for the extraction of mineral absorption features from high spectral resolution remote sensing data', *Photogrammetric Engineering and Remote Sensing* 57, 1303-1309.

Ridd, M.K., Ritter, N.D., Bryant, N.A. and Green, R.O. (1992) 'AVIRIS data and neural networks applied to an urban ecosystem', *in* R.O. Green (ed.) *Third Airborne Geoscience Workshop*, National Aeronautics and Space Administration, Jet Propulsion Laboratory, Pasadena, Ca., 1, pp. 129-131.

Riggs, G.A. and Running, S.W. (1991) 'Detection of canopy water stress in conifers using the airborne imaging spectrometer', *Remote Sensing of Environment* 35, 51-68.

Rivard, B. and Arvidson, R.E. (1992) 'Utility of imaging spectrometry for lithologic mapping in Greenland', *Photogrammetric Engineering and Remote Sensing* 58, 945-949.

Roberts, D.A., Smith, M.O. and Adams, J.B. (1993) 'Green vegetation, non-photosynthetic vegetation and soils in AVIRIS data', *Remote Sensing of Environment*, 44, 255-269.

Rock, B.N., Hoshizaki, T. and Miller, J.R. (1988) 'Comparison of *in situ* and airborne spectral measurements of the blue shift associated with forest decline', *Remote Sensing of Environment* 24, 109-127.

Rothery, D.A. and Oppenheimer, C.M.M. (1990) 'The potential of imaging spectrometry for measuring surface temperatures and energy budgets of volcanoes', *in* S.E. Plummer (ed.) *Applications and Developments in Imaging Spectrometry,* Remote Sensing Society, Nottingham. pp. 70-74.

Salomonson, V.V., Barnes, W.L., Maymon, P.W., Montgomery, H.E. and Ostrow. H. (1989) 'MODIS: Advanced Facility Instrument for studies of the Earth as a system', *IEEE Transactions on Geoscience and Remote Sensing* 22, 145-153.

Schanzer, D. and Staenz, K. (1992) 'Discussion of band selection and methodologies for the estimation of precipitable water vapour from AVIRIS data', *in* R.O. Green (ed.) *Third Airborne Geoscience Workshop*, National Aeronautics and Space Administration, Jet Propulsion Laboratory Publication, Pasadena, Ca., 1, pp. 135-137.

Slater, P.N. (1985) 'Survey of multispectral imaging systems for Earth observation', *Remote Sensing of Environment* 17, 85-102.

Smith, M.O., Adams, J.B., and Sabol, D.E. (1994a) 'Spectral mixture analysis - new strategies for the analysis of multispectral data', *in* J. Hill and J. Mégier (eds.) *Imaging Spectrometry - a Tool for Environmental Observations,* Kluwer Academic Publishers, Dordrecht (this volume).

Smith, M.O., Adams, J.B., and Sabol, D.E. (1994b) 'Mapping sparse vegetation canopies', *in* J. Hill and J. Mégier (eds.) *Imaging Spectrometry - a Tool for Environmental Observations,* Kluwer Academic Publishers, Dordrecht (this volume).

Ustin, S.L., Smith, M.O., Roberts, D., Gammon, J.A. and Field, C.B. (1992) 'Using AVIRIS images to measure temporal trends in abundance of photosynthetic and non-photosynthetic canopy components', *in* R.O. Green (ed.) *Proceedings, Third JPL Airborne Geoscience Workshop*, National Aeronautics and Space Administration, Jet Propulsion Laboratory Publication, Pasadena, Ca., 1, pp. 5-10.

Vane, G. ed. (1987) *Proceedings, Third Airborne Imaging Spectrometer Data Analysis Workshop,* National Aeronautics and Space Administration, Jet Propulsion Publication, Pasadena, Ca.,

Vane, G. ed. (1988) *Proceedings, Airborne Visible/Infrared Imaging Spectrometer (AVIRIS) Performance Evaluation Workshop*, National Aeronautics and Space Administration, Jet Propulsion Laboratory Publication, Pasadena, Ca.

Vane, G. and Goetz, A.F.H. eds (1985) *Proceedings, Airborne Imaging Spectrometer Data Analysis Workshop,* National Aeronautics and Space Administration, Jet Propulsion Laboratory Publication, Pasadena, Ca.

Vane, G. and Goetz, A.F.H. eds (1986) *Proceedings, Second Airborne Imaging Spectrometer Data Analysis Workshop*, National Aeronautics and Space Administration, Jet Propulsion Laboratory Publication, Pasadena, Ca.

Vane, G. and Goetz, A.F.H. (1988) 'Terrestrial imaging spectrometry', *Remote Sensing of Environment* 24, 1-29.

Verstraete, M.M. and Pinty, B. (1992) 'Extracting surface properties from satellite data in the visible and near-infrared wavelengths', *in* P.M. Mather (ed.) *TERRA-1 Understanding the Terrestrial Environment*, Taylor and Francis, London, pp. 203-209.

Wessman, C.A. (1994a) 'Remote sensing and the estimation of ecosystem parameters and functions', *in* J. Hill and J. Mégier (eds.) *Imaging Spectrometry - a Tool for Environmental Observations,* Kluwer Academic Publishers, Dordrecht (this volume).

Wessman, C.A. (1994b) 'Estimating canopy biochemistry through imaging spectrometry', *in* J. Hill and J. Mégier (eds.) *Imaging Spectrometry - a Tool for Environmental Observations,* Kluwer Academic Publishers, Dordrecht (this volume).

Wessman, C.A., Aber, J.D., Peterson, D.L. and Melillo, J.M. (1988a) 'Foliar analysis using near infrared reflectance spectroscopy', *Canadian Journal of Forest Research* 18, 6-11.

Wessman, C.A., Aber, J.D., Peterson, D.L. and Melillo, J.M. (1988b) 'Remote sensing of canopy chemistry and nitrogen cycling in temperate forest ecosystems', *Nature* 335, 154-156.

Wilson, A.K. (1990) 'The NERC 1989 Compact Airborne Spectrographic Imager (CASI) Campaign', *Proceedings, NERC Workshop on Airborne Remote Sensing*, Natural Environment Research Council, Swindon, pp. 259-283.

SCIENTIFIC ISSUES AND INSTRUMENTAL OPPORTUNITIES IN REMOTE SENSING AND HIGH RESOLUTION SPECTROMETRY

MICHEL M. VERSTRAETE
Institute for Remote Sensing Applications
Commission of the European Communities
Joint Research Centre
I-21020 Ispra (Va), Italy

ABSTRACT. The effective use of remote sensing techniques requires a basic understanding of the fundamental processes that affect radiation during its transport between the source of light, the target of interest, and the detector. The principles of radiation emission and scattering in the optical domain are outlined, paying particular attention to the spatial, temporal, spectral, and directional sources of variability in the data. The problems of measuring and interpreting these observations are addressed, and the specifications of existing and planned space-borne instruments are discussed.

1. Principles of remote sensing

Strictly speaking, remote sensing does not exist: it is not possible to 'sense' remotely. What is intended by this misnomer is the local sensing of a signal which may carry information about the properties of remote objects by virtue of having interacted with them in the past. The remote object of interest is often called the target. The messenger or carrier of meaningful information about this target (the signal) serves as a proxy for *in situ* measurements of target properties. The technical difficulty in remote sensing is to detect and measure a signal with appropriate accuracy, reliability, and resolution. The scientific challenge is to interpret these measurements in terms of the properties of the target.

Our own senses of vision, hearing or smell constitute our personal remote sensing devices. When photographic plates were invented, at the end of last century, they allowed the objective observation of remote scenes and the storage of that information on a medium characterized by a prolonged lifetime. Historical accounts of the scientific and technological developments of remote sensing are available in the literature (e.g. Simonett, 1983; Estes and Consentino, 1989).

In scientific and industrial applications, the signal is usually carried by an electromagnetic radiation or waves, but it could as well be a sound wave or any other system capable of transporting information. Electromagnetic waves (including light), have at least three advantages: first, they travel very fast, thereby minimizing the time needed to carry the information. Second, electromagnetic and quantum theories provide a firm basis for the understanding of these waves and their interactions with various media. And last but not least, advances in electronics allow the conversion of electromagnetic waves (or photons) into measurable electrical quantities which can be digitized and processed by computers.

J. Hill and J. Mégier (eds.), Imaging Spectrometry - a Tool for Environmental Observations, 25–38.

Remote sensing instruments can be classified in two broad categories, depending on the source of radiation. Active sensors emit their own electromagnetic waves. These are directed towards the target of interest, which scatters a fraction of this incoming radiation back to the instrument. Examples of active sensing systems include radars and lasers; they tend to require very high power supplies. On the other hand, passive sensors are those that rely on the detection of a signal emitted by an independent and uncontrolled source. Most of the space-based sensors operating in the optical range are of the passive type, because the Sun provides a reasonably well known, stable, reliable and cheap source of light. In that case, it is the reflectance of a target which is measured. In the thermal infrared range or in passive microwave applications, it is the source itself which is being observed, and the relevant properties are its temperature and emissivity. The rest of this chapter focuses exclusively on passive instruments in the solar spectral range.

No useful information can be retrieved about the target if the latter does not influence the radiation in any way: it is because the observed radiation is modified (e.g. by absorption or scattering) by the target that we can infer some information about it. Of course, no information could be gathered if the target perfectly absorbed all the incoming radiation, since no signal would reach the detector. By the same token, other objects not necessarily of interest but located on the path of the radiation may have been interacting with it also. Finally, no information can be deduced on the target unless the characteristics of the source of radiation are precisely known; in fact, it is the difference between the signal that would be received directly from the source in the absence of target and that actually received that provides information about the target. A major issue in remote sensing therefore consists in identifying the various factors that affect radiation along its path from the source to the detector, and assessing their relative impacts. Clearly, this implies that only those objects or phenomena which directly affect the transmission of light between its source and the detector can be studied by remote sensing.

The fundamental problem associated with the use of remote sensing in practical applications is to quantitatively interpret local observations of a signal that has interacted with remote objects, in terms of the properties of these remote objects. This interpretation must account for signal variations that may be due to changes in the source of light, to the interaction of the signal with multiple objects during its brief lifetime, and to the particular conditions of observation. Some of these issues will be considered below.

2. From the Sun to the detector

The light we receive from the Sun is actually emitted in its photosphere, the visible outer layer of the star. The Earth revolves around the Sun at a mean distance of $149.68 \cdot 10^9$ m; and at this distance the flux of radiant energy is currently around 1370 W m^{-2}. This value is known as the 'solar constant', because it is a very stable value, even though it is not absolutely constant. This value can be converted in an equivalent brightness temperature as follows: Since the Sun's disk subtends an angle of 32", its radius is 696 10^6 m, and the power emitted at the surface amounts to 63.5 10^6 W m^{-2}. An ideal black body at a temperature of about 5800 K would emit thermal radiation at the same rate. (Rozelot, 1973; Houghton, 1986). Solar photons take about 500 seconds (8.3 minutes) to reach the Earth, but they cross the atmosphere, interact with the surface, traverse the atmosphere again, and reach the detector on board a polar-orbiting satellite such as NOAA in only 3.3 milliseconds.

For all practical purposes, the bulk of solar radiation is emitted at wavelengths between 0.3 and 3.0 micron, as shown in figure 1. About half of the total energy flux is emitted at wavelengths below 0.7 micron, which is the upper limit of visible radiation. The other half is emitted in the near infrared spectral region. The theoretical distribution of spectral radiation emitted by a perfect black body, in this case at the slightly different temperature of 6000 K, has also been drawn on this figure. Significant departures from this theoretical curve occur mostly at the shortest wavelengths (ultra-violet), either because the Sun is not a perfect emitter, because of absorption lines in the Sun's outer layers, or because of absorption and scattering by interplanetary dust and particles.

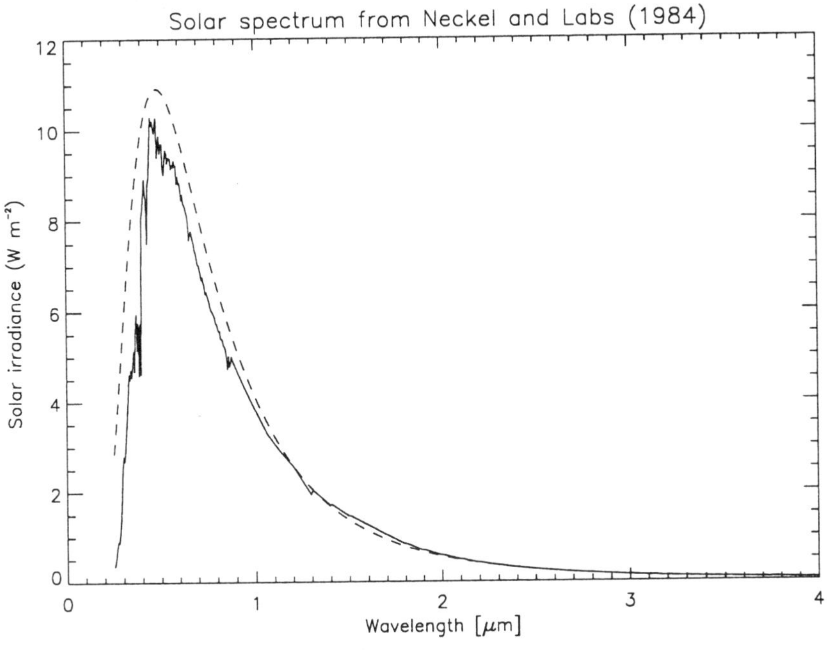

FIGURE 1. The solar spectrum (after Neckel and Labs, 1984).

Much more important for our purpose is the absorption and scattering of solar radiation within the atmosphere, both before and after the interaction of the radiation with the surface. The Earth atmosphere is a gaseous layer principally composed of molecular nitrogen (78% per volume) and oxygen (21%). Some trace gases occur in relatively minor concentrations, but nevertheless play a dominant role in the climate system, mainly because of their capacity to absorb radiation in specific spectral bands (wavelength intervals). The most notable are well-mixed compounds such as carbon dioxide (about 350 ppm), methane (2 ppm) and nitrous oxide (0.3 ppm). The most important variable constituents are water vapor and ozone. The former is characterized by a highly variable spatial and temporal distribution over a wide range of scales, while the latter is smoothly varying with latitude, altitude, and season. (Houghton, 1986). Clearly, the position and width of these absorption bands define the spectral intervals in which the surface can effectively be observed from space.

Other atmospheric constituents present additional challenges: aerosols originating at the surface modify the scattering properties of the atmosphere, and a variety of clouds either prevent (the most common case) or contaminate (in the case of thin cirrus veils or subpixel clouds) surface observations. At any particular time, clouds occupy, on average, about half of the planet's surface.

As a result, the observation of the Earth surface from space with optical instruments is only feasible part of the time, and the detailed interpretation of the results requires complex algorithms. This state of affairs is both an advantage and an inconvenience: on the positive side, these atmospheric effects on radiation transfer can be used to gain information on the atmosphere itself (e.g. its composition). The drawback is that atmospheric effects increase the complexity of the algorithms used to interpret the data, and that any residual effect not accounted for becomes a source of noise in the signal, as far as surface studies are concerned.

3. From photons to reflectances

The fundamental event in remote sensing observations is the arrival of one or more photons at a detector. These photons are absorbed by the instrument, where they deposit a small but measurable quantity of energy. A variety of devices and techniques exist to detect photons, including photographic plates, photoconductive and photodiode detectors, or charge-coupled devices (CCD). Photographic techniques had a major impact on the development of remote sensing, and are still used extensively in aerial surveys, but this approach has been completely superseded by electronic devices for space-borne instruments. The fundamental process exploited by electronic detectors is the photoelectric effect, in which incoming photon displace electrons in the semi-conductor of the detector. The resulting current (or potential difference) is then measured and transformed into a digital count with an analog/digital converter.

Different detectors are sensitive to radiation at different wavelengths, but it is feasible to design detectors sensitive to a broad range of wavelengths, covering a large part of the solar spectrum. An optical subsystem, often including mirrors and lenses, collects the incoming light from a small solid angle (the instantaneous field of view, or IFOV) pointing at the target, and focuses it on the detector. If spectral information is desired, it is necessary to use filters to screen access to the detector, or to disperse the incoming light before it reaches the detector with optical prisms or diffraction gratings. In both cases, an array of detectors is used to measure the incoming radiation in multiple spectral bands simultaneously.

To spectrally image a scene, it is possible to use a single detector array and point it towards the various locations of interest, as is done in scanning instruments. In practice, the detector array is kept fixed, but a mirror system is installed in front of the optical system, such that different points on the surface of the Earth are imaged successively as the mirror rotates. A half rotation of the mirror results in the observation of a line of points at the surface. The scanning direction is oriented across the direction of motion of the satellite, so that successive lines of pixels can be used to image the scene. The total field of view (FOV) of such an instrument is the angle between the nadir and the most extreme direction for which a measurement is actually obtained.

An alternative design, called pushbroom, consists in building a two-dimensional array of detectors: in this case, all points within the total FOV are observed simultaneously, in all spectral bands. This approach poses significant technical challenges in terms of data acquisition and processing, since a large number of individual detectors are active at the same time, but present other advantages, such as the absence of moving parts. Additional introductory information on the design of detectors can be found in Slater (1983), Robinson and DeWitt (1983), Norwood and Lansing (1983), or Rees (1990).

The processes involved in the conversion of photons to digital counts take place within the observing instrument itself, or its associated electronic boxes. The resulting digital data are then temporarily stored and transmitted back to receiving stations on the ground, together with other engineering data (such as clock marks) for further processing. These raw data are known as Level 0 data. Level 1 data results from a first stage of processing, where the time-referenced raw data at full resolution is complemented by various ancillary information, including radiometric and geometric calibration coefficients and georeferencing parameters. Different agencies may use slightly different terminologies. For example, at NASA, Level 1B data has been processed into sensor units, i.e. the calibration has been applied (Butler, 1987), while AVHRR products at Level 1B from NOAA include ancillary data about calibration, but the latter has not been applied (Kidwell, 1991). In any case, the calibration of raw data yields radiance measurements, in W m^{-2} sr^{-1} $micron^{-1}$.

The incoming solar radiation flux at the top of the atmosphere varies seasonally because of the ellipticity of the orbit of the Earth. The Earth is closest to the Sun around January 3 and farthest on July 5 (Sellers, 1965). This results in a predictable 3.5% variation in light intensity, and it is usual to normalize the measured radiances into reflectances by dividing them by the instantaneous incoming solar flux.

A Level 2 data set contains environmental parameters derived from Level 1 data, at the same spatial and temporal resolution. It is often useful or necessary to analyze data on a regular grid, for example to compare data taken at different times, or from different instruments. Gridded data is known as Level 3 data. Their creation involves the mapping of Level 2 data, accounting for the curvature of the Earth, the motion of the satellite platform, and the varying geometry of observation. Missing observations may be interpolated, or compositing techniques may be used to generate data sets with desirable properties (e.g., mostly cloud free scenes). Finally, Level 4 applies to data not directly measured but produced by models or resulting from analyses, on the basis of data at lower levels (NASA, 1986).

A computer software known as a 'pre-processing chain' is used to generate data sets at Levels 1, 2, or 3. These products can be displayed on video screens or printed as images. At Level 2, and at full resolution, each individual picture element (pixel) represents a single measurement in a given spectral band. Sophisticated computer programs are available to manipulate these images, for instance to enhance the contrast, combine measurements in different spectral bands, manipulate colours, or perform statistical operations.

4. From reflectances to the desired information

The basic data set used in remote sensing is a collection of numbers representing calibrated radiances or reflectances, in one or more spectral bands, for a particular location and time. Additional data sets may be available for the same location at different times, for other places, or from a different perspective. In this section, we investigate the nature of the information that can be retrieved from these data sets.

4.1 SOURCES OF VARIABILITY

Since remote sensing measurements are repeated for different places and times, in various spectral bands and for various geometrical configurations, these observations can be considered implicit functions of the corresponding independent variables. Formally, a reflectance ρ at the level of the satellite is a function of a set of intrinsic parameters y_i of the media with which the radiation has interacted, and these parameters are in turn functions of independent variables such as space, time, spectral band location, and illumination and observation geometry:

$$\rho = \rho(y_1, y_2, \ldots, y_n) \tag{1}$$

where

$$y_i = y_i(x, t, \lambda, \Theta) \qquad i = 1, 2, \ldots, n \tag{2}$$

In this formula, x stands for spatial coordinates, such as the longitude and latitude of the observed location, t represents time, λ indicates that the parameters y_i are spectrally dependent, and θ reminds us that the measurements may depend on the particular position of the Sun and the observer at the time of the measurement.

This implies that there are four main sources of variation, and therefore of information, in such a data set. The spatial variability is normally preserved as faithfully as possible throughout the processing chain, as it is used to represent the spatial distribution of the objects in the scene. Similarly, the temporal variations are used to describe the dynamic evolution of these objects. Spectral contrasts serve to identify the nature or composition of the target being observed. Finally, signal variations due to changes in the conditions of illumination or observation constitute a very promising (and little used) source of signal variability. Advanced mathematical models must be used to exploit this latter feature, as will be seen in a later chapter.

Often, different sources of variability can be combined to improve a particular product; for example, the temporal evolution of the signal over a small homogeneous region can complement the information obtained from its spectral signature, and help discriminate surface types. The basic types of data variability are now reviewed.

4.2 SPATIAL VARIABILITY - MAPPING

The simplest application of remote sensing consists in producing maps of the observed areas. In this case, the spatial variation of the reflectance of the surface can be exploited directly, even in a single broad channel, to produce maps showing the location of objects at the surface, provided the effects of the Earth's surface curvature and of the geometry of observation are taken into consideration. This application has mostly exploited very high resolution systems, such as aerial photography and space instruments such as HRV on SPOT and MSS or TM on Landsat. But even when lower resolution instruments are used, there is almost always a need to refer observations to a particular location on the ground.

Beyond the localization of the information, the spatial variability of the signal itself can be used. For example, the spatial structure of a scene has been used to describe the heterogeneity of the environment. Advanced statistical tools such as semi-variograms and local variance analysis have been combined with principal components and clustering methods to improve the characterization and classification of surface types (e.g. Vogt, 1992).

4.3 TEMPORAL VARIABILITY - EVENT DETECTION

The next simplest application is to detect temporal changes in the environment, such as the construction of a new road, the expansion of urbanization or agriculture, deforestation, etc. Basically, any environment modification that results in a significant change of the optical properties of the surface can be detected by processing two or more scenes from the same region, taken some time apart. The detailed interpretation of non-anthropogenic changes must rely on local observations ('ground truth'), but when these changes are drastic enough (as is the case for deforestation, which replaces deep forest by essentially bare ground), the effect is so obvious, even in the raw data, that many authors have forgone many steps in the pre-processing, save the geometric correction, to quickly arrive at a usable product. Obviously, the ability to detect and interpret these changes depends on the capacity of the analyst to remove extraneous effects and perturbations such as those resulting from atmospheric contamination.

4.4 SPECTROMETRY - TARGET IDENTIFICATION

As hinted above, spectral variability in the data (i.e. the information coming from different spectral channels for the same region) can be used to identify the nature of the observed target. This is because most natural objects in the environment have distinctive optical properties (in particular absorptance). This is shown very clearly in figure 2, which exhibits the spectral reflectance of a few typical natural surfaces (soils, leaves, cloud, water). It can be seen for instance, that the reflectance of water is very low throughout the solar spectrum, while clouds are very bright in the visible but progressively darker in the near-infrared regions. While the reflectance of soils typically increases with wavelength, that of green leaves demonstrates a bimodal behavior, where solar radiation is strongly absorbed in the visible part of the spectrum and strongly reflected at wavelengths above 0.7 micrometer.

The simplest way to exploit this differential response is to measure the light reflected by these surfaces in two or more spectral bands, and to combine these measurements in simple indices. A number of vegetation indices, which are typical examples of this approach, have been developped to date (Curran, 1981; Pinty and Verstraete, 1992; Kaufman and Tanré, 1992), and their use will continue for the foreseeable future. The main advantage of these indices is that they may provide useful information at a very low cost. They suffer a number of drawbacks, however, including relatively high sensitivity to soils, atmospheric conditions, or illumination and viewing geometry. Some of these indices have been optimized to reduce one of these sources of errors, but often at the expense of a larger sensitivity to other types of contamination.

The advent of true spectrometers (especially airborne instruments) has stimulated a flurry of activities to extract information from multiple contiguous narrow bands. For example, the position and slight displacement of the red edge (the jump in reflectance of leaves around 0.7 micrometer may be linked to the state and health of the plant. More information on these developments will be found elsewhere in this volume.

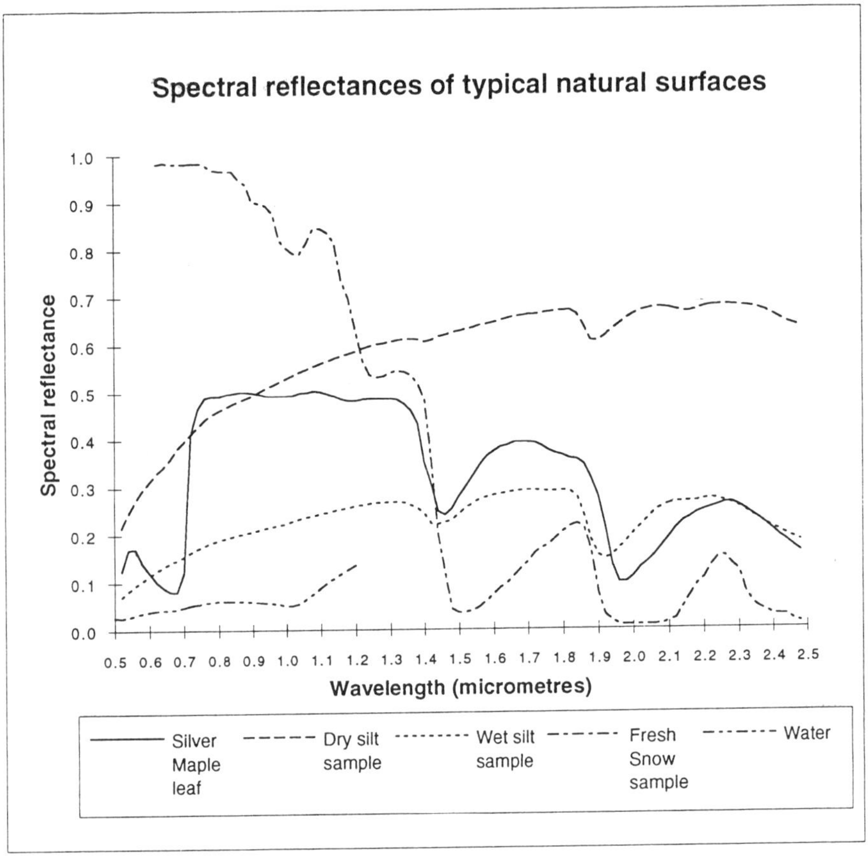

FIGURE 2. Spectral reflectances of typical natural surfaces.

4.5 ANISOTROPY - OPTICAL AND STRUCTURAL PROPERTIES OF THE SURFACE

All natural surfaces are anisotropic, in the sense that they reflect solar radiation differently in different directions. The most obvious examples are the near total specular reflection of a calm water surface, compared to the otherwise rather dark appearance of water at other observation angles, or the increased reflectance of vegetated surfaces in the direction of illumination (hot spot), which results from the absence of apparent shadows. Since all remote sensing measurements are made with instruments characterized by a relatively small instantaneous field of view, they depend strongly on the angular position of the Sun and of the observer with respect to the target, at the time of the measurement. For this reason, observations are called 'bidirectional reflectances'.

Directional variability in the data is often considered a drawback, because observations of the same area from different directions will not yield the same values, even if the surface and the atmosphere are identical, but it is a source of additional information about the surface, particularly about its optical and structural properties. Advanced models must be inverted against remote sensing data to yield the desired information, as will be seen in a later chapter.

5. Review of existing and planned optical instruments

There is an extensive literature on the design and operation of both existing and planned space instruments. The purpose of this section is to summarize in a few paragraphs the characteristics of the principal instruments currently used to monitor terrestrial surfaces, and to provide an overview of sensors under development which will become operational in the next decade or so.

5.1 AVHRR

The US National Oceanic and Atmospheric Administration (NOAA) operates a series of space platforms known as the NOAA satellites. These platforms support a number of instruments, mostly designed for operational meteorology, of which the Advanced Very High Resolution Radiometer (AVHRR) is the best known for surface studies. This instrument acquires measurements in five spectral bands, shown in Table 1.

Channel number	Central Wavelength [nm]	Width [nm]
1	635	130
2	850	260
3	3730	400
4	10790	920
5	11900	1000

TABLE 1. AVHRR (NOAA-9) spectral bands (Kidwell, 1986)

Data are collected at the base spatial resolution of about 1.1 km at nadir. Observations at this resolution are being permanently broadcast as they are acquired, and can be received in High Resolution Picture Transmission (HRPT) mode if and while a ground receiving station is within sight of the platform. A limited amount of data at full resolution can be recorded on board the satellite, for later downloading in selected locations (Local Area Coverage, or LAC data). The standard product available globally from this instrument series, however, is known as the Global Area Coverage (GAC) data. This data set is obtained in averaging the values of consecutive sets of four pixels in a scan line, skipping one pixel between each set, and skipping two scan lines before processing one again. The effective spatial resolution of the GAC data is therefore of the order of 4-5 km at nadir.

The principal advantage of this instrument as far as surface observations are concerned results from the fact that the two optical channels, one in the visible and one in the near-infrared, are located on either side of the 0.7 micrometer threshold in reflectance for green leaves (figure 2). The simultaneous availability of thermal channels provides a means to estimate brightness surface temperature at the same resolution, location, and time. The availability of over 10 years of GAC data globally is also a major advantage. However, this instrument suffers from severe shortcomings, in particular because it is not calibrated in flight, and because of the rather peculiar subsampling strategy used to produce the GAC data set. Navigation is poorly controlled, because the on-board clock drifts in time, and pixel registration is difficult to achieved at better than one or even a few pixels. The spectral channels are very wide, and the data are therefore significantly contaminated by atmospheric effects.

5.2 SPOT AND LANDSAT

While AVHRR provides daily global coverage, its spatial resolution (1 km or higher) is too coarse for many applications, such as land management, mapping, soil degradation monitoring, etc. This capability is offered by the High Resolution Visible (HRV) instruments on the French SPOT satellites, or by the Multispectral Scanning (MSS) or Thematic Mapper (TM) instruments on the American Landsat platforms. These instruments, operated by for-profit companies, also have more spectral bands than the AVHRR, as can be seen from Table 2 in the case of TM. Data are expensive, and not collected systematically. In the case of SPOT, for example, the sensor can be pointed sideways to provide stereoscopic views of an area of interest, but this implies that other regions are not observed while the pointing mechanism changes the orientation of the instrument, or while the sensor is pointed in another direction.

Channel number	Central Wavelength [nm]	Width [nm]
1	485	70
2	560	80
3	660	60
4	830	140
5	1650	200
6	11450	2100
7	2215	270

TABLE 2. Landsat-TM solar spectral bands (Freden & Gordon, 1983)

These instruments are designed to operate at very much higher spatial resolution than the AVHRR, typically 10 m for the panchromatic mode of SPOT-HRV, or 30 m for the Landsat-TM instrument (The older Landsat-MSS instrument had a spatial resolution of 80 m). The penalty for using these instruments is that global coverage can only be acquired in 16 days with the Landsat system, and every 26 days with SPOT. The cost of acquiring and processing these data is currently so high as to be prohibitive for all but the smallest investigations, or for all but the wealthiest customers.

5.3 MODIS AND MERIS

The MODerate resolution Imaging Spectrometer (MODIS) is a new instrument currently being built for launch around 1998, as part of the first Earth Observing System (EOS) platform of NASA. This is not truly a spectrometer, since it features a set of 36 non contiguous spectral bands, but it will nevertheless provide pretty good coverage of the response of the surface over a wide spectrum. Table 3 shows the 19 spectral bands in the solar part of the spectrum.

The European Space Agency (ESA) is planning to launch the MEdium Resolution Imaging Spectrometer (MERIS) on its ENVISAT-1 platform, also around 1998. This instrument will actually measure the incoming radiation over the entire range from 400 to 1050 nm, in 260 contiguous spectral bands of 2.5 nm each. However, technological limitations in the rate of data transmission will restrict access to 15 channels at any one time. Each channel is programmable in flight and can include multiple elementary spectral bands. To achieve adequate signal to noise ratio, it is currently planned to download most channels at 10 nm, i.e. as averages of four individual bands. Of course, any change in the selection of bands to download will affect the period of time for which continuous data are available. Table 4 shows a typical selection of bands for MERIS.

Channel number	Central Wavelength [nm]	Width [nm]	IFOV [m]	Main Mission
8	415	15	1000	Chlorophyll (Ocean)
9	443	10	1000	Chlorophyll (Ocean)
3	470	20	500	Soil, Veg. Differences
10	490	10	1000	Chlorophyll (Ocean)
11	531	10	1000	Chlorophyll (Ocean)
12	555	10	1000	Sediments
4	555	20	500	Green Vegetation
1	659	50	250	Veg., Chloro., Land Cover
13	667	10	1000	Sediments, Atmosphere
14	681	10	1000	Chlorophyll Fluorescence
15	750	10	1000	Aerosol Properties
16	865	15	1000	Aerosol/Atm. Properties
2	865	40	250	Cloud, Veg., Land Cover
17	905	30	1000	Cloud/Atm. Properties
18	936	10	1000	Cloud/Atm. Properties
19	940	50	1000	Cloud/Atm. Properties
5	1240	20	500	Leaf/Canopy Differences
6	1640	20	500	Snow/Cloud Differences
7	2130	50	500	Land & Cloud Properties

TABLE 3. MODIS solar spectral bands (Pagano & Young, 1992)

5.4 HIRIS, AVIRIS, AND HRIS

NASA had initiated the design of a very ambitious high resolution imaging spectrometer, code-named HIRIS, as part of its EOS strategy. This instrument has now been abandoned, in part because of the technical difficulty of building an instrument to the very stringent specifications selected in its design, and in part because of its excessive cost. Nevertheless, an airborne version of this instrument, AVIRIS, has been implemented and has been flown in a number of missions on board the ER-2 plane of NASA, both in the US and in Europe. Selected data sets are available for test applications.

ESA also has a high resolution imaging spectrometer (PRISM) under development. This instrument is currently intended to fly on the ENVISAT-2 platform around 2005. It will cover the 450 to 2350 nm spectral regions, with a spectral resolution of 10 nm and a spatial resolution of 40 m. The absolute calibration accuracy is expected to be very tight. The overall instrument concept is already in place, and the mission objective will focus on terrestrial applications.

Channel number	Central Wavelength [nm]	Width [nm]	IFOV [m]	Main Mission
1	410.0	10.0	300	Open & Coastal Ocean
2	445.0	10.0	300	Open & Coastal Ocean, Veg.
3	490.0	10.0	300	Open & Coastal Ocean
4	520.0	10.0	300	Open & Coastal Ocean, Veg.
5	565.0	10.0	300	Open & Coastal Ocean, Veg.
6	620.0	10.0	300	Open & Coastal Ocean
7	665.0	10.0	300	Open & Coastal Ocean, Veg.
8	682.5	5.0	300	Open & Coastal Ocean, Veg.
9	710.0	10.0	300	Open & Coastal Ocean
10	755.0	10.0	300	Open & Coastal Ocean
11	762.5	2.5	300	Atmos. Properties
12	765.0	2.5	300	Atmos. Properties
13	767.5	2.5	300	Atmos. Properties
14	880.0	10.0	300	Ocean, Atm., Veg.
15	900.0	10.0	300	Atmos. Properties
16	1022.5	25.0	300	Atmos. & Ocean

TABLE 4. MERIS spectral bands (ESA, 1992)

6. Outstanding issues

There is little doubt that remote sensing can provide useful information to deal with the long suite of climatic and environmental problems that need to be addressed. There is also a general consensus on the fact that instruments with multiple narrow spectral bands would contribute a lot to our understanding of the relevant mechanisms. This is because narrow channels may be less sensitive to the absorption bands of atmospheric constituents, and because additional channels allow a more reliable identification of the nature and state of the targets.

But imaging spectrometers that span the entire solar spectrum with a spectral resolution of 10 nm or better, and a high spatial resolution, are very expensive to design, build and operate. The user community therefore needs to provide a compelling scientific justification for these instruments. For example, while the first two bands of AVHRR are clearly not enough to properly characterize the surface and address Global Change issues, how many more are needed? One approach to answer this question would be to recursively evaluate the scientific return of one additional band, in other words to estimate the marginal return on investment. Since different applications need different spectral bands, this may be a complex but nevertheless necessary exercise.

Another issue that will need to be addressed soon is the analysis of the very large data sets that will be produced by the instruments currently under developments. This is not only a technical problem (archival, distribution of gigabytes of data over international computer networks), but also a scientific problem. We must start to develop now the models and approaches that will be needed to address the most pressing issues, so that we can not only diagnose the problems after the fact (retrospective studies) or witness the issues in real time (monitoring), but also predict with some confidence the likely evolution of our environment, and permit preventive actions to be taken.

7. Acknowledgments

The comments of Alan Belward, Stéphane Flasse, and Philippe Martin on earlier drafts of this document are greatly appreciated.

8. References

Butler, D. (editor) (1987) 'From pattern to process: The strategy of the Earth Observing System', Report of the Eos Science Steering Committee, Vol II.

Curran, P. J. (1981) 'Multispectral remote sensing for estimating biomass and productivity', in: Smith (ed.) *Plants and the daylight spectrum*, Academic Press, London, 65-96.

ESA (1992) 'The Medium Resolution Imaging Spectrometer (MERIS)', Draft internal document, European Space Agency, 92 pp.

Estes, J. and M. Consentino (1989) 'Remote sensing of vegetation', *in* M. Rambler, L. Margulis, and R. Fester (eds.) *Global ecology*, Academic Press, New York, 75-111.

Freden, S. and F. Gordon, Jr. (1983) 'Landsat satellites', *in* R. Colwell (ed.) '*Manual of Remote Sensing, Vol. 1* ', American Society of Photogrammetry, Falls Church, 517-570.

Houghton, J. (1986) 'The Physics of Atmospheres', Cambridge University Press, Cambridge, 271 pp.

NASA (1986) 'Data and information system: Report of the EOS Data Panel', *in Earth Observing System, Vol IIa*, NASA Technical Memorandum 87777, Washington, 49 pp.

Kaufman, Y. and D. Tanré (1992) 'Atmospherically resistant vegetation index (ARVI) for EOS-MODIS', *IEEE Transactions on Geoscience and Remote Sensing* 30, 261-270.

Kidwell, K. (1991) 'NOAA Polar Orbiter Data Users Guide', US Department of Commerce, NOAA, Washington.

Neckel, H. and D. Labs (1984) 'The solar radiation between 3300 and 12500 A', *Sol. Phys.*, 90, 205-258.

Norwood, V. and J. Lansing (1983) 'Electro-optical imaging sensors', *in* R. Colwell (ed.), *Manual of Remote Sensing, Vol. 1*, American Society of Photogrammetry, Falls Church, 335-367.

Pagano, T. and J. Young (1992) 'MODIS-N Instrument Status', Santa Barbara Research Center, Document 92-0257-1, Hughes Corporation.

Rees, W. (1990) 'Physical Principles of Remote Sensing', Cambridge University Press, Cambridge, 247 pp.

Robinson, B. and D. DeWitt (1983) 'Electro-optical non-imaging sensors', in R. Colwell (ed.), *Manual of Remote Sensing, Vol. 1*, American Society of Photogrammetry, Falls Church, 293-333.

Rozelot, J.-P. (1973) 'La Couronne Solaire', Doin, Paris, 144 pp.

Sellers, W. (1965) 'Physical Climatology', Chicago University Press, Chicago, 272 pp.

Simonett, D. (1983) 'The development and principles of remote sensing', *in* R. Colwell (ed.), *Manual of Remote Sensing, Vol. 1*, American Society of Photogrammetry, Falls Church, 1-35.

Slater, P. (1983) 'Photographic systems for remote sensing', in R. Colwell, (ed.), *Manual of Remote Sensing, Vol. 1*, American Society of Photogrammetry, Falls Church, 231-291.

Pinty, B. and M. M. Verstraete (1992) 'GEMI: A non-linear index to monitor global vegetation from satellites', *Vegetatio,* 101, 15-20.

Vogt, J. (1992) 'Characterizing the spatio-temporal variability of surface parameters from NOAA AVHRR data: A case for Southern Mali', Ph.D. Thesis, Trier University, 216 pp.

REMOTE SENSING AND THE ESTIMATION OF ECOSYSTEM PARAMETERS AND FUNCTIONS

CAROL A. WESSMAN
Environmental, Population, and Organismic Biology
Cooperative Institute for Research in Environ. Sciences (CIRES)
University of Colorado
Boulder, Colorado 80309-0449

ABSTRACT. Remote sensing provides the synoptic views needed to study ecosystem dynamics occurring at landscape, regional and global scales. The type of biophysical attributes sensible from space will determine how widely ecological extrapolations can be implemented. Monitoring large scale biological processes is contingent on two factors. First, key structural and physiological characteristics of ecosystems must be identified which can be directly linked to underlying processes. Second, the capability to measure these characteristics remotely must exist. Successful estimation of parameters such as absorbed photosynthetically active radiation (APAR), canopy chemistry, and canopy water content will provide insights on changes in seasonal photosynthesis and photosynthate allocation, the susceptibility of plants to disease, rate of litter decomposition, and other changes related to environmental stress. Linkage between small-scale ecological understanding and large-scale dynamics is important and can be accomplished through characterization of reflectance properties at the leaf, canopy, and landscape scales.

1. Introduction

The capability to predict the response of ecosystems to environmental and climate change relies on our ability to understand and model the effective functioning of biotic processes at regional and global scales. Terrestrial carbon dynamics, specifically, have presented significant challenges to global carbon cycle research. The magnitude of the terrestrial carbon pools vary in time and space; and the balance between the principal acting processes, namely carbon assimilation by photosynthesis, respiration rates and carbon turnover time, determine the net exchange between the biosphere and the atmosphere. Since our study of ecosystem functioning has been concentrated at local scales for logistical reasons, we are handicapped by a poor understanding of how these processes scale from the level of our field studies to larger regions. Remote sensing technology provides the temporal and spatial information that can aid our extrapolations of ecosystem functions such as flux rates of carbon dioxide, nutrients, and water, thereby enhancing our understanding of ecosystem dynamics within the global system.

The contribution of remote sensing to ecosystem studies ranges from empirically-based classification and mapping of land cover types to quantitative characterization of radiative transfer and energy balance. Statistical classification of digital imagery is used to describe spatial patterns in land cover types, their location, area, and change over time. Process-level questions require explicit linkages between the ecosystem function under study and the structure of the landscape in

J. Hill and J. Mégier (eds.), Imaging Spectrometry - a Tool for Environmental Observations, 39–56.

space and time. Quantiative remote sensing of parameters that represent such links provides information on dynamics at spatial and temporal scales previously inaccessible to study.

This paper is a review of ecosystem parameters that are currently and potentially retrievable from remote sensing data. Three areas important to large-scale ecology are covered: (1) biophysical processes linking the biosphere to the atmosphere, (2) biogeochemical cycles, and (3) ecosystem dynamics and change over time.

2. Remote Sensing of Ecosystem Structure and Function

2.1 BIOPHYSICAL PROCESSES

The application of remote sensing as a tool for estimating biophysical rates of photosynthesis and transpiration draws from the simple thesis that plant growth is related to the fraction of incident radiation absorbed by the canopy and the dry matter:radiation quotient (an "efficiency" coefficient defining the carbon fixed per radiation intercepted) (Monteith 1972, 1977). Radiation interception properties of plants are strongly influenced by chlorophyll; its unique absorption of energy in the red (R) spectral region relative to the highly reflected near-infrared (NIR) region distinguishes live vegetation from soil and other non-photosynthetic materials. Since observations in the red and near-infrared regions are indicative of factors related to chlorophyll density, which in turn relates to carbon fixation rates, these observations should provide information regarding photosynthetic capacity (Tucker and Sellers 1986). In this context, photosynthetic capacity specifies the upper limit of the photosynthetic rate for a given PAR flux; i.e. the gross photosynthetic rate that occurs under no environmental stress. Rates of transpiration can be derived from this value of photosynthetic capacity since water vapor diffuses out of leaves via the stomatal pores which open for the influx of atmospheric carbon dioxide.

Early field studies investigated the near-linear relationships between spectral reflectance indices based on red and NIR reflectance (e.g., a simple ratio NIR/R or Normalized Difference Vegetation Index (NDVI = (NIR-R)/(NIR+R)) and standard measurements of the canopy properties of biomass, leaf area and photosynthetically active radiation (PAR) absorbed by the canopy (Tucker 1979, Hipps et al. 1983, Asrar et al. 1984, Hatfield et al. 1984). Remote sensing of the amount of leaf surface area available for gas and moisture exchange (described by leaf area per ground area -- leaf area index (LAI)) has been of particular interest to the ecological community. Vegetation indexes (VI) are asymptotic with respect to LAI as the signal saturates (see Asrar et al. 1989, Peterson and Running 1989), but linearity can extend from LAIs of 2 to 6 for crop and grassland canopies (Tucker 1977, Asrar et al. 1984, Ripple 1985) and up to approximately 8 for coniferous forests (Peterson et al. 1987, Running et al. 1989). However, changes in background reflectance, canopy closure, and understory vegetation introduce inconsistencies in prediction capabilities even within the range of the linear response (Baret and Guyot 1991). Ground measurements of canopy transmitted light have gained in importance for rapid characterization of canopy leaf area and architecture (see Norman and Campbell 1991). These measurements greatly enhance capabilities to acquire adequate ground calibrations for satellite measurements (Pierce and Running 1988, Gower and Norman 1991, Lathrop and Pierce 1991).

Strong relationships have been demonstrated between time integrals of satellite-derived VIs and net primary production (NPP) (Goward et al. 1985, Fung et al. 1987), the geography and seasonality of vegetative cover (Justice et al. 1985, Tucker et al. 1985), and simulated photosynthesis and transpiration (Running and Nemani 1988). Theoretical analyses by Sellers (Sellers 1985, 1987) examined the links between spectral vegetation indexes and canopy properties of LAI, absorbed PAR, photosynthetic capacity, and canopy resistance to water vapor efflux. A mechanistic basis for the observed correlations (given a horizontally uniform canopy) was demonstrated with a two-stream approximation model of radiative transfer and simple leaf models of photosynthesis and stomatal resistance. These results suggest that indexes such as the simple ratio and NDVI are indicative of instantaneous biophysical rates of photosynthesis and conductance, but are not reliable estimators for any state (leaf area, biomass) associated with vegetation. Furthermore, they are related to the maximum photosynthetic output of the vegetation; the actual rates being determined by the PAR flux and environmental factors.

Sellers model assumed that all leaves in the canopy had the same light response curve, i.e. all leaves throughout the depth of the canopy responded identically to the flux of PAR. With increasing PAR flux above the canopy, the upper-most leaves would saturate and leaves lower down would remain below saturation, resulting in an increasingly non-linear relationship between photosynthesis and the fraction of absorbed PAR. In reality, photosynthetic capacity changes in parallel with the depth-distribution of PAR (Björkman 1981). By modifying the physiological model to incorporate change in photosynthetic response with depth in the canopy, Sellers et al. (1992) produced a stronger theoretical relationship between canopy biophysical rates (photosynthesis, conductance) and spectral vegetation indexes because the contributions of canopy structure and environmental forcings could be separated from those of leaf physiology and radiation flux.

Conditions still constraining the predictive powers of vegetation indexes include those that affect the photosynthesis/PAR relationship such as environmental stress and different photosynthetic pathways (C3, C4), and conditions that may influence spectral estimates of absorbed PAR such as contributions from background soil and litter reflectance. Biological processes and their respective sensitivity to VIs and environmental variables must be considered for different vegetation types (Bartlett et al. 1990). For example, land cover should be stratified according to ecosystem or biome type before relationships are established between PAR and a vegetation index. Fung et al. (1987) determined global net primary production from NDVI using an empirically-derived scaling factor that essentially accounted for Monteith's conversion efficiency for each biome type. Prince (1991) has cited efficiency factors converting annual APAR energy in megajoules (MJ) to NPP in grams for different biome types.

The relationships between NDVI, absorbed PAR and photosynthetic capacity are highly linear in spatially heterogeneous (but physiologically uniform) canopies (Asrar et al. 1992) and when background reflectance (soil, rocks, litter) is minimal (Sellers 1987, Sellers et al. 1992). However, it is questionable whether a measurement in two spectral bands can provide an unambiguous measure of vegetation when background reflectance is a significant component of the total surface reflectance. Confounding influences from background variation, atmospheric attenuation and off-nadir viewing cannot all be accounted for using a two-band ratio (Choudhury 1987, Huete and Jackson 1988, Baret and Guyot 1991, Goward et al. 1991, Middleton 1991). Modifications to NDVI have been suggested to account for first-order soil-vegetation interactions (i.e. soil brightness effects) (Huete 1988, Baret et al. 1989). However, secondary soil variations due to soil optical properties can only be addressed, using multiple spectral bands, through (i) factor-analytic inversion models which allow composite plant-soil mixtures to be separated into component spectra (Huete 1986, Huete and Escadafal 1991), or (ii) selection of spectral regions where soils reflectance varies linearly.

Data from high spectral resolution instruments may yield more information on biophysical and biochemical processes. Variables of spectral shape such as width, depth, skewness, and symmetry of absorption features are more directly indicative of biochemical state and canopy physiology than broad-band measurements made with current operational sensors (see Wessman 1990). Studies relating chlorophyll content with the location of the inflection point of the long wavelength edge of the feature have met with varied success (Schutt et al. 1984, Rock et al. 1988, Milton and Mouat 1989, Curran et al. 1990, Miller et al. 1991); but it appears that the wavelength of the inflection point in the red-edge region is less dependent than broad band VIs on soil optical properties, atmospheric effects and irradiance conditions (Baret et al. 1992). Pigments other than chlorophyll have been identified that are more directly indicative of actual photosynthetic rates (as opposed to photosynthetic capacity) (Demmig-Adams 1990). Wavelength-specific absorption differences among the variety of photosynthetic pigments may permit quantification of their concentrations. Light-induced changes in a xanthophyll pigment closely linked to changes in photosynthetic activity have been related to spectral changes in green reflectance at 531 nm (Gamon et al. 1990, 1992). Multiple spectral bands will permit separation of scene components through spectral mixture analysis (Adams et al. 1989, Smith et al. 1990a, 1990b, Ustin et al. 1992) and derivative spectroscopy (Wessman 1990, Demetriades-Shah et al. 1990). Second derivatives of high spectral resolution reflectance data in the visible and near infrared regions appear to be strongly related to absorbed PAR and relatively insensitive to the reflectance of non-photosynthetically active materials such as litter and soils (Hall et al. 1990). However, derivative techniques are likely to be problematic due to their sensitivity to noise.

2.2 Biogeochemical Cycles

Remote sensing of photosynthesis, as described above, can provide substantial information for modeling aboveground carbon pools and their dynamic interaction with the global system. Additional properties of carbon and other elemental cycles must also be quantified for better understanding of regional ecosystem dynamics and global atmosphere-biosphere exchanges.

2.2.1 Canopy Chemistry Some of the terms used to calculate carbon turnover time, nutrient availability and soil respiration may be provided by new techniques in imaging spectrometry that offer the possibility for determining the chemical composition of vegetation canopies (Waring et al. 1986, Peterson et al. 1988). These ecosystem processes are intimately linked with rates of decomposition, which are strongly regulated by the chemical quality of the organic matter (Melillo et al. 1982, Meentemeyer and Berg 1986, Aber et al. 1990a). Remotely sensed estimations of lignin (the most recalcitrant material in litter), canopy nitrogen, or other constituents related to C:N ratios may serve to constrain decomposition submodels in ecosystem simulations, thus stabilizing model inversions (Aber et al. 1990a, Schimel et al. 1991).

Analytical spectroscopy of organic mixtures in the shortwave infrared region (0.7 to 2.5 μm) is a well established technique for biochemical analyses in agricultural forage assessment and the food industry (e.g. Barton and Burdick 1979, Shenk et al. 1981, Wetzel 1983, Marten et al. 1985). Reflectance spectra of organic mixtures in this region consist of harmonic overtones and combinations due mainly to stretching and bending vibrations of strong molecular bonds between atoms of low weight, primarily C-H, O-H, and N-H (Wetzel 1983, Weyer 1985). Changes in the concentrations of these bonds in an organic mixture will induce shifts in amplitude and frequency within the region of 0.7 to 2.5 μm. The nature of these absorptions within organic mixtures (such as a leaf or canopy) are weak and complex since they consist of overlapping overtone and combination bands. The origins of the observed vibrations are limited and they are all associated with primary constituents of vegetation. Knowledge of absorption characteristics of each of the major leaf constituents (e.g., cellulose, starch and protein) may permit remote assessment of canopy level concentrations if high spectral resolution reflectance information is acquired.

Spectroscopy applications to analyses of foliar biochemistry of native species has strengthened sampling strategies for ecosystem studies; the rapidity of the method enables processing of large numbers of samples (Wessman et al. 1988a, McLellan et al. 1992). Application of these techniques to imaging spectrometer data over temperate forests yielded strong relationships with ground measurements of canopy lignin concentrations that in turn allowed the mapping of nitrogen mineralization for the study site (Wessman et al. 1988b, 1989). Significant correlations have also been noted between imaging spectrometer data and canopy nitrogen content across a range of coniferous forest stands in Oregon (Peterson and Running 1989) and fertilization plots of Douglas-fir (*Pseudotsuga menzeisii*) in New Mexico (Swanberg and Matson 1987). Gao and Goetz (1990) demonstrated that canopy water content can be retrieved, using spectral curve fitting techniques, from canopy reflectance acquired by imaging spectrometers. Further studies on the question of remote sensing of canopy chemistry are currently being pursued (Goetz et al. 1990, Martin and Aber 1990, Curran et al. 1992).

The application of analytical spectroscopy to remotely sensed data is still early in its development. Detection of minor absorption characteristics will rely on high spectral resolution sampling, sufficient characterization of atmospheric conditions, and high signal-to-noise sensors. Certainly, integrating spectrometry studies at the leaf, canopy and landscape will enhance our understanding of vegetation optical properties and the transfer of spectral information with increasing scale and landscape complexity. Importantly, these investigations into the question of canopy chemistry have led us to consider the use of remote sensing in extrapolation models of biogeochemistry. Here we must utilize the concept of surrogates since belowground processes so significant to biogeochemical cycling are invisible to the sensor (Wessman 1991). This amplifies our need to better understand how properties such as plant physiology and biochemistry reflect the balance between factors limiting to the system (Aber et al. 1990, Schimel et al. 1991).

2.2.2 Trace Gas Exchange Attempts to develop regional and global budgets of biogenically released trace gases have been made (Khalil and Rasmussen 1990, Fung et al. 1991), but source/sink strengths and the processes which control flux rates remain areas of great uncertainty. This is largely due to the complexity of the biological, physical, and chemical systems that are involved and the difficulty of measuring exchange fluxes in the field (see Andreae and Schimel 1989). Quantification of flux processes at large-scales requires an understanding of the variability within and among vegetation types as well as variation with time at any given site (e.g. Sebacher et al. 1986, Bartlett et al. 1989, Desjardins et al. 1989, Harris et al. 1988a, Mosier et al. 1991).

Regional and global trace gas budgets have been largely based on area-weighted extrapolation of *in situ* flux measurements considered representative of a given land category. In other words, the region is first stratified into areas considered homogeneous and likely to have lower within-class variance in flux rates. Matthews and Fung (1987) demonstrated the importance of wetlands to the global methane budget using a stratification of global wetlands based on environmental characteristics governing methane emissions. The original wetland data base at 1° resolution was developed from global digital data of vegetation, innundation characteristics and soil properties.

The five derived strata were then assigned typical methane fluxes integrated over the methane-production season as defined by latitude.

While useful for first-cut global and regional estimates and for targeting major contributors, area-weighted field estimates can incur significant error from the high spatial and temporal variability in point measurements and the subsequent process of aggregation. An assumption of "representativeness" for the land cover type may fail to account for sources of spatial variation. Matson and Vitousek (1987) suggest that the high variability in tropical forests will result in a parallel variability in trace gas fluxes; i.e., there is no representative site for the general category of "tropical forest". Gradients of factors (e.g. soil fertility) that control both fluxes and ecosystem properties and processes in tropical forests may be more useful for extrapolating fluxes and for calculating budgets of nitrous oxide and other trace gases. In fact, stratification of tropical forest types that reflect soil characteristics reveal consistently higher nitrous oxide fluxes in forests on acid clay soils (terra firme) than other forest types within a Brazilian study area, even though within-type variation is significant (Matson et al. 1990). Estimates of nitrous oxide emission rates in moist tropical forests were extrapolated by areal estimates of forests stratified by soil fertility to derive a tropical contribution to the global nitrous oxide source (Matson and Vitousek 1990). The total flux, calculated as 2.4 Tg/y, was considerably lower than that based on a general lumping of tropical and subtropical forests and woodlands (7.4 Tg/yr), but remaining very significant in the global budget. Combination of these types of data with remote sensing-based areal extrapolations may be particularly useful in refining regional flux measurements or locating "hot spots" worthy of further investigation.

Single physical characteristics of the surface in wetlands (water and soil depth, soil temperature) do not appear to be quantitatively associated with the variability of methane flux rates within a single regional wetland system (Bartlett et al. 1989) nor have quantitative relationships been found to link wetlands in different physiographic and climatic regimes (Matthews and Fung 1987). In the absence of parameters which offer predictive relationships with fluxes, vegetation community distribution can provide a relevant stratification for area-weighted regional flux inventory. An emission inventory of the Everglades was substantially improved by using Landsat Thematic Mapper data to direct *in situ* sampling efforts in important habitats and by providing a means for calculating area-weighted mean fluxes for the system as a whole (Bartlett et al. 1989). Reiners et al. (1989) followed a similar strategy to predict nitrogen mineralization rates over sagebrush steppe landscape using an ecosystem simulation model parameterized for ecosystem types defined by Landsat Thematic Mapper data.

Aircraft flux measurements provide independent data for testing regional and global extrapolations of trace gas fluxes (Desjardins and MacPherson 1989). Regional tower and aircraft flux measurements which integrate gaseous exchange over large areas can provide impetus to mechanistically relating ground-based measurements to regional scales. The Amazon Boundary Layer Experiment (ABLE 2A and 2B) was designed to quantitatively characterize the spatial and temporal variability of trace gases and aerosols over the Brazilian Amazon during wet and dry season conditions (Harriss et al. 1988b, 1990). Continuous ground-based sampling served to characterize temporal variability with aircraft and satellite observations providing spatial sampling capabilities. The success of these experiments suggests the use of airborne eddy correlation flux surveys in future research programs to select representative ground sites for continuous tower measurements.

The great uncertainty in estimating global trace gas budgets arises from the heterogeneity of source distributions (Harriss et al. 1988a). Difficulties in producing consistent estimates result from i) high variability in a single habitat and (ii) uncertainty in geographical extent and seasonal variability in extent of wetland environments. Despite an extensive field sampling program, large sampling errors surrounding ground-based estimates of trace gas emissions occur if the stratification of the area does not consider the controlling factors. Combined remote sensing of landscape biophysical and ecological characteristics and trace gas measurements are needed to generalize to regional and global flux models. Approaches to extrapolations of local to regional measurements should include process-level modeling, stratified sampling by the most relevant landscape units, and synthesis with the aid of geographic information systems (Harriss 1989, Harriss et al. 1988a).

2.3 TEMPORAL DYNAMICS

Regional biogeochemical flux estimates and atmosphere-biosphere interactions are significantly influenced by the type and successional stage of ecosystems within a landscape. The rapid rate of land-use changes occuring in many parts of the world contribute directly to perturbations in those dynamics. Successional patterns reflect local variations in resource availability and linked carbon and nitrogen cycles. Effects of climate change or human disturbance will, in turn, be modified by the stage and pattern of succession within the landscape (Pastor and Post 1986). Large-scale spatial heterogeniety and long-term patterns of successional dynamics have prevented past extrapolations of ecosystem research from local to regional scales (Hall et al. 1991). Remote sensing and ground-based evaluations provide the most promising tools for compiling geographical information on the stage and condition of ecosystems over time.

The ability to detect long-term change in ecosystems requires that we are able to detect conditions in the static situation, e.g. health, structure, and seasonal productivity (Hobbs 1990). We can consider several remotely sensible variables that, when monitored over time, will lead us to deeper insights on ecosystem dynamics. Seasonally integrated vegetation indexes and canopy chemistry are variables that will be affected by and respond to environmental change. Information on canopy and landscape structure and their change over time can be derived from studies of image texture (e.g. Otterman 1981, Franklin and Peddle 1990, Briggs and Nellis 1991). Image spatial variance has been used to quantify the number and spacing of forest trees (Franklin et al. 1986, Li and Strahler 1986, 1988) which, when monitored over time, could be used to track forest stand dynamics such as gap formation and regrowth. Spectral mixture analysis also provides a means to estimate the spatial cover of vegetation in a sparse community, independent of the spectral characteristic of the substrate (Ustin et al. 1986, Smith et al. 1990a, 1990b). Rapid dynamics occurring in the spatial pattern of Minnesota boreal forest ecological states and their transition rates were tracked using a ten-year time-series of Landsat Multispectral Scanner data (Hall et al. 1991).

Reliable information on global land cover/land use maps is a top priority for global change research (IGBP 1990). Existing maps of global vegetation are generally compiled from disparate sources at varying scales and contain many inconsistencies (Townshend et al. 1991). Data from NOAA's Advanced Very High Resolution Radiometer (AVHRR) has been applied to monitoring regional land use change (Tucker et al. 1984, Justice and Hieurnaux 1986) and classifying land cover at continental scales (Goward et al. 1985, Tucker et al. 1985, Townshend et al. 1987), suggesting that global land cover classification by remote sensing is a real possibility (Townshend et al. 1991). Operational provision of global land cover data will require systems with consistent internal (instrument) calibration, and appropriate temporal and spatial characteristics.

3. Linking Remote Sensing and Simulation Models

Progress in addressing questions of interactions among ecosystems and between ecosystems and the atmosphere is achieved by (1) direct observation via ground and satellite platforms and (2) linking observations to process simulation models for calibration and validation of model runs. Remote observations will be critical for describing the current state of the earth, monitoring near-term change and estimating initial conditions for predictive modeling; models extend observations for long-term prognoses. The use of remote sensing in the extrapolation of process models will require rethinking of traditional modeling approaches and the use of inverse modeling techniques. This, in turn, will be directed by the understanding of how remote measurements of plant physiology and biochemistry can reflect the balance between above and belowground limiting factors (Schimel et al. 1991). Identification of important phenomena sensible from space, as discussed in the above sections and summarized in table 1, will assist in defining the role of remote sensing. Ecosystem simulation models that recognize the relationship between fluxes and controlling factors will be best positioned to link to spatial data bases provided by remote sensing.

Relatively few examples exist where remote observations are incorporated as model drivers for flux rate calculations. Fung et al. (1987) integrated global NDVI values with field data on soil respiration and climate data to obtain global distributions of monthly atmosphere-biosphere exchange of CO_2 exchange. Satellite-derived LAI has been used to drive simulations of regional evapotranspiration and photosynthesis in a forest ecosystem (Running et al. 1989). NASA has conducted two intensive field and remote sensing experiments in forests with the intention of interfacing such data with forest ecosystem models (e.g. Ranson and Smith 1990, Willams and Walthall 1990, Johnson and Peterson 1991). The goal of the Oregon Transect Ecosystem Research (OTTER) project is to examine canopy and landscape characteristics along climate and fertility gradients that may be indicators of ecosystem processes and physiological dynamics and to integrate remotely sensed observations of these gradients with the Forest BGC model. The Forest Ecosystem Dynamics (FED) project in Maine, supports a hierarchy of submodels describing forest growth and development, soil processes, radiative transfer, and establishing functional linkages between them. Multi-scale remote sensing acquisitions will serve as inputs to the models, as well as provide, through independent remote observations, validation for the sub-models and integrated model.

Plant Ps/Respiration	Carbon Storage Vegetation and Soil	Decomposition (Soil respiration)	Trace Gases
Photosynthetic capacity	Biomass	Litter input	Land cover type
Leaf area index	Land cover type	Foliar chemistry	Photosynthesis
Greenness	Vegetation height		
APAR	Vegetation spatial distribution		

TABLE 1. Ecosystem parameters sensible from space. Attributes of landscapes demonstrated to be sensible from space or, from limited studies, show strong potential for direct observation.

The dominant use of remote sensing with respect to simulation models has been as an independent validation tool. Simulation models may be useful for developing methods to extrapolate across scales because they can test the implications of various scaling rules, but ground validations for regional and global simulations are a fundamental problem. Regional geographic-based simulations have provided significant insight into storage and fluxes of C and N within grassland ecosystems (Burke et al. 1990, 1991), and seasonal patterns of net primary production in South America (Raich et al. 1991). In each of these studies, remote sensing has the potential to provide temporal driving variables and to validate spatial and temporal predictions.

4. Concluding Remarks

Remote sensing provides the synoptic views needed to study ecosystem dynamics occurring at landscape, regional and global scales. The type of biophysical attributes sensible from space will determine how widely ecological extrapolations can be implemented. Monitoring large scale biological processes is contingent on two factors. First, key structural and physiological characteristics of ecosystems must be identified which can be directly linked to underlying processes. Second, the capability to measure these characteristics remotely must exist. Reflectance data from vegetation are indicative of the biophysical processes of photosynthesis and transpiration, and have been shown to correlate with biomass and productivity. Imaging spectrometry offers improved characterization of reflectance properties, leading to spectral models that aid in better defining landscape complexity. Successful estimation of parameters such as absorbed photosynthetically active radiation (APAR), canopy chemistry, and canopy water content will provide insights on changes in seasonal photosynthesis and photosynthate allocation, the susceptibility of plants to disease, rate of litter decomposition, and other changes related to environmental stress. Coupling these data to ecosystem simulation models and geographic information systems will add substantially to the scaling of ecological information from local to global systems.

5. References

Aber, J.D., Melillo, J.M. and McClaugherty, C.A. (1990a) 'Predicting long-term patterns of mass loss, nitrogen dynamics, and soil organic matter formation from initial fine litter chemistry in temperate forest ecosystems', *Canadian Journal of Botany*, 68(10), 2201-2208.

Aber, J.D., Wessman, C.A., Peterson, D.L., Melillo, J.M. and Fownes, J.H. (1990b) 'Remote sensing of litter and soil organic matter decomposition in forest ecosystems', in R. J. Hobbs and H. A. Mooney (eds), *Remote Sensing of Biosphere Functioning*, Springer-Verlag, New York, 87-103.

Adams, J.B., Smith, M.O. and Gillespie, A.R. (1989) 'Simple models for complex natural surfaces: a strategy for the hyperspectral era of remote sensing', *Proc. IGARSS'89 12th Canadian Symposium on Remote Sensing*, 1, 16-21.

Andreae, M.O. and Schimel, D.S., ed. (1989) *Exchange of Trace Gases between Terrestrial Ecosystems and the Atmosphere*, Wiley and Sons, Chichester.

Asrar, G., Fuchs, M., Kanemasu, E.T. and Hatfield, J.L. (1984) 'Estimating absorbed photosynthetic radiation and leaf area index from spectral reflectance in wheat', *Agronomy Journal*, 76, 300-306.

Asrar, G., Myneni, R.B. and Choudhury, B.J. (1992) 'Spatial heterogeneity in vegetation canopies and remote sensing of absorbed photosynthetically active radiation: A modeling study', *Remote Sensing of Environment*, 41, 85-103.

Asrar, G., Myneni, R.B., and Kanemasu, E.T. (1989) 'Estimation of plant-canopy attributes from spectral reflectance measurements', in G. Asrar (ed.), *Theory and Applications of Optical Remote Sensing*, Wiley Pub, New York, pp. 252-296.

Baret, F. and Guyot, G. (1991) 'Potentials and limits of vegetation indices for LAI and APAR assessment', *Remote Sensing of Environment*, 35, 161-173.

Baret, F., Guyot, G., and Major, D. (1989) 'TSAVI: a vegetation index which minimizes soil brightness effects on LAI and APAR estimation', in *12th Canadian Symp. on Remote Sesning and IGARSS'90*, Vancouver, Canada, 4 pp.

Baret, F., Jacquemoud, S., Guyot, G. and Leprieur, C. (1992) 'Modeled analysis of the biophysical nature of spectral shifts and comparison with information content of broad bands', *Remote Sensing of Environment*, 41, 133-142.

Bartlett, D.S., Bartlett, K.B., Hartman, J.M., Harriss, R.C., Sebacher, D.I., Pelletier-Travis, R., Dow, D.D. and Brannon, D.P. (1989) 'Methane emissions from the Florida Everglades: patterns of variability in a regional wetland ecosystem', *Global Biogeochemical Cycles*, 3(4), 363-374.

Bartlett, D.S., Whiting, G.J. and Hartman, J.M. (1990) 'Use of vegetation indices to estimate intercepted solar radiation and net carbon dioxide exchange of a grass canopy', *Remote Sensing of Environment*, 30, 115-128.

Barton, F.E. and Burdick, D. (1979) 'Preliminary study on the analysis of forages with a fliter-type near-infrared reflectance spectrometer', *Journal of Agriculture and Food Chemistry*, 27(6), 1248-1252.

Barton, F.E., Himmelsbach, D.S., Duckworth, J.H. and Smith, M.J. (1992) 'Two-dimensional vibration spectroscopy: correlation of mid- and near-infrared regions', *Applied Spectroscopy*, 46, 420-429.

Björkman, O. (1981) 'Responses to different quantum flux densities', in O.L. Lange, P.S. Nobel, C.B. Osmond, H. Ziegler (eds.), *Encyclopedia of Plant Physiology, Vol. 12A. Plant Physiological Ecology*, Springer-Verlag, Berlin, pp 57-107.

Briggs, J.M. and Nellis, M.D. (1991) 'Seasonal variation of heterogeneity in the tallgrass prairie: A quantitative measure using remote sensing', *Photogrammetric Engineering and Remote Sensing*, 57(4), 407-411.

Burke, I.C., Kittel, T.G.F., Lauenroth, W.K., Snook, P., Yonker, C.M. and Parton, W.J. (1991) 'Regional analysis of the Central Great Plains', *BioScience*, 41(10), 685-692.

Burke, I.C., Schimel, D.S., Yonker, C.M., Parton, W.J., Joyce, L.A. and Lauenroth, W.K. (1990) 'Regional modeling of grassland biogeochemistry using GIS', *Landscape Ecology*, 4(1), 45-54.

Choudhury, B.J. (1987) 'Relationships between vegetation indices, radiation absorption, and net photosynthesis evaluated by a sensitivity analysis', *Remote Sensing of Environment*, 22, 209-233.

Curran, P.J., Dungan, J.L. and Gholz, H.L. (1990) 'Exploring the relationship between reflectance red edge and chlorophyll content in slash pine', *Tree Physiology*, 7, 33-48.

Curran, P.J., Dungan, J.L., Macler, B.A., Plummer, S.E. and Peterson, D.L. (1992) 'Reflectance spectroscopy of fresh whole leaves for the estimation of chemical composition', *Remote Sensing of Environment*, 39(2), 153-166.

Demetriades-Shah, T.H., Steven, M.D. and Clark, J.A. (1990) 'High resolution derivative spectra in remote sensing', *Remote Sensing of Environment*, 33, 55-64.

Demmig-Adams, B. (1990) 'Carotenoids and photoprotection in plants: A role for the xanthophyll zeaxanthin', *Reviews on Biochimica et Biophysica Acta*, 1020, 1-24.

Desjardins, R.L. and MacPherson, J.I. (1989) 'Aircraft-based measurements of trace gas fluxes', in M. O. Andreae and D. S. Schimel (eds), *Exchange of Trace Gases between Terrestrial Ecosystems and the Atmosphere*, Wiley and Sons, Chichestor, 135-152.

Desjardins, R.L., MacPherson, J.I., Schuepp, P.H. and Karanja, F. (1989) 'An evaluation of aircraft flux measurements of CO2, water vapor and sensible heat', *Boundary-Layer Meteorology*, 47, 55-70.

Franklin, J., Logan, T.L., Woodcock, C.E. and Strahler, A.H. (1986) 'Coniferous forest classification and inventory using Landsat and digital terrain data', *IEEE Transactions in Geoscience and Remote Sensing*, GE-24(1), 139-149.

Franklin, S.E. and Peddle, D.R. (1990) 'Classification of SPOT HRV imagery and texture features', *International Journal of Remote Sensing*, 11(3), 551-556.

Fung, I., John, J., Matthews, E., Prather, M., Steele, L.P. and Fraser, P.J. (1991) 'Three-dimensional model synthesis of the global methane cycle', *Journal of Geophysical Research*, 96, 13033-13065.

Fung, I.Y., Tucker, C.J. and Prentice, K.C. (1987) 'Application of Advanced Very High Resolution Radiometer vegetation index to study atmosphere-biosphere exchange of CO2', *Journal of Geophysical Research*, 92(D3), 2999-3015.

Gamon, J.A., et al. (1990) 'Remote sensing of the xanthophyll cycle and chlorophyll fluorescence in sunflower leaves and canopies', *Oecologia*, 85, 1-7.

Gamon, J.A., Penuelas, J. and Field, C.P. (1992) 'A narrow-waveband spectral index that tracks diurnal changes in photosynthetic efficiency', *Remote Sensing of Environment*, 41, 35-44.

Gao, B.C. and Goetz, A.F.H. (1990) 'Column atmospheric water vapor and vegetation liquid water retrievals from airborne imaging spectrometer data', *Journal of Geophysical Research*, 95(D4), 3549-3564.

Goetz, A.F.H., Gao, B.C., Wessman, C.A. and Bowman, W.D. (1990) 'Estimation of biochemical constiuents from fresh, green leaves by spectrum matching techniques', *Proceedings International Geocience and Remote Sensing Symposium*, 2, 971-974.

Goward, S.N., Markham, B., Dye, D.G., Dulaney, W. and Yang, J. (1991) 'Normalized difference vegetation index measurements from the Advanced Very High Resolution Radiometer', *Remote Sensing of Environment*, 35, 257-277.

Goward, S.N., Tucker, C.J. and Dye, D.G. (1985) 'North american vegetation patterns observed with the NOAA-7 advanced very high resolution radiometer', *Vegetatio*, 64, 3-14.

Gower, S.T. and Norman, J.M. (1991) 'Rapid estimation of leaf area index in conifer and broad-leaf plantations', *Ecology*, 72(5), 1896-1900.

Hall, F.G., Botkin, D.B., Strebel, D.E., Woods, K.D. and Goetz, S.J. (1991) 'Large-scale patterns of forest succession as determined by remote sensing', *Ecology*, 72(2), 628-640.

Hall, F.G., Huemmrich, K.F. and Goward, S.N. (1990) 'Use of narrow-band spectra to estimate the fraction of absorbed photosynthetically active radiation', *Remote Sensing of Environment*, 32(1), 47-54.

Harriss, R.C. (1989) 'Experimental design for studying atmosphere-biosphere interactions', in M. O. Andreae and D. S. Schimel (eds), *Exchange of Trace Gases between Terrestrial Ecosystems and the Atmosphere*, Wiley and Sons, Chichester, 291-301.

Harriss, R.C., Garstang, M., Wofsy, S.C., Beck, S.M., Bendura, R.J., Coelho, J.R.B., Drewry, J.W., Hoell, J.M., Jr., Matson, P.A., McNeal, R.J., et al. (1990) 'The Amazon Boundary Layer Experiment: Wet Season 1987', *Journal of Geophysical Research*, 95(D10), 16,721-16,736.

Harriss, R.C., Sebacher, C.D., Bartlett, K.B., Bartlett, D.S. and Crill, P.M. (1988a) 'Sources of atmospheric methane in the South Florida environment', *Global Biogeochemical Cycles*, 2, 231-243.

Harriss, R.C., Wofsy, S.C., Garstang, M., Browell, E.V., Molion, L.C.B., NcNeal, R.J., Hoell, J.M., Jr., Bendura, R.J., Beck, S.M., Navarro, R.L., et al. (1988b) 'The Amazon boundary layer experiment (ABLE 2A): Dry season 1985', *Journal Article of Geophysical Research*, 93(D2), 1477-1486.

Hatfield, J.L., Kanemasu, E.T., Asrar, G., Jackson, R.D., Pinter, P.J., Jr., Reginato, R.J. and Idso, S.B. (1984) 'Leaf area estimates from spectral reflectance measurements over various planting dates of wheat', *International Journal of Remote Sensing*, 46, 651-656.

Hipps, L.E., Asrar, G., and Kanemasu, E.T. (1983) 'Assessing the interception of photosynthetically active radiation in winter wheat', *Agricultural Meteorology*, 28, 253-259.

Hobbs, R.J. (1990) 'Remote sensing of spatial and temporal dynamics of vegetation', in R. Hobbs and H. Mooney (eds), *Remote Sensing of Biosphere Functioning*, Springer-Verlag, New York, 203-219.

Huete, A.R. (1986) 'Separation of soil-plant spectral mixtures by factor analysis', *Remote Sensing of Environment*, 19, 237-251.

Huete, A.R. (1988) 'A soil-adjusted vegetation index (SAVI)', *Remote Sensing of Environment*, 25, 295-309.

Huete, A.R. and Escadafel, R. (1991) 'Assessment of biophysical soil properties through spectral decomposition techniques', *Remote Sensing of Environment*, 35, 149-159.

Huete, A.R. and Jackson, R.D. (1988) 'Soil and atmosphere influences on the spectra of partial canopies', *Remote Sensing of Environment*, 25, 89-105.

IGBP (International Geosphere-Biosphere Programme) (1990) *The International Geosphere-Biosphere Programme: A Study of Global Change*, Report No. 12, The Initial Core Projects, Stockholm: IGBP Secretariat.

Johnson, L.F. and Peterson, D.L. (1991) 'AVIRIS observations of forest ecosystems along the Oregon Transect', *Proc. 3rd Airborne Visible/Infrared Imaging Spectrometer (AVIRIS) Workshop*, JPL Pub. 91-28, 190-196.

Justice, C.O. and Hieurnaux, P. (1986) 'Monitoring the grasslands of the Sahel using NOAA/AVHRR data: Niger 1983', *International Journal of Remote Sensing*, 7, 1475-1498.

Justice, C.O., Townshend, J.R.G., Holben, B.N. and Tucker, C.J. (1985) 'Analysis of the phenology of global vegetation using meteorological satellite data', *International Journal of Remote Sensing*, 6(8), 1271-1318.

Khalil, M.A.K. and Rasmussen, R.A. (1990) 'Atmospheric methane: recent global trends', *Environmental Science and Technology*, 24, 549-553.

Lathrop Jr., R.G. and Pierce, L.L. (1991) 'Ground-based canopy transmittance and satellite remotely sensed measurements for estimation of coniferous forest canopy structure', *Remote Sensing of Environment*, 36, 179-188.

Li, X. and Strahler, A.H. (1986) 'Geometric-optical bidirectional reflectance modeling of a coniferous forest canopy', *IEEE Transactions in Geoscience and Remote Sensing*, GE-24, 906-919.

Li, X. and Strahler, A.H. (1988) 'Modeling the gap probability of a discontinuous vegetation canopy', *IEEE Transactions in Geoscience and Remote Sensing*, GE-26, 161-170.

Marten, G.C., Shenk, J.S. and Barton, F.E. (1985) 'Near infrared reflectance spectroscopy (NIRS): analysis of forage quality', *Agric. Handbook No. 643*, 1-95.

Martin, M.E. and Aber, J.D. (1990) 'Effects of moisture content and chemical composition on the near infrared spectra of forest foliage', 171-177 in *Proceedings, Imaging Spectroscopy of the Terrestrial Environment, SPIE.* vol.1298.

Matson, P.A. and Vitousek, P.M. (1987) 'Cross-system comparisons of soil nitrogen transformations and nitrous oxide flux in tropical forest ecosystems', *Global Biogeochemical Cycles*, 1, 163-170.

Matson, P.A. and Vitousek, P.M. (1990) 'Ecosystem approach to a global nitrous oxide budget', *BioScience*, 40(9), 667-672.

Matson, P.A., Vitousek, P.M., Livingston, G.P. and Swanberg, N.A. (1990) 'Sources of variation in nitrous oxide flux from Amazonian ecosystems', *Journal of Geophysical Research*, 95(D10), 16,789-16,798.

Matthews, E. and Fung, I. (1987) 'Methane emission from natural wetlands: Global distribution, area, and environmental characteristics of sources', *Global Biogeochemical Cycles*, 1(1), 61-86.

McLellan, T., Martin, M.E., Aber, J.D., Melillo, J.M., Nadelhoffer, K.J. and Dewey, B. (1991) 'Comparison of wet chemistry and near infrared reflectance measurements of carbon-fraction chemistry and nitrogen concentration of forest foliage', *Canadian Journal of Forest Research*, 21, 1689-1693.

Meentemeyer, V. and Berg, B. (1986) 'Regional variation in rate of mass loss of Pinus sylvestris needle in Swedish pine forests as influenced by climate and litter quality', *Scandinavian Journal of Forest Research*, 1, 167-180.

Melillo, J.M., Aber, J.D. and Muratore, J.F. (1982) 'Nitrogen and lignin control of hardwood leaf litter decomposition dynamics', *Ecology*, 63(3), 621-626.

Middleton, E.M. (1991) 'Solar zenith angle effects on vegetation indices in tallgrass prairie', *Remote Sensing of Environment*, 38,45-62.

Miller, J.R., Wu, J. and Boyer, M.G. (1991) 'Seasonal patterns in leaf reflectance red edge characteristics', *International Journal of Remote Sensing*, 12(7), 1509-1524.

Milton, N.M. and Mouat, D.A. (1989) 'Remote sensing of vegetation responses to natural and cultural environmental conditions', *Photogrammetric Engineering and Remote Sensing*, 55, 1167-1174.

Monteith, J.L. (1972) 'Solar radiation and productivity in tropical ecosystems', *Journal of Applied Ecology*, 9, 747-766.

Monteith, J.L. (1977) 'Climate and the efficiency of crop production in Britain', *Philosophical Transactions of the Royal Society, London, Series*, B281, 277-294.

Mosier, A., Schimel, D., Valentine, D., Bronson, K. and Parton, W. (1991) 'Methane and nitrous oxide fluxes in native, fertilized and cultivated grasslands', *Nature*, 350(6316), 330-332.

Norman, J.M. and Campbell, G.S. (1991) 'Canopy structure', in R.W. Pearcy, J.R. Ehleringer, H.A. Mooney, P.W. Rundel (eds.), *Plant Physiological Ecology: Field Methods and Instrumentation*, Chapman and Hall Publishers, London, pp 301-325.

Otterman, J. (1981) 'Satellite and field studies of man's impact on the surface in arid regions', *Tellus*, 33, 68-77.

Pastor, J. and Post, W.M. (1986) 'Influence of climate, soil moisture, and succession on forest carbon and nitrogen cycles', *Biogeochemistry*, 2, 3-27.

Peterson, D.L. and Running, S.W. (1989) 'Applications in forest science and management', in G. Asrar (ed.), *Theory and Applications of Optical Remote Sensing*, Wiley Pub, New York, pp. 429-473.

Peterson, D.L., Aber, J.D., Matson, P.A., Card, D.H., Swanberg, N., Wessman, C. and Spanner, M. (1988) 'Remote sensing of forest canopy and leaf biochemical contents', *Remote Sensing of Environment*, 24, 85-108.

Peterson, D.L., Spanner, M.A., Running, S.W. and Teuber, K.B. (1987) 'Relationship of Thematic Mapper Simulator data to leaf area index of temperate coniferous forest', *Remote Sensing of Environment*, 22, 323-341.

Pierce, L.L. and Running, S.W. (1988) 'Rapid estimation of coniferous forest leaf area index using a portable integrating radiometer', *Ecology*, 69(6), 1762-1767.

Prince, S.D. (1991) 'A model of regional primary production for use with coarse resolution satellite data', *International Journal of Remote Sensing*, 12, 1313-1330.

Raich, J.W., Rastetter, E.B., Melillo, J.M., Kicklighter, D.W., Steudler, P.A., Peterson, B.J., Grace, A.L., Moore III, B. and Vorosmarty, C.J. (1991) 'Potential net primary productivity in South America: Application of a global model', *Ecological Applications*, 1(4), 399-429.

Ranson, K.J. and Smith, J.A. (1990) 'Airborne SAR experiment for forest ecosystems research: Maine 1989 experiment', *Proc. 10th Annual International Geoscience and Remote Sensing Symposium*, University of Maryland, Maryland, 1, 861-864.

Reiners, W.A., Strong, L.L., Matson, P.A., Burke, I.C. and Ojima, D.S. (1989) 'Estimating biogeochemical fluxes across sagebrush-steppe landscapes with Thematic Mapper imagery', *Remote Sensing of Environment*, 28, 121-129.

Ripple, W.J. (1985) 'Landsat Thematic Mapper bands for characterizing fescue grass vegetation', *International Journal of Remote Sensing*, 6, 1373-1384.

Rock, B.N., Hoshizaki, T. and Miller, J.R. (1988) 'Comparison of in situ and airborne spectral measurements of the blue shift associated with forest decline', *Remote Sensing of Environment*, 24, 109-127.

Running, S.W. and Nemani, R.R. (1988) 'Relating seasonal patterns of the AVHRR vegetation index to simulated photosynthesis and transpiration of forests in different climates', *Remote Sensing of Environment*, 24, 347-367.

Running, S.W., Nemani, R.R., Peterson, D.L., Band, L.E., Potts, D.F., Pierce, L.L. and Spanner, M.A. (1989) 'Mapping regional forest evapotranspiration and photosynthesis by coupling satellite data with ecosystem simulation', *Ecology*, 70(4), 1090-1101.

Schimel, D.S., Kittel, T.G.F. and Parton, W.J. (1991) 'Terrestrial biogeochemical cycles: global interactions with the atmosphere and hydrology', *Tellus*, 43(AB), 188-203.

Schutt, J.B., Rowland, R.R. and Heartly, W.H. (1984) 'A laboratory investigation of a physical mechanism for the extended infrared absorption ('red shift') in wheat', *International Journal of Remote Sensing*, 5(1), 95-102.

Sebacher, D.I., Harriss, R.C., Bartlett, K.B., Sebacher, S.M. and Grice, S.S. (1986) 'Atmospheric methane sources: Alaskan tundra bogs, an alpine fen, and a subarctic boreal marsh', *Tellus*, 38B, 1-10.

Sellers, P.J. (1985) 'Canopy reflectance, photosynthesis and transpiration', *International Journal of Remote Sensing*, 6(8), 1335-1372.

Sellers, P.J. (1987) 'Canopy reflectance, photosynthesis, and transpiration. II. The role of biophysics in the linearity of their interdependence', *Remote Sensing of Environment*, 21, 143-183.

Sellers, P.J., Berry, J.A., Collatz, G.J., Field, C.B. and Hall, F.G. (1992) 'Canopy reflectance, photosynthesis and transpiration III. A reanalysis using improved leaf models and a new canopy integration scheme', *Remote Sensing of Environment*, in press,

Shenk, J.S., Landa, I., Hoover, M.R. and Westerhaus, M.O. (1981) 'Description and evaluation of a near infrared reflectance spectro-computer for forage and grain analysis', *Crop Science*, 21(3), 355-358.

Smith, M.O., Ustin, S.L., Adams, J.B. and Gillespie, A.F. (1990a) 'Vegetation in deserts: I. A regional measure of abundance from multispectral images', *Remote Sensing of Environment*, 31, 1-26.

Smith, M.O., Ustin, S.L., Adams, J.B. and Gillespie, A.F. (1990b) 'Vegetation in deserts: II. Environmental influences on regional abundance', *Remote Sensing of Environment*, 29, 27-52.

Swanberg, N.A. and Matson, P.A. (1989) 'Determining experimentally induced variation in coniferous canopy chemistry with airborne imaging spectrometer data', *Proceedings International Geoscience and Remote Sensing Symposium,* vol.IEEE No. 89CH2768-0, Vancouver, B.C., Canada.

Townshend, J., Justice, C. and Kalb, V. (1987) 'Characterization and classification of South American land cover types using satellite data', *International Journal of Remote Sensing*, 8, 1189-1207.

Townshend, J., Justice, C., Li, W., Gurney, C. and McManus, J. (1991) 'Global land cover classification by remote sensing: present capabilities and future possibilities', *Remote Sensing of Environment*, 35, 243-255.

Tucker, C.J. (1977) 'Spectral estimation of grass canopy variables', *Remote Sensing of Environment*, 6, 11-26.

Tucker, C.J. (1979) 'Red and photographic infrared linear combinations for monitoring vegetation', *Remote Sensing of Environment*, 8, 127-150.

Tucker, C.J. and Sellers, P.J. (1986) 'Satellite remote sensing of primary production', *International Journal of Remote Sensing*, 7(11), 1395-1416.

Tucker, C.J., Holben, B.N. and Goff, T.E. (1984) 'Intensive forest clearing in Rondonia, Brazil, as detected by satellite remote sensing', *Remote Sensing of Environment*, 15, 255-261.

Tucker, C.J., Townshend, J.R.G. and Goff, T.E. (1985) 'African land-cover classification using satellite data', *Science*, 227(4685), 369-375.

Ustin, S.L., Adams, J.B., Elvidge, C.D., Rejmanek, M., Rock, B.N., Smith, M.O., Thomas, R.W. and Woodward, R.A. (1986) 'Thematic Mapper studies of semiarid shrub communities', *BioScience*, 36(7), 446-452.

Ustin, S.L., Smith, M.O. and Adams, J.B. (1992) 'Remote sensing of ecological processes: A strategy for developing and testing ecological models using spectral mixture analysis', in J. Ehlringer and C. Field (eds), *Scaling Ecological Processes between Leaf and Landscape*, Academic Press, in press.

Waring, R.H., Aber, J.D., Melillo, J.M. and Moore III, B. (1986) 'Precursors of change in terrestrial ecosystems', *BioScience*, 36(7), 433-438.

Wessman, C.A. (1990) 'Evaluation of canopy biochemistry', in R. J. Hobbs and H. A. Mooney (eds), *Remote Sensing of Biosphere Functioning*, Springer-Verlag, New York, 135-156.

Wessman, C.A. (1991) 'Remote sensing of soil processes', *Agriculture, Ecosystems and Environment*, 34, 479-493.

Wessman, C.A. (1992) 'Spatial scales and global change: bridging the gap from plots to GCM grid cells', *Annual Review of Ecology and Systematics*, 23, 175-200.

Wessman, C.A., Aber, J.D. and Peterson, D.L. (1989) 'An evaluation of imaging spectrometry for estimating forest canopy chemistry', *International Journal of Remote Sensing*, 10(8), 1293-1316.

Wessman, C.A., Aber, J.D., Peterson, D.L. and Melillo, J.M. (1988a) 'Foliar analysis using near infrared spectroscopy', *Canadian Journal of Forest Research*, 18, 6-11.

Wessman, C.A., Aber, J.D., Peterson, D.L. and Melillo, J.M. (1988b) 'Remote sensing of canopy chemistry and nitrogen cycling in temperate forest ecosystems', *Nature*, 335, 154-156.

Wetzel, D.L. (1983) 'Near-infrared reflectance analysis', *Analytical Chemistry*, 55, 1165a-1171a.

Weyer, L.G. (1985) 'Near-infrared spectroscopy of organic substances', *Applied Spectroscopy Reviews*, 21(1&2), 1-43.

Williams, D.L. and Walthall, C.L. (1990) 'Helicopter-based multispectral data collection over the Northern Experimental Forest. Preliminary results for the 1989 field season', *Proc. 10th Annual International Geoscience and Remote Sensing Symposium*, University of Maryland, Maryland, 1, 875-878.

ESTIMATING CANOPY BIOCHEMISTRY THROUGH IMAGING SPECTROMETRY

CAROL A. WESSMAN
Environmental, Population, and Organismic Biology (CIRES)
Cooperative Institute for Research in Environmental Sciences
University of Colorado
Boulder, Colorado 80309-0449

ABSTRACT. Broad-band reflectance measurements of vegetation have been widely applied in the form of indexes based on the unique differential between chlorophyll absorption in the red wavelengths and reflectance in the near infrared region. Background and atmospheric effects also have an influence on the measured signal and are only partially removed through ratioing of wavebands. High spectral resolution data acquired by imaging spectrometers provides information on absorption feature characteristics. Spectral shape parameters such as width, depth, skewness, and symmetry are more indicative of biochemical state and canopy physiology than average reflectance measured over relatively broad spectral regions. Variations in spectrum shape in the visible region relate to chlorophyll concentration, chlorophyll degradation, and other pigment activity. Other canopy biochemical constituents, such as cellulose and lignin, influence reflectance in the shortwave infrared and can potentially be quantified using imaging spectrometry. Capability to estimate biochemical properties in terrestrial ecosystems would aid in the assessment of carbon fixation/allocation patterns, nutrient availability and soil respiration.

1. Introduction

Quantitative information on seasonal activity of vegetation at large scales is a basic requirement to the understanding of the dynamics of major ecosystems and for the development of accurate global biogeochemical and climate models. Remote sensing has demonstrated capability in monitoring coarse-scale changes in vegetation type and architecture, as well as phenological changes associated with absorbed photosynthetically active radiation. However, remote sensing of ecosystem processes with current operational sensors is limited by two significant factors: (1) not all variation in important ecosystem processes is accompanied by large-scale change; and (2) broad-scale measurements of vegetation reflectance are irretrievably convolved with atmospheric and background (e.g. soil) effects. Spatial resolution will certainly play a role in both cases; but, in a more direct sense, spectral resolution will constrain interpretations of current remotely sensed data. Two-band indices currently in use are influenced by background variation, atmospheric attenuation and off-nadir viewing. Given the multiple factors affecting surface reflectance, these indexes cannot provide an unambiguous measure of vegetation.

Advances in imaging spectrometry suggest that measurements of land surface reflectance at sufficient spectral detail and quality will permit the detection of spectral shape characteristics indicative of vegetation biochemical state and will enable the reduction of atmosphere and

J. Hill and J. Mégier (eds.), Imaging Spectrometry - a Tool for Environmental Observations, 57–69.

background effects. This paper presents an overview of the remote sensing of vegetation chemistry. The discussion includes current capabilities and limitations to assess ecosystem processes using broad-band instruments; and the capabilities of imaging spectrometry for estimating canopy chemistry through high spectral resolution measurements of the visible and shortwave infrared region.

1.1 BIOCHEMICAL INDICATORS OF ECOSYSTEM PROCESSES

Characteristics of ecosystem functions such as carbon fixation, nutrient flux and heat exchange are reflected in vegetation canopy morphology and biochemistry. Alterations in large-scale ecosystem processes that result from changes in canopy morphology through disturbance or succession generally operate on relatively coarse temporal and/or spatial scales. Subtle changes in ecosystem functioning are often expressed in the canopy biochemistry as a result of altered carbon allocation patterns, metabolic processes and nutrient availability. These changes may occur over long time scales in response to change in environmental factors; or they may occur over relatively short time periods, e.g. normal phenological changes over a growing season. Their spatial scale of operation will be largely dependent on the heterogeneity of environmental resources within the landscape.

The capability to detect changes in canopy biochemistry using remote sensing would provide a means to assess spatial extent and variation of carbon/nutrient sources and sinks crucial to understanding gas exchange between vegetation and the atmosphere. Present knowledge is limited to small-scale, site-specific studies. Relative concentrations of carbohydrates and nitrogenous compounds in plant tissue often reflect the partitioning of carbon resources between roots and shoots (Lainson 1982, Chapin et al. 1987); these relationships are dependent on the plant's relative growth rate. Foliar nitrogen content, related to both chlorophylls and protein, is related to photosynthetic capacity, nitrogen uptake and primary productivity (Vitousek 1982, Birk and Vitousek 1986, Binkley and Hart 1989). Reduction in nitrogen supply can promote secondary wall thickening and lignification (Gartlan et al. 1980, Waring et al. 1985, Chapin et al. 1986). Storage carbohydrates such as starch, and defensive compounds, such as polyphenols and fibers, also vary predictably with resource availability (Mooney and Gulmon 1982, Bryant et al. 1983, Coley et al. 1985); and seasonal variations in starch content have been related to allocation patterns of labile carbohydrates in pine (Gholz et al. 1985).

Ecosystem carbon and nitrogen cycles are mutally linked because the quality (organic chemical composition) and quantity of litter supplied by the canopy modulate the processes of decomposition, mineralization and nitrification (Meentemeyer 1978, Melillo et al. 1982, Pastor and Post 1986, Aber et al. 1990). These processes, in turn, strongly regulate system productivity through their influence on nitrogen availability (Vitousek 1982, Pastor et al. 1984). Significant changes in foliar lignin to nitrogen ratios may indicate corresponding changes in decomposition rates affecting nutrient cycling and trace gas fluxes (Goodroad and Keeney 1984). Biochemical deposition may alter these processes through their action on foliar chemistry. Chronic deposition of nitrogen acid aerosols into northeastern forests of the United States have been shown to lead to excess nitrogen accumulation in soils, increased nitrogen uptake by trees, and decreased lignin content (McNulty et al. 1990). Several studies have identified pre-visual biochemical changes as well as acute responses to chronic ozone exposure (e.g. Ustin and Curtiss 1990, Beyers et al. 1991).

2. Spectrometry of Foliar Chemistry

The chemical constituents of leaves affect their spectral properties in the visible (400 - 700 nm) and shortwave infrared (700 - 2500 nm) regions. Absorption by photosynthetic pigments (chlorophyll, xanthophyll, and carotene) dominates the visible wavelengths. Each of the pigments has absorption maxima in the 300-500 nm region, however only chlorophyll absorbs in the red wavelengths (Salisbury and Ross 1969). Principal absorption peaks of extracted chlorophyll a occur at 430 and 660 nm and those of chlorophyll b at 455 and 640 nm (figure 1). When measured *in vivo*, these peaks shift approximately 20 nm towards the longer wavelengths due to the difference in refractive indices between the extract solvent and leaf water (Mackinney 1938, Setak et al. 1971). Changes in chlorophyll concentration with phenological development produce apparent spectral shifts (on the order of 5 to 20 nm) of the absorption edge near 700 nm (Gates et al. 1965, Horler et al. 1983). Environmental stresses which result in chlorophyll loss cause narrowing of the absorption band in the red region and a shift of the red edge to shorter wavelengths. This shift has been reported for studies of vegetation exposed to heavy metal stress (Horler et al. 1980, Collins et al. 1983), ozone (Westman and Price 1988, Ustin and Curtiss 1990), and acidic deposition (Rock et al. 1988, Buschmann et al. 1991).

Reflectance characteristics of vegetation in the region from 700 to 2500 nm exhibit high reflectance in the near infrared (700 - 1300 nm) and high absorption in the middle infrared (1300 - 2500 nm). The near IR wavelengths are greatly influenced by cellular structure and refractive index discontinuities within the leaf (Knipling 1970). Minor water absorption features near 960 and 1200 nm vary significantly in shape and depth and may be related to both cellular arrangement within the leaf and hydration state (Gausman et al. 1978, Goetz et al. 1983). The mid-IR region is dominated by leaf water absorption and has been related to plant water status through indices combining these and near-IR bands (Tucker 1980, Hunt and Rock 1989, Hunt 1991). The region intermediate to the water absorption maxima at 1450 and 1940 nm may be strongly influenced by cell structure, morphology and tissue constituents (Kleman and Fagerlund 1987; Wessman et al. 1988a).

Research in analytical chemistry has demonstrated that concentrations of constituents within organic mixtures can be evaluated from reflectance in the SWIR (Wetzel 1983; Weyer 1985, McDonald 1986). Near infrared reflectance spectroscopy (0.7 to 2.5 μm) has been used to successfully predict protein, lignin, fiber fractions, and *in vitro* or *in vivo* digestible dry matter of forage (e.g. Norris et al. 1976, Shenk et al. 1981, Barton et al. 1992) and nitrogen and lignin content in foliage of native forest and prairie species (Wessman et al. 1988a, McLellan et al. 1991). The physical basis for the extraction of biochemical information is the absorption of radiation by the molecular functional groups of C-H, O-H and N-H found within all foliar material. Overtone and combination bands specific to these compounds occur in the near infrared region and are particularly sensitive to changes in chemical concentrations. Reviews of spectral characterization of foliar constituents can be found in Himmelsbach et al. (1988), Curran (1979) and Wessman (1990).

Near infrared spectroscopy (NIRS) analysis relies on high instrumental signal-to-noise ratios and wavelength reproducibility, one or both of which are difficult conditions to meet in remote sensing instrumentation. *In situ* canopy reflectance has high variablity due to environmental and sensor effects which may introduce enough noise to limit interpretation of the SWIR signal. Nevertheless, advances in NIRS have stimulated work with whole leaf reflectance that promises improved understanding of foliar optical characteristics. The approach suggests that, while the individual spectra of pure leaf components such as cellulose, lignin, and protein are not immediately apparent in a composite vegetation spectrum, they are decidedly important factors in its shape. Remotely-sensed measurements of vegetation canopy reflectance will certainly not be as sensitive as those of a laboratory spectrophotometer to foliar chemical constituents, but, if sampled at sufficient spectral detail, can indicate those constituents that strongly influence the shape of the spectra. The potential to estimate canopy constituents remotely rests on: i) the influence of individual or functional groups of foliar constituents on the overall canopy reflectance curve; and ii) the development of high spectral resolution instruments which will measure the reflectance signal at sufficient detail and quality to document subtle changes in spectral shape and allow reduction of background and atmospheric effects.

3. Spectral Analysis of Organic Mixtures

The Beer-Lambert law is the fundamental relationship upon which virtually all spectrometric techniques are based, and is stated here as a function of transmission

$$-\log T(\lambda) = A(\lambda) = \varepsilon(\lambda)\, b(\lambda)\, c \tag{1}$$

The term $T(\lambda)$ is the transmission as a function of wavelength, $A(\lambda)$ is the absorbance as a function of wavelength, $\varepsilon(\lambda)$ is the absorption coefficient as a function of wavelength, the term $b(\lambda)$ is the absorption path length as a function of wavelength, and the term c is the concentration of the absorbing compound. Solving for the concentration gives

$$c = A(\lambda)/\varepsilon(\lambda)\, b(\lambda) \tag{2}$$

Beer-Lambert's law describes pure absorption and serves as an adequate model for spectrophotometer data acquired in the laboratory. However, it does not include the appropriate scattering and absorption coefficients which are necessary when modeling radiation propagating through a canopy. Scattering caused by the architecture of the canopy as well as the internal structure of each leaf will lead to increased optical depth and, as a result, increased absorption (Allen et al. 1973). Pure scattering will linearly attenuate light with depth, unlike the exponential effect of absorption (Fukshansky 1981). Additional complications in the relationship between foliar constituent concentrations and canopy reflectance will result from background spectral contributions, atmospheric effects and sun-sensor geometry.

The spectral behavior of mixtures, either of single surfaces with inherent complexity (e.g., a leaf) or of a complex of many surfaces (e.g., a landscape), is a function of the type and quantity of reflecting components and their relative influence on the measured response. Absorption bands due to electronic transitions and bond vibrations may assist in identifying concentrations of foliar constituents using local (derivatives) or global (curve fitting) analyses of laboratory-acquired spectral data (e.g. Card et al. 1988; Wessman et al. 1988a). Such information, uncomplicated by the atmosphere and illumination geometry, can be used to interpret more complex spectral mixtures acquired with airborne and satellite sensors. Tracking spectral features in reflectance measurements made in the laboratory up to those made at the pixel level should provide some indication of the transfer of spectral information across scales.

3.1 Derivative Spectrometry

Derivative spectrometry is commonly employed to resolve or enhance absorption features that are masked by interfering background absorptions and/or by noise (Talsky et al. 1978; Dixit and Ram 1985). The technique aids in separating overlapping bands and isolating shoulders and weak signals from unwanted background. For a constant intensity I_O over the whole wavelength range (as measured by a spectrophotometer), the first derivative is obtained as:

$$(dI/d\lambda)/I = -C\ell\,(d\alpha/d\lambda) \tag{3}$$

and will be linearly proportional to concentration (Talsky et al. 1978). The sensitivity of the measurements will be high in inflection areas. The second derivative reads as:

$$\left(d^2 I/d\lambda^2\right)/I = C^2\,\ell^2\,(d\alpha/d\lambda)^2 - C\ell\left(d^2\alpha/d\lambda^2\right) \tag{4}$$

where direct proportionality to concentration exists only if $d\alpha / d\lambda$ equals zero. If $d^2\alpha / d\lambda^2$ has an extremum value, then sensitivity is very high.

Derivative transformations can be applied to remotely sensed data to reduce baseline shifts (albedo variations) resulting from surface topography, illumination conditions, and/or lack of appropriate calibration information (Dixit and Ram 1981, Wessman et al. 1989, Demetriades-Shah et al. 1990). Derivative spectra from lab- and field-acquired measurements are useful for the characterization of shifts in the chlorophyll absorption edge (e.g. Horler et al. 1983, Rock et al. 1988; Ustin and Curtiss 1990) and reduction of background soil reflectance (Hall et al. 1990). A derivative transformation of Airborne Imaging Spectrometer (AIS) imagery over temperate deciduous and coniferous forests reduces apparent brightness differences due to canopy architecture and shifts in albedo between flight lines (Wessman et al. 1989). Correlative analysis of the transformed spectral data and canopy lignin concentrations suggests that absorption characteristics of lignin or a closely associated canopy property influences reflectance in a predictable fashion (Wessman et al. 1988b).

3.2 Spectral Curve-Fitting

An alternative approach to derivative spectroscopy was proposed by Goetz et al. (1990) who showed that leaf water could be modeled by fitting a liquid water absorption curve of known concentration to the leaf spectrum. Gao and Goetz (1990) initially developed this approach to model the atmospheric water vapor column in high spectral resolution images, where they noted that liquid water absorption bands were offset up to 60 nm from vapor absorptions and could be spectrally modeled.

Using this technique, Goetz et al. (1990) demonstrated that the reflectance spectra of fresh green leaves are in fact a combination of spectral components contributed by the dry leaf material and liquid water. The relative dominance of water absorption beyond 1 μm tends to obscure other leaf optical characteristics in that spectral region. However, spectral reflectance of dried leaves exhibit diagnostic absorption features of major foliar chemical constituents such as cellulose, starch and lignin. By use of spectrum matching techniques originally used to quantify whole column water abundance in the atmosphere and equivalent liquid water thickness in leaves, the

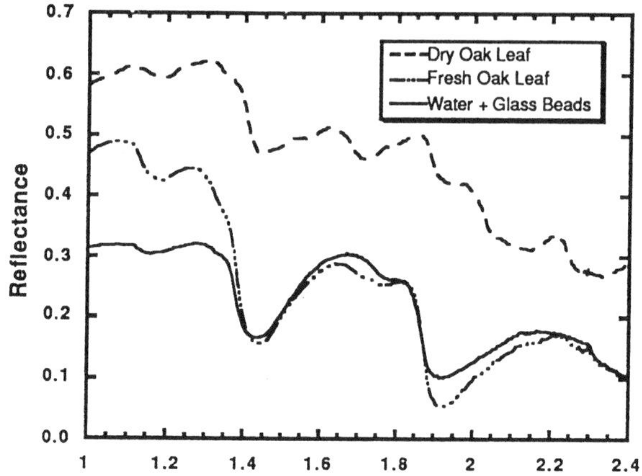

FIGURE 1. Reflectance spectra of fresh and dried oak leaves and glass beads plus water (Goetz et al., 1990).

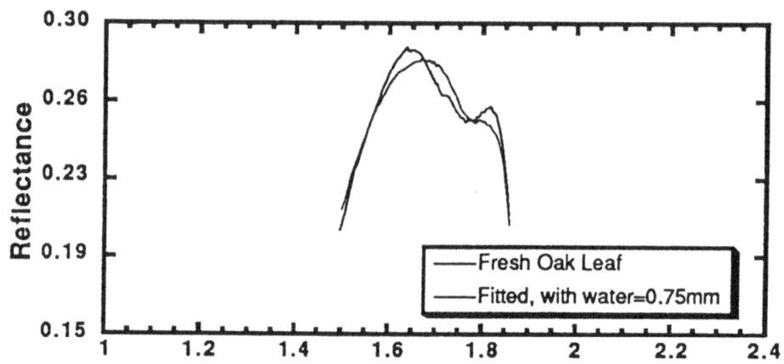

FIGURE 2. Fit of the fresh leaf spectrum with the dry leaf and water spectral reflectance components (Goetz et al., 1990).

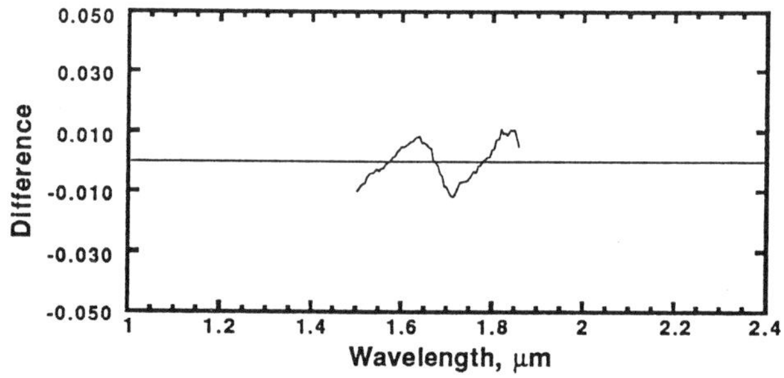

FIGURE 3. Difference between fresh leaf spectrum and water spectrum (Goetz et al., 1990).

liquid water contribution was removed from spectra of fresh, green oak leaves. Figure 1 shows measured reflectance spectra of fresh oak leaves ($R_{FM}(\lambda)$), dry oak leaves ($R_{DM}(\lambda)$), and glass beads mixed with liquid water ($R_{WM}(\lambda)$). Because the glass beads alone have no absorption bands in the 1.0-2.5 μm region, $R_{WM}(\lambda)$ contains only liquid water absorption features center at approximately 1.18, 1.43, and 1.92 μm.

The calculated fresh leaf spectrum ($R_{FC}(\lambda)$) is derived according to

$$R_{FC}(\lambda) = (A + B\lambda)\cdot \exp\{-[C_w K_w(\lambda) + C_d K_d(\lambda)]\} \tag{5}$$

where the approximate absorption coefficients of dry leaves ($K_d(\lambda)$) and liquid water ($K_w(\lambda)$) are derived from the reflectance spectra $R_{DM}(\lambda)$ and $R_{WM}(\lambda)$ according to the equations $K_d(\lambda) = -\log(R_{DM}(\lambda))$ and $K_w(\lambda) = -\log(R_{WM}(\lambda))$. The residual remaining from curve fitting the fresh oak leaf ($R_{FM}(\lambda)$) with the dry leaf plus water spectral reflectance spectra ($R_{FC}(\lambda)$) (figure 2) in the region of 1.5 - 1.7 μm (where lignin is an important constituent) resembled mixtures of reflectance spectra from pure leaf components (figure 3). This experiment establishes that the absorption features of chemical constituents clearly observed in the dry leaf spectrum influence the fresh leaf spectrum. By first removing the effects of liquid water, these features are distinguished and may be quantified.

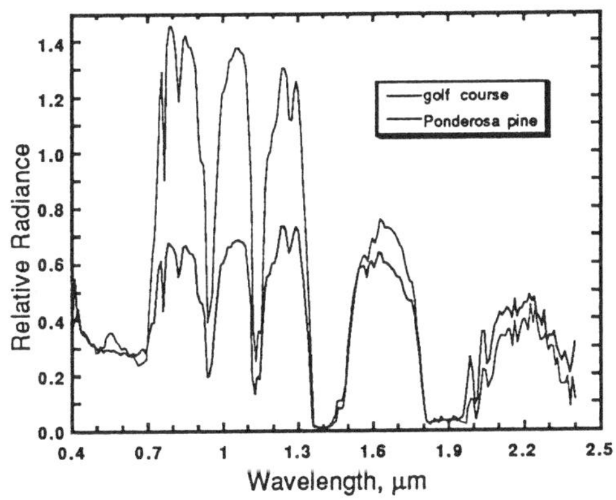

FIGURE 4. Relative radiance spectra of a golf course and a stand of Ponderosa pine from AVIRIS data over Oregon (B.-C. Gao, unpublished data).

The same water extraction procedure was applied to Airborne Visible/Infrared Imaging Spectrometer (AVIRIS) data in order to observe differential absorption characteristics of grass (golf course) and pine (*Pinus ponderosa*) in the region of lignin absorption, 1.5 to 1.7 μm (B.-C. Gao, unpublished data). Radiance data from the two vegetation types (figure 4) were normalized to one another (figure 5) to remove albedo differences. A liquid water spectrum was fit to each of the two AVIRIS spectra and the residual spectra (figure 6) showed a lignin absorption feature more intense in the spectrum of the pine stand than in that of the grass. Pine will have considerably higher lignin concentrations (15 - 30%) than grasses (3 - 15%), however a quantitative estimate of lignin values from these data would require consideration of path length through the canopy.

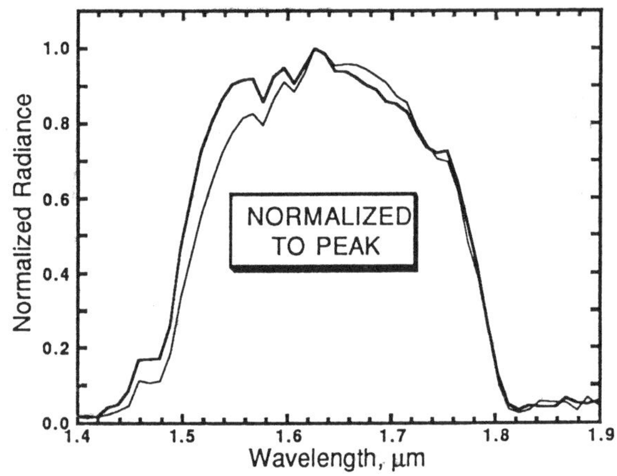

FIGURE 5. Radiance data from golf course and ponderosa pine normalized to the peak radiance in the region of 1.5 to 1.7 μm. (B.-C. Gao, unpublished data)

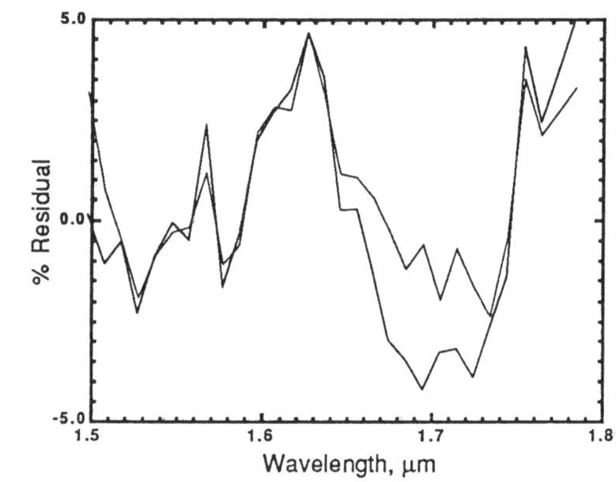

FIGURE 6. Residual from removal of liquid water spectrum from AVIRIS radiance spectra of golf course and ponderosa pine. (B.-C. Gao, unpublished data)

These techniques are still under development. While spectral curve-fitting techniques can differentiate between levels of absorption, estimates can only be relative. Further modeling of the canopy pathlength is required to provide quantitative estimates and several validated experiments will be required before these methods can be confirmed.

4. Concluding Remarks

The nature of absorptions within organic mixtures are weak and complex since they consist of overlapping overtone and combination bands. The origins of the observed vibrations are limited and they are all associated with primary constituents of vegetation. Knowledge of absorption characteristics of each of the major leaf constituents (e.g., cellulose, starch and protein) may permit remote assessment of canopy level concentrations if high spectral resolution reflectance information is acquired. Direct assessment of low level constituents is unlikely; chances increase with the predominant materials such as cellulose, lignin and protein. It may be necessary to create a simpler taxonomy of constituents and their effective spectral and ecological combinations. For example, structural materials such as cellulose and lignin may influence the canopy spectrum in such similar ways as to be indifferentiable. Nonetheless, an estimate of their combined concentration may be very useful for large scale ecological applications. High spectral resolution measurements will be needed to study characteristics of the canopy reflectance curve and to better separate background factors from the vegetation response.

We already recognize the need for temporal and spatial information on biospheric functioning, and the prohibitive logistics of acquiring such information other than through the use of remote sensing technology. Moreover, we recognize that patterns of regional and global interactions will most likely be imperceptible from our present vantage point. Assessments of global primary productivity using remote sensing (e.g. Justice et al. 1986; Tucker et al. 1986) will increase in usefulness as we become more capable of utilizing such information for modeling and monitoring global-level processes. Estimates of canopy biochemistry through direct assessment of spectral reflectance features or inferred through relationships with other factors contributing to canopy reflectance (e.g. water content) will provide further insights on the nature of biosphere function response to environmental change.

5. References

Aber, J.D., Wessman, C.A., Peterson, D.L., Melillo, J.M. and Fownes, J.H. (1990) 'Remote sensing of litter and soil organic matter decomposition in forest ecosystems', in R. J. Hobbs and H. A. Mooney (eds), *Remote Sensing of Biosphere Functioning*, Springer-Verlag, New York, 87-103.

Allen, W.A., Gausman, H.W. and Richardson, A.J. (1973) 'Willstatter-Stoll theory of leaf reflectance evaluated by ray tracing', *Applied Optics*, 12(10), 2448-2453.

Barton, F.E., Himmelsbach, D.S., Duckworth, J.H. and Smith, M.J. (1992) 'Two-dimensional vibration spectroscopy: correlation of mid- and near-infrared regions', *Applied Spectroscopy*, 46, 420-429.

Beyers, J.L., Reichers, G.H. and Temple, P.J. (1991) 'Photosynthetic capacity of ponderosa pine seedlings exposed to different levels of atmospheric ozone and drought stress', *Bulletin of the Ecological Society of America*, 72(2), 69.

Binkley, D. and Hart, S.C. (1989) 'The components of nitrogen availability assessments in forest soils', *Advances in Soil Science*, 10, 57-112.

Birk, E.M. and Vitousek, P.M. (1986) 'Nitrogen availability and nitrogen use efficiency in Loblolly pine stands', *Ecology*, 67(1), 69-79.

Buschmann, C., Rinderle, U. and Lichtenthaler, H.K. (1991) 'Detection of stress in coniferous forest trees with the VIRAF spectrometer', *IEEE Transactions on Geoscience and Remote Sensing*, 29(1), 96-100.

Card, D.H., Peterson, D.L., Matson, P.A. and Aber, J.D. (1988) 'Prediction of leaf chemistry by the use of visible and near infrared reflectance spectroscopy', *Remote Sensing of Environment*, 26, 123-147.

Chapin III, F.S., Bloom, A.J., Field, C.B. and Waring, R.H. (1987) 'Plant responses to multiple environmental factors', *BioScience*, 37(1), 49-57.

Chapin III, F.S., McKendrick, J.D. and Johnson, D.A. (1986) 'Seasonal changes in carbon fractions in Alaskan tundra plants of differing growth form: implications for herbivory', *Journal of Ecology*, 74, 707-731.

Collins, W., Chang, S.H., Canney, F. and Ashley, F. (1983) 'Airborne biogeochemical mapping of hidden mineral deposits', *Economic Geology*, 78, 737-749.

Curran, P.J. (1989) 'Remote sensing of foliar chemistry', *Remote Sensing of Environment*, 30, 271-278.

Demetriades-Shah, T.H., Steven, M.D. and Clark, J.A. (1990) 'High resolution derivative spectra in remote sensing', *Remote Sensing of Environment*, 33, 55-64.

Dixit, L. and Ram, S. (1985) 'Quantitative analysis by derivative electronic spectroscopy', *Applied Spectroscopy Reviews*, 21(4), 311-418.

Fukshanski, L. (1981) 'Optical properties of plants', in H. Smith (eds), *Plants and the Daylight Spectrum*, Academic Press, London, 21-40.

Gartlan, J.S., McKey, D.B., Waterman, P.G., Mbi, C.N. and Struhsaker, T.T. (1980) 'A comparative study of the phytochemistry of two African rain forests', *Biochem. Syst. and Ecology*, 8, 401-422.

Gates, D.M., Keegan, H.J., Schleter, J.C. and Weidner, V.R. (1965) 'Spectral properties of plants', *Applied Optics*, 4(1), 11-20.

Gausman, H.W., Escobar, D.E., Everitt, J.H., Richardson, A.J. and Rodriguez, R.R. (1978) 'Distinguishing succulent plants from crop and woody plants', *Photogrammetric Engineering and Remote Sensing*, 44(4), 487-491.

Gholz, H.L., Perry, C.S., Cropper, W.P., Jr. and Henry, L.C. (1985) 'Litterfall, decomposition, and nitrogen and phosphorus dynamics in a chronosequence of slash pine (*Pinus elliotii*) plantations', *Forest Science*, 31, 463-478.

Goetz, A.F.H., Gao, B.C., Wessman, C.A. and Bowman, W.D. (1990) 'Estimation of biochemical constiuents from fresh, green leaves by spectrum matching techniques', *Proceedings International Geocience and Remote Sensing Symposium*, 2, 971-974.

Goetz, A.F.H., Rock, B.N. and Rowan, L.C. (1983) 'Remote sensing for exploration: an overview', *Economic Geology*, 78(4), 573-590.

Goodroad, L.L. and Keeney, D.R. (1984) 'Nitrous oxide emission from forest, marsh, and prairie ecosystems', *Journal of Environmental Quality*, 13(3), 448-452.

Hall, F.G., Huemmrich, K.F. and Goward, S.N. (1990) 'Use of narrow-band spectra to estimate the fraction of absorbed photosynthetically active radiation', *Remote Sensing of Environment*, 32(1), 47-54.

Himmelsbach, D.S., Boer, H., Akin, D.E. and Barton II, F.E. (1988) 'Solid-state carbon-13 NMR, FTIR and NIR spectroscopic studies of ruminant silage digestion', in A. M. C. Davies (eds), *Analytical Applications of Spectroscopy*, Royal Society of Chemistry, London.

Horler, D.N.H., Barber, J. and Barringer, A.R. (1980) 'Effects of heavy metals on the absorbance and reflectance spectra of plants', *International Journal of Remote Sensing*, 1(2), 121-136.

Horler, D.N.H., Dockray, M. and Barber, J. (1983) 'The red edge of plant leaf reflectance', *International Journal of Remote Sensing*, 4(2), 273-288.

Hunt, E.R., Jr. and Rock, B.N. (1989) 'Detection of changes in leaf water content using near- and middle-infrared reflectances', *Remote Sensing of Environment*, 30, 43-54.

Hunt, J., E.R. (1991) 'Airborne remote sensing of canopy water thickness scaled from leaf spectrometer data', *International Journal of Remote Sensing*, 12(3), 643-649.

Justice, C.O., Townshend, J.R.G., Holben, B.N. and Tucker, C.J. (1985) 'Analysis of the phenology of global vegetation using meteorological satellite data', *International Journal of Remote Sensing*, 6(8), 1271-1318.

Kleman, J. and Fagerlund, E. (1987) 'Influence of different nitrogen and irrigation treatments on the spectral reflectance of barley', *Remote Sensing of Environment*, 21, 1-14.

Knipling, E.B. (1970) 'Physical and physiological basis for the reflectance of visible and near-infrared radiation from vegetation', *Remote Sensing of Environment*, 1, 155-159.

Lainson, R.A. (1982) 'A model for leaf expansion in cucumber', *Ann. Bot.*, 50, 407-425.

Mackinney, G. (1938) 'Applicability of Kundt's rule to chlorophyll', *Plant Physiology*, 13, 427-430.

McDonald, R.S. (1986) 'Review: infrared spectrometry', *Analytical Chemistry*, 58, 1906-1925.

McLellan, T., Martin, M.E., Aber, J.D., Melillo, J.M., Nadelhoffer, K.J. and Dewey, B. (1991) 'Comparison of wet chemistry and near infrared reflectance measurements of carbon-fraction chemistry and nitrogen concentration of forest foliage', *Canadian Journal of Forest Research*, 21, 1689-1693.

McNulty, S. (1990) 'Effects of chronic nitrogen and sulfur deposition on the nutrient cycling, productivity, trace gas production and canopy chemistry of northeast U.S. forest ecosystems', *Proc. IV International Congress of Ecology*, Yokohama, Japan, 190.

Meentemeyer, V. (1978) 'Macroclimate and lignin control of litter decomposition rates', *Ecology*, 59, 465-472.

Melillo, J.M., Aber, J.D. and Muratore, J.F. (1982) 'Nitrogen and lignin control of hardwood leaf litter decomposition dynamics', *Ecology*, 63(3), 621-626.

Norris, K.H., Barnes, R.F., Moore, J.E. and Shenk, J.S. (1976) 'Predicting forage quality by infrared reflectance spectroscopy', *Journal of Animal Science*, 43(4), 889-897.

Pastor, J. and Post, W.M. (1986) 'Influence of climate, soil moisture, and succession on forest carbon and nitrogen cycles', *Biogeochemistry*, 2, 3-27.

Pastor, J., Aber, J.D., McClaugherty, C.A. and Melillo, J.M. (1984) 'Aboveground production and N and P cycling along a nitrogen mineralization gradient on Blackhawk Island, Wisconsin', *Ecology*, 65(1), 256-268.

Rock, B.N., Hoshizaki, T. and Miller, J.R. (1988) 'Comparison of in situ and airborne spectral measurements of the blue shift associated with forest decline', *Remote Sensing of Environment*, 24, 109-127.

Salisbury, F.B. and Ross, C. (1969) *Plant Physiology*. Wadsworth, San Francisco.

Setak, S., Catsky, J. and Jarvis, P.G. (1971) 'Determination of chlorophylls a and b', in (eds), *Plant Photosynthetic Production: Manual of Methods*, Junk, The Hague, 673-701.

Shenk, J.S., Landa, I., Hoover, M.R. and Westerhaus, M.O. (1981) 'Description and evaluation of a near infrared reflectance spectro-computer for forage and grain analysis', *Crop Science*, 21(3), 355-358.

Talsky, G., Mayring, L. and Kreuzer, H. (1978) 'High-resolution, higher-order UV/VIS derivative spectrophotometry', *Angew. Chem. Int. Engl.*, 17, 785-799.

Tucker, C.J. (1980) 'Remote sensing of leaf water content in the near infrared', *Remote Sensing of Environment*, 10, 23-32.

Tucker, C.J., Fung, I.Y., Keeling, C.D. and Gammon, R.H. (1986) 'Relationship between atmospheric CO2 variations and a satellite-derived vegetation index', *Nature*, 319(6050), 1-5.

Ustin, S.L. and Curtiss, B. (1990) 'Spectral characteristics of ozone-treated conifers', *Environmental and Experimental Botany*, 30(3), 293-308.

Vitousek, P.M. (1982) 'Nutrient cycling and nutrient use efficiency', *American Naturalist*, 119(4), 553-572.

Waring, R.H., McDonald, A.J.S., Larsson, S., Ericsson, T., Wiren, A., Arwidsson, E. and al., e. (1985) 'Differences in chemical composition of plants grown at constant relative growth rates with stable mineral nutrition', *Oecologia*, 66, 157-160.

Wessman, C.A. (1990) 'Evaluation of canopy biochemistry', in R. J. Hobbs and H. A. Mooney (eds), *Remote Sensing of Biosphere Functioning*, Springer-Verlag, New York, 135-156.

Wessman, C.A., Aber, J.D. and Peterson, D.L. (1989) 'An evaluation of imaging spectrometry for estimating forest canopy chemistry', *International Journal of Remote Sensing*, 10(8), 1293-1316.

Wessman, C.A., Aber, J.D., Peterson, D.L. and Melillo, J.M. (1988a) 'Foliar analysis using near infrared spectroscopy', *Canadian Journal of Forest Research*, 18, 6-11.

Wessman, C.A., Aber, J.D., Peterson, D.L. and Melillo, J.M. (1988b) 'Remote sensing of canopy chemistry and nitrogen cycling in temperate forest ecosystems', *Nature*, 335, 154-156.

Westman, W.E. and Price, C.V. (1988) 'Spectral changes in conifers subjected to air pollution and water stress: Experimental studies', *IEEE Transactions in Geoscience and Remote Sensing*, GE-26(1), 11-21.

Wetzel, D.L. (1983) 'Near-infrared reflectance analysis', *Analytical Chemistry*, 55, 1165a-1171a.

Weyer, L.G. (1985) 'Near-infrared spectroscopy of organic substances', *Applied Spectroscopy Reviews*, 21(1&2), 1-43.

SOIL SPECTRAL PROPERTIES AND THEIR RELATIONSHIPS WITH ENVIRONMENTAL PARAMETERS - EXAMPLES FROM ARID REGIONS

RICHARD ESCADAFAL
Mission ORSTOM
B.P. 434
1004 El Menzah
Tunisia

ABSTRACT. The development of global studies on terrestrial environments has led to a renewed interest in soils. In addition to supporting and nourishing the biomass, soils play a very important interface and buffer function in the ecosystems. For instance, large parts of the carbon, water and gases fluxes are controlled by soils which can be referred as the 'skin' of terrestrial ecosystems.

Remote sensing techniques are essential in the global approach and links between soil properties and spectral features have been investigated. New instruments currently enable to measure these features with high resolution. In this paper we will first summarise the techniques used to measure soil spectra in the laboratory, in the field and remotely. Then, we will discuss the relationships between the shapes of the spectral reflectance curves observed and the soil composition (organic carbon, iron oxides, and hydrous minerals, particularly). The spectral characteristics of aridic soils and their use as 'desertification' indicators will illustrate the potential of imaging spectrometry for the assessment of environmental changes through the detection of soil surface spectral variations. Finally, the example of simulated impact of soils on the remote measurement of vegetation parameters will enable to conclude on the necessity of giving more consideration to soils and their spectral variability.

1. Introduction

In classical soil science, soils have been studied rather independently as natural phenomena. Focus was on establishing soil types and classification schemes, on inventorying the distribution of soils and on determining their suitability for crop production. This approach of considerable interest and impact when soils were still to be discovered, seems to have lost its attraction nowadays while the inventory phase is mostly achieved.

In the very recent years however, the rapid environmental changes we are facing have enlightened the necessity of studying the processes globally, and particularly of understanding the soils in interaction with the other ecosystems components. When considered from a global point of view, soils play a very important part as interface between the biosphere and the geosphere ('pedosphere', Arnold et al., 1990). The soil mantle at the Earth surface supports the biota, serves as the source of nutrients for the functioning and for the production of biomass.

J. Hill and J. Mégier (eds.), Imaging Spectrometry - a Tool for Environmental Observations, 71–87.

Thus, the pedosphere is obviously the important natural resource for growing food, feed, timber and fibres. But it has also very important atmospheric, hydrospheric and lithospheric functions. For instance, it can be reminded that large quantities of gases are exchanged between the soil surface and the atmosphere (O_2, CO_2, CH_4), whereas water fluxes are strongly modulated by soils (i.e. infiltration, runoff, and percolation rates vary largely among soils). From this interface position, the pedosphere has an important buffer rule and can be considered as the 'skin' of our planet.

This global approach of the ecosystems has led to the need of appropriate tools for assessing the various parameters. Large hopes have been put in remote sensing techniques and several of them can be used for soil characteristics and properties measurements. In this paper we will summarise recent studies on the relationships between soil spectral properties and soil composition, and their potential use in imaging spectrometry. As an example, soil surface spectral features related with arid environment changes and desertification processes will be discussed.

2. Measuring soil spectral properties

The generic term of 'Soils' refers to a three-dimensional continuum at the Earth surface. In this 'mantle' lateral and vertical differenciations are observed in pits and trenches dug to carry out observations. These field observations usually reveal a characteristically layered pattern. These layers, known as *horizons*, are mainly distinguished through differences in colour, texture (particle size distribution), and structure (soil particles arrangement). Several types of horizons have been recognised and defined by soil scientists, forming the basis for systems of soil classifications (or 'taxonomy').

Thus, in a first approach, the spectral characterisation of a given soil can be performed by measuring the spectral reflectance of its different horizons. It should be noted that the first criterion used to discriminate the horizons is of spectral nature, as colour is basically the visual sensation related to an uneven reflection of the light (see below for discussion on the relationships between soil spectral reflectance and colour). When dealing with applications for remote sensing, we generally focus on the first horizon, the upper layer of the soil and its surface which can be observed from space. However, knowing the spectral properties of lower horizons is essential when looking at processes such as soil erosion. The lower soil materials progressively appearing at the soil surface during the denudation process, may have different spectral properties enabling the remote detection of erosion (see, for instance, the simulation of Latz et al., 1984).

2.1 LABORATORY MEASUREMENTS

Spectrophotometers are largely used in various laboratory analysis techniques, e.g. for measuring light transmission through solutions. With the help of an integrating sphere and of a calibrated white standard, the spectral reflectance of small samples can be easily computed with a spectrophotometer. Soil samples may be scanned after a simple air drying and sieving, but a more rigorous approach requires grinding down to a standard size (see Fernandez and Schulze, 1987; Bedidi et al., 1992 for more details).

This technique enables precise and standardised measurements of soil diffuse reflectance, with a high spectral resolution (down to 1 nm) and within a wide spectral range (typically 350 to 2500 nm). However, the whole process of sample collection, preparation and scanning is rather long, and these measures deal only with the fine-sized fraction of small soil samples.

2.2 Field Measurements

Recently developed portable spectroradiometers enable the recording of soil surface radiances with increasing precision and spectral range width as the technology evolves. Compact hand-held devices currently commercialised record spectra in a few seconds typically over 1 m^2 surface samples. Bi-directional reflectance factors (BRF) are computed by ratioing the radiances measured over soils by the one measured over a white reference panel (Escadafal et al., 1990). Technically, the varying illumination conditions, the differences in viewing geometry and the non-lambertian behaviour of the reference targets have to be taken into account when comparing field measured reflectances with laboratory measurements (Jackson et al., 1990). In the case of arid soil surfaces, we have shown that the variations in geometrical condition do not alter the shape of soil spectra (Escadafal and Huete, 1991).

The main advantage of the field spectroradiometers is that they measure the spectral properties of the soil surface as a whole, including coarse elements (as stones) and debris or litter. Moreover, individual surface elements of patchy soil surfaces, such as those encountered in arid regions, can be measured separately, which enables further spectral modelling of mixed surfaces .

2.3 Measurements from Aircraft and Spacecraft

While soils can easily be observed and scanned on the ground, they are less easily remotely sensed as they are very often - at least partially - covered by vegetation. Two favourable situations enable direct optical remote sensing of soils: in cultivated areas after ploughing or after other cultural practices removing the vegetation, and under dry climates where the vegetation is scarce. In the other cases soils are part of mixed pixels ('mixels') and the measurements made from satellite are a combination of soil and vegetation signals. Although most of the attention has been focused on plant spectral properties, the soil spectral variability has a strong impact on the signal recorded over those incomplete canopies (Huete, 1987; Escadafal and Huete, 1992). Very interesting recent examples of spectra recorded from airborne imaging spectrometers can be found in Hill and Mégier (1991) and Hill et al. (1994).

3. Main types of soil spectra

Laboratory measurements on soil samples have been reported in a few publications, among these two concern sets of data covering a significantly large number of soils (Condit, 1970; Stoner et al., 1980). The latter is the only published data set available for further analysis, with spectra recorded in the 550 to 2320 nm range over 564 wetted soil samples collected from the USA and from Brazil. More recently, outdoors spectroradiometric measurements have been carried out on soils (e.g. Escadafal and Huete, 1991; Hill and Mégier, 1991) but currently, there is no global data base on soil spectral properties available for the scientific community. However, the main types of spectra associated to broad soil groups have been progressively recognised and described in the literature (Baumgardner et al., 1985; Courault et al., 1988).

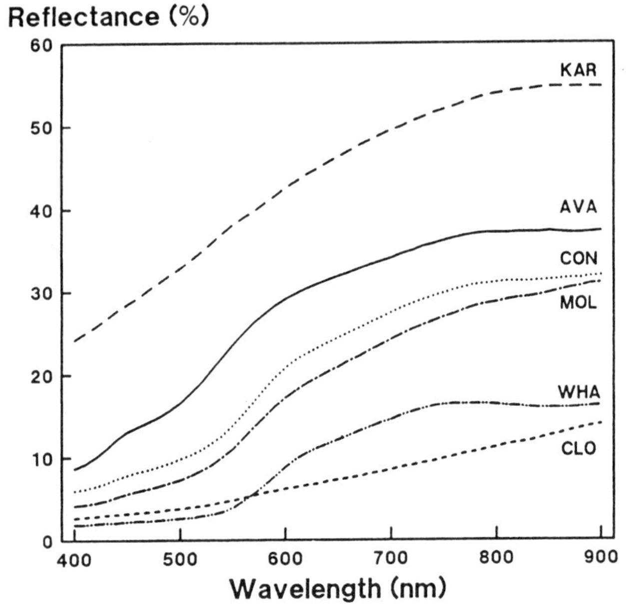

FIGURE 1. Examples of soil spectra illustrating the three main types observed in the Visible-NIR-region (see table 1 for legend).

Symbol	Series	Soil Taxonomy	Munsell Color(*)	%Clay	%Sand	%Fe	%Carb
AVA	Ava	mesic typic fragiudalf	5.1 YR 6.0/4.0	31	9	1.25	0.880
CLO	Cloversprings	cumulic cryoborolls	4.5 YR 2.7/1.5	21	40	2.00	0.930
COM	Comoro	thermic typic torrifluvent	4.9 YR 3.6/1.9	5	84	1.60	0.240
CON	Contine	hyperthermic typic haplargids	3.8 YR 4.5/3.7	26	52	0.70	0.970
COR	Cornutt	mesic ultic haloxeralfs				24.80	0.005
DAV	Davidson	thermic rhodic kandiudults	2.9 YR 3.3/4.2	52	25	10.20	0.580
HOL	Holtville	hyperthermic typic torrifluvent	4.4 YR 4.4/2.1	41	9	0.90	0.380
KAR	Karro	thermic ustollic calciorthids	9.9 YR 6.9/2.3			0.19	1.198
LAV	Laveen	hyperthermic typic calciorthids	4.1 YR 4.7/3.0	19	46	0.80	0.005
MOH	Mohave	thermic typic haplargids	3.8 YR 4.8/3.6	23	59	0.60	0.630
MOL	Molokai	isohyperthermic typic torrox	2.2 YR 2.7/4.0	52	23	12.50	0.740
NIC	Nicholson	mesic typic fragiudalf	5.2 YR 5.5/4.0	49	4	2.80	0.005
PIN	Pinaleno	thermic typic haplargids	4.2 YR 4.7/3.4	7	71	0.90	0.890
RED	Red Cinders		1.4 YR 3.1/2.7			2.25	0.110
SUP	Superstition	hyperthermic typic calciorthids	4.6 YR 5.7/3.1	2	96	0.47	0.300
VIN	Vint	hyperthermic typic torrifluvent	4.3 YR 4.9/2.9	4	82	0.80	0.840
WHA	White house	thermic ustollic haplargids	3.9 YR 4.1/4.0	7	79	1.50	0.890
YUM	Yuma		4.8 YR 5.5/3.1	30	2	0.90	1.000

(*) Munsell color computed from reflectance spectra (see Escadafal et al., 1989 for details on technique)

TABLE 1. Characteristics of the soil samples referred to in this paper.

The first striking feature of soil reflectance is the high variability in brightness: soils can reflect only a small percentage of the incoming light ('black' soils), as well as up to 70 percent ('whitish' soils). One of the first consequences of this observation is the impact of soil albedo on the Earth energy balance (e.g. Wilson and Henderson-Sellers, 1987). But soils are not only dark or bright, their reflectance curves also vary in shape and hence, they have different colour appearances. The three main types of spectral shapes recognised in previous work (Condit, 1970) can be clearly distinguished on figure 1.

3.1 SOIL SPECTRAL FEATURES IN THE VISIBLE-NEAR INFRARED RANGE (VIS-NIR)

As an illustration of the different types of curves frequently observed in the visible-NIR range, figure 1 shows a series of soil spectra recently obtained with a spectroradiometer by Huete and Escadafal (1991). The composition of the soils cited as examples varies largely as reported in table 1.

Soils have a regularly increasing reflectance, forming a simple convex curve in the first case (soil sample KAR of figure 1). This type can be described as 'featureless' giving a pale colour (the flatter the curve, the more greyish the soil appears). On the contrary, the second type shows strong absorption features in the blue-green part of the spectrum and can be described as sigmoïdal (soil sample AVA, CON and MOL of figure 1). As a result of this uneven light reflection, these soils appear coloured yellowish to red. The more pronounced the curve, the more vivid the colour. The third type has a very low reflectance associated with a convex shape, and is typical of organic matter-rich soils (soil sample CLO of figure 1).

3.2 SOIL SPECTRA AND SOIL COLOUR

Soil colour is a variable of great importance in soil science, used for soil characterisation in the field, as well as for soil classification. Since the fifties, an internationally adopted method has been developed to normalise soil colour description. A given soil sample is compared with a special set of colour chart, the Munsell atlas of soil colour. Colour is a visual sensation related to the spectral properties of the observed objects. In controlled conditions, laws have been established linking the spectral reflectance and the composition of the incoming light to the colour stimuli (Wyszecki and Stiles, 1982).

This colourimetric approach has been recently used to establish the relationships between Munsell colour estimated by the soil scientists and soil spectral properties measured with laboratory and field instruments, and satellite sensors. Thanks to the simple shape of soil spectra and using some simplifications in colour theory, we have shown that i) soil spectra can be simulated from Munsell colour data, ii) soil colour can be remotely sensed using a limited number of measurements in the visible domain (Escadafal et al., 1988 and Escadafal, 1992)

3.3 SOIL SPECTRAL FEATURES IN THE SHORT WAVE INFRARED RANGE (SWIR)

Less data are available on soil features in this spectral domain, field instruments are less common and less easy to use, and hand-held devices still under development. Here again, the only published large series of data on soils in this domain is the above cited 'atlas' of Stoner et al. However, as the soils have been wetted prior to scanning, the dominant feature is the large absorbing band of water potentially masking more subtle phenomena.

A more recent study, based on laboratory spectrophotometric measurements of a series of 83 air-dried soil samples, shows that the soil spectra in the SWIR are largely dependent on the soil mineralogy (Courault et al., 1988). Laboratory spectral data on minerals have been extensively studied (particularly for ore exploration) and are largely available in the geological literature (see

Hunt et al, 1970, 1971, quoted in Curran, 1994). For our purpose they will be referred to, since minerals in soils are similar to the pure minerals described in those publications. The fact that soils are mixtures may however dampen the typical absorption features.

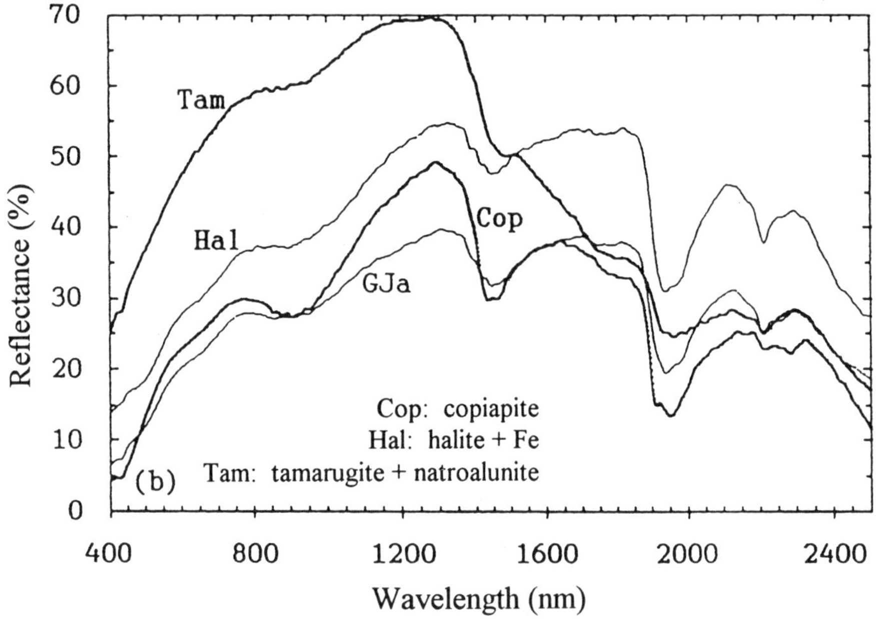

FIGURE 2. Examples of soil spectra with strong features in the SWIR region (salt affected soils of Casamance, Senegal) (after Mougenot, 1991).

Still, when compared to geological applications, remote sensing of soil mineralogy is somewhat easier. Soils are generally rather intensively homogenised by the natural fauna and/or human activity, so that their surface often reflects the inner composition of soil, contrary to exposed rocks whose patina or surface alterations may present different minerals. Hydrous minerals or OH-bearing minerals have the most spectacular absorption features as it can be seen on Figure 2 displaying spectra of minerals found at the surface of salt affected soils in Casamance, Senegal (Mougenot, 1991). These absorption bands are harmonics of main bands observed at longer wavelengths (Mulders, 1987). However, the strong water absorption bands of wet soils are the same as those found in the atmosphere. They are not usable in remote sensing of soil water content.

In a first approach, soil organic matter has no peculiar absorption features in this part of the spectrum. But, thanks to the strong interest of a large number of scientists on detecting and quantifying carbon in the soils, new subtle features may be detected in the near future and lead to spectroradiometric applications.

4. Soil spectrometry

4.1 Spectroscopy of Soil Iron Oxides

Spectrometry enables a more quantitative approach of the relationships between spectral shape and soil composition. The sigmoïdal shape of coloured soils is due to the presence of iron oxides. Although usually present only in limited amounts, these alteration products are very common in soils. They reflect the type and degree of evolution of the soil and therefore soil colour is a very important diagnosis variable. Subtle colour variations are often related to changes in soil texture and soil water regime, for instance.

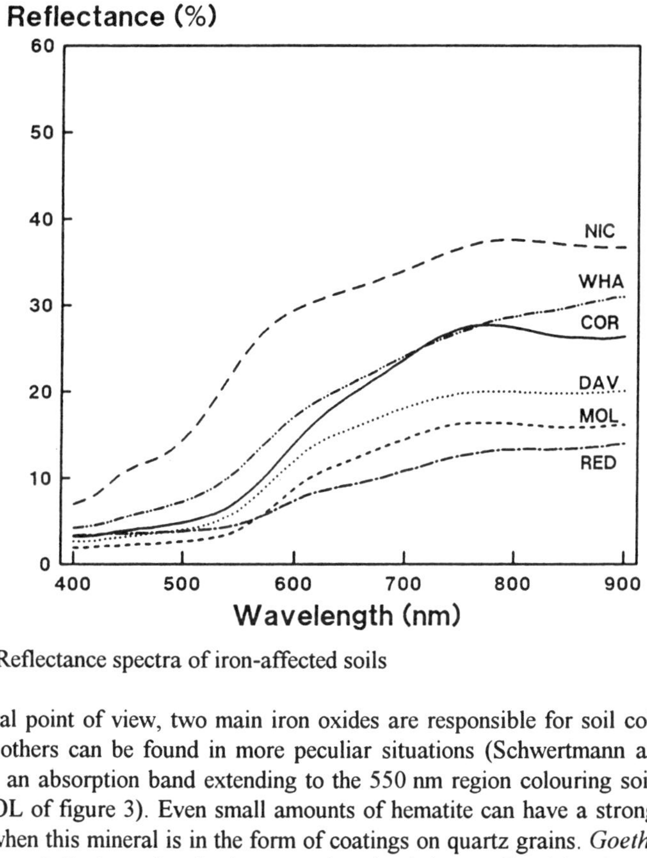

FIGURE 3. Reflectance spectra of iron-affected soils

From a general point of view, two main iron oxides are responsible for soil colour, hematite and goethite, but others can be found in more peculiar situations (Schwertmann and Taylor, 1977). *Hematite* has an absorption band extending to the 550 nm region colouring soils in red (samples COR and MOL of figure 3). Even small amounts of hematite can have a strong colouring effect, for instance when this mineral is in the form of coatings on quartz grains. *Goethite* has less strong features with two inflexion points in short wavelength, giving a yellowish colour to soils.

Derivative spectroscopy is an interesting technique for detecting subtle spectral differences (as discussed in detail in previous lectures of this course). Figure 4 displays the first derivative of iron oxides affected soil spectra, where goethite and hematite signature can clearly be distinguished. By analogy with laboratory titration techniques, the size of the peaks observed is related to the 'concentration' of the iron oxide considered. But in the case of soils this parameter expresses only the apparent content, since the size and distribution of oxides grains vary largely among soils

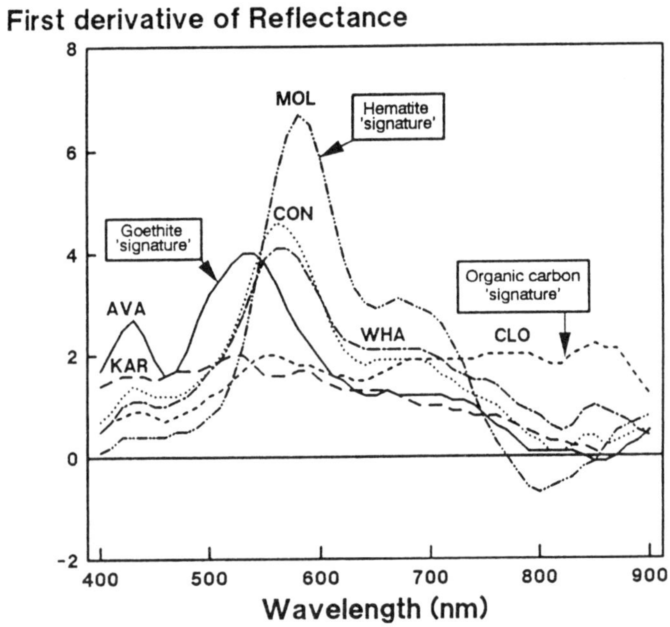

FIGURE 4. First derivative of iron- and organic matter-affected soil spectra.

However, when considering a limited range of soils, quantitative relationships can be established. Based on results of Torrent et al. (1983) relating colour derived indices to hematite in soils, Madeira (1991) has recently proposed a spectral index to determine the hematite content of tropical soils from Brazil ('latosols'). This index is based on a combination of reflectances measured in the three visible bands of Landsat Thematic Mapper, characterising the shape of the spectrum. Higher spectral resolution data from spectroimagers will certainly enable to refine this approach.

4.2 POTENTIAL OF SPECTROMETRY FOR SOIL ORGANIC CARBON

As it appears on figure 5, soils containing around 1 % of organic carbon have rather flat spectra, only some of them present a slightly concave shape (samples COM and CLO). This feature will be detectable with spectrometric data, but has not been used so far for remote sensing of organic carbon in soils. However, attempts have been made to correlate the organic carbon content of soils with their brightness. This technique has gained some success when restricted to the field or farm level. In the same manner as iron oxides, the darkening efficiency of the organic matter varies largely among soils, so that the 'spectral shape' approach maybe more successful in the future.

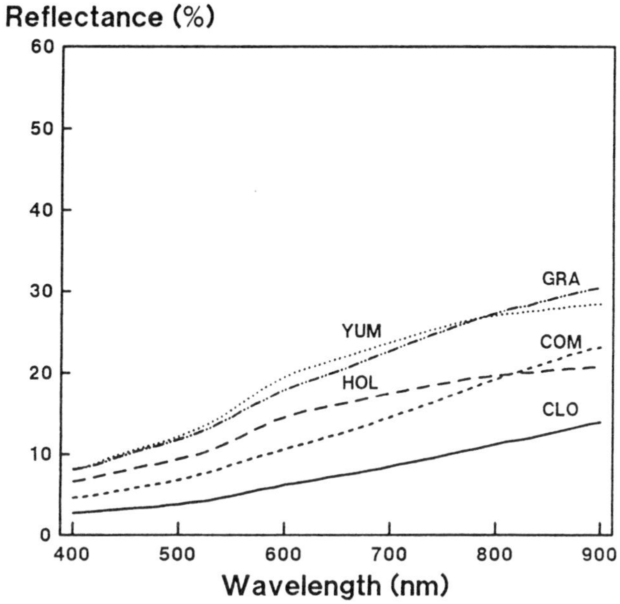

FIGURE 5. Examples of organic matter-affected soil spectra

4.3 Influence of Soil Water Content

Since the first observations made in the field and the first spectra recorded in the laboratory, it has been recognised that wetted soils are darker than dry ones. Bowers and Hanks (1965) have published curves illustrating the lowering of the overall reflectance of a soil sample with increasing water content. This same effect is observed with increasing organic matter content or roughness. Thereby, soil water content cannot generally be obtained from visible-NIR spectral data. However, in the case of red soils from Brazil, Bedidi et al. (1992) have recently found that spectra of wetted soils are not homothetic with the ones of the same soils in dry condition. These iron-rich soils present a spectral 'shift' of the inflexion point of the reflectance curve when the water content increases (figure 6). This is another example of foreseeable application of imaging spectrometry to soil parameters assessment, although more performing techniques are concurrently developed to measure soil water content (e.g. microwaves).

4.4 Soils Are a Mixture: Examples of 'Unmixing'

As an attempt to summarise, in the visible-NIR spectral range, the spectral properties of a given soil can be considered as a combination of individual spectra of a few main components: skeletal (e.g.: quartz) and matrix (e.g.: clay) minerals, iron oxides (goethite and/or hematite), organic matter, water. Surface roughness variations alter the overall brightness but do not modify the spectral shape (Escadafal and Huete, 1991a). Thus, although the statistical dimensionality of soil spectra is generally low (Price, 1990), spectral unmixing techniques can be applied, since the number of components to extract is also limited (see Smith, 1993, this volume). As an example, Huete and Escadafal (1991) have shown for a series of 46 different soils that any of the studied spectra can be reconstructed by linear mixing of four 'basis' curves (eigenspectra). The latter are related to the spectral features of the major soil components (e.g., 'average' skeleton, organic carbon, iron oxides, etc.). The contribution of each of these eigenspectra is an estimation of the *apparent* soil content in those components.

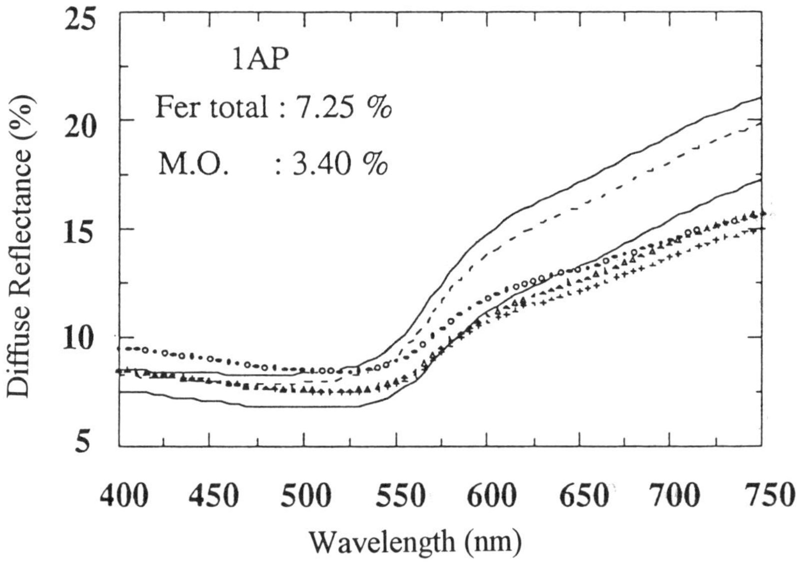

FIGURE 6. Diffuse reflectance of an iron-rich soil sample with increasing water content (after Bedidi et al., 1991).

5. Case study: applications of arid soils spectrometry

As stressed in the introduction, in arid regions soils can be easily observed from space, vegetation cover is low and atmospheric conditions are often favourable for optical remote sensing. Moreover in those environments, since soils are in direct contact with the atmosphere, their impact on the water and plant cycles is essential. Recent studies in different arid regions (Sahel, Northern Africa) have shown that infiltration is controlled by the soil surface composition and features such as sand sheets, crusting etc. (Casenave et Valentin, 1989; Escadafal, 1989).

In the northern fringe of the Sahara, these surface characteristics were found to condition as well the spontaneous emergence and development of new plants (Asseline et al., 1989) and the dust production during wind storms (Escadafal et Callot, 1991). Surface characteristics are often related to specific soil types, so that remote sensing of surface condition is a powerful technique for arid soils mapping (Escadafal et Pouget, 1989).

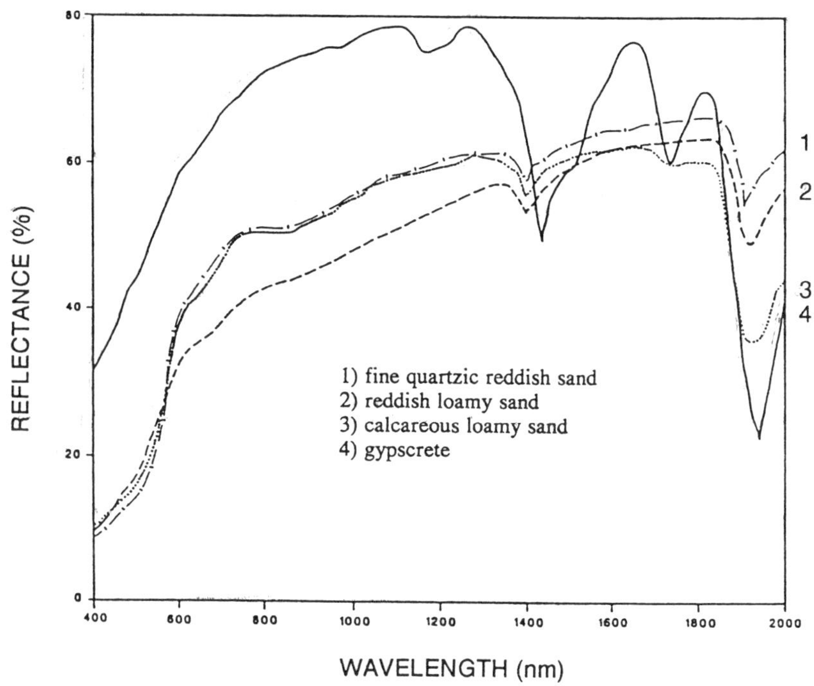

FIGURE 7. Examples of laboratory arid soil surface spectra (spectral range: Visible- SWIR)

5.1 ARIDIC SOILS SPECTRAL PROPERTIES

When observed from above, arid land surfaces are composed by 70 to 100 percent of soils (and rocks). These soils have a very low organic matter content, are often coloured by iron oxides and maybe salt affected (see example figure 7). Therefore, measured form space, their brightness is generally high, but will decrease with increasing roughness, water content and steppic vegetation cover (Escadafal, 1989 ; Epema, 1990).

5.2 'SPECTRAL' DESERTIFICATION INDICATORS (EXAMPLE OF NORTHERN AFRICA)

Classical techniques of monitoring the environment with satellites are often based on assessing the changes in biomass through vegetation indices. These indices are based on the high infrared reflectance of green vegetation (relatively to the soil). However, in arid regions this approach encounters severe limitations, the vegetation signal is very low since the vegetation is scarce and the steppic plants are only partially and temporarily green.

Considering that changes in arid environments are also affecting the soil surface, in an undergoing project in Southern Tunisia we are currently investigating potential surface spectral modifications which might indicate land degradation phenomena ('desertification').

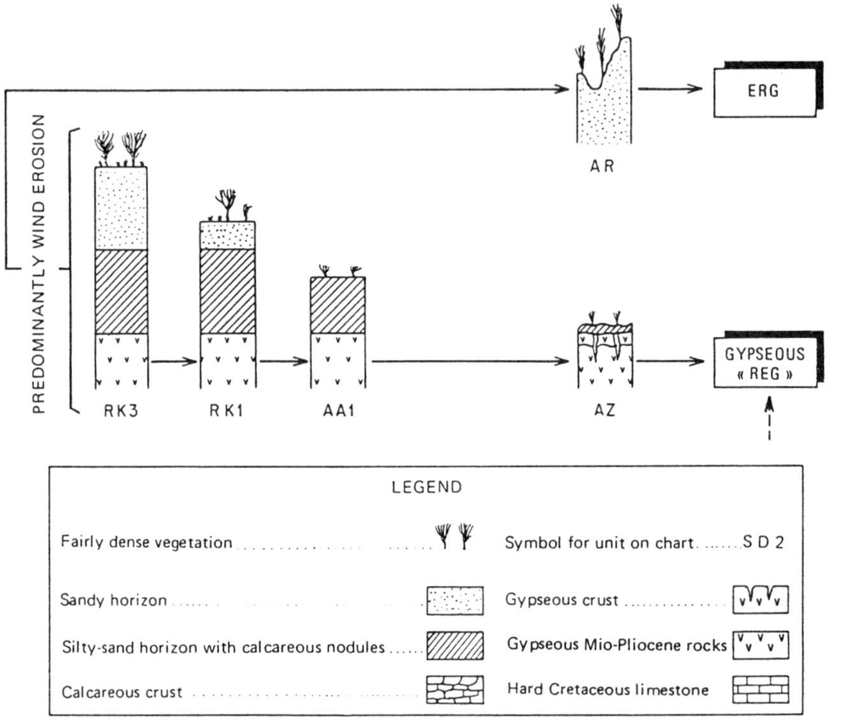

FIGURE 8. Degradation sequence of sandy ecosystems in arid Tunisia (after Mabbutt and Floret, 1980)

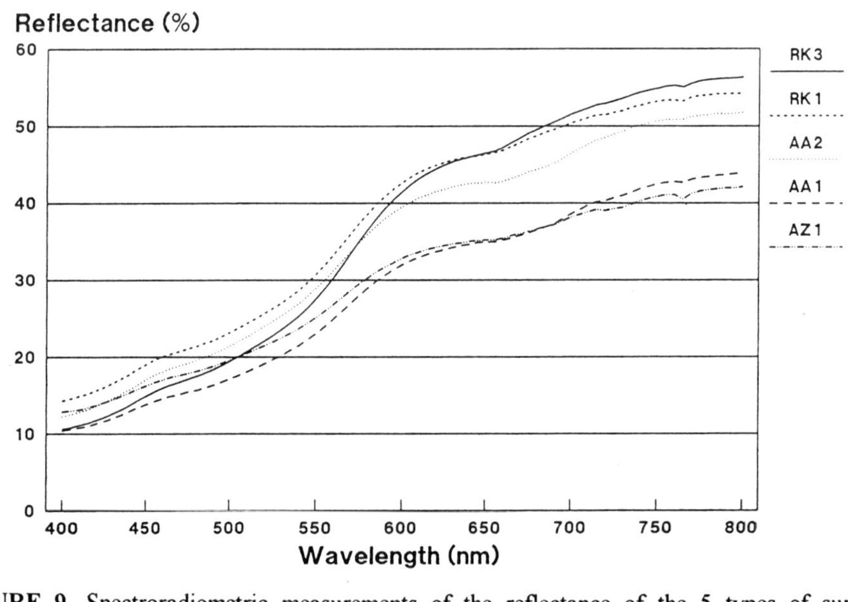

FIGURE 9. Spectroradiometric measurements of the reflectance of the 5 types of surfaces (ecosystem degradation sequence, see also figure 8)

Degradation scenarii in this area have already been studied in the field for many years by ecologists. Here we have used as an example the sequence of the sandy ecosystems described in Mabbut and Floret (1980, figure 8). Soils surface spectra have been acquired and processed over surfaces characterised by an increasing degradation level. Figure 9 displays the reflectance spectra (averages of transects) of the five types of surfaces recognised by the ecologists as affected by an increasing desertification level. The spectra are very similar, differences in shapes appear however when computing the first derivative of these curves (figure 10). Clearly, the surface degradation correspond to a flattening of the curve (lower values of the derivative). This spectral phenomena is independent of the brightness and could then be detected even with varying surface roughness or humidity (contrary to the albedo).

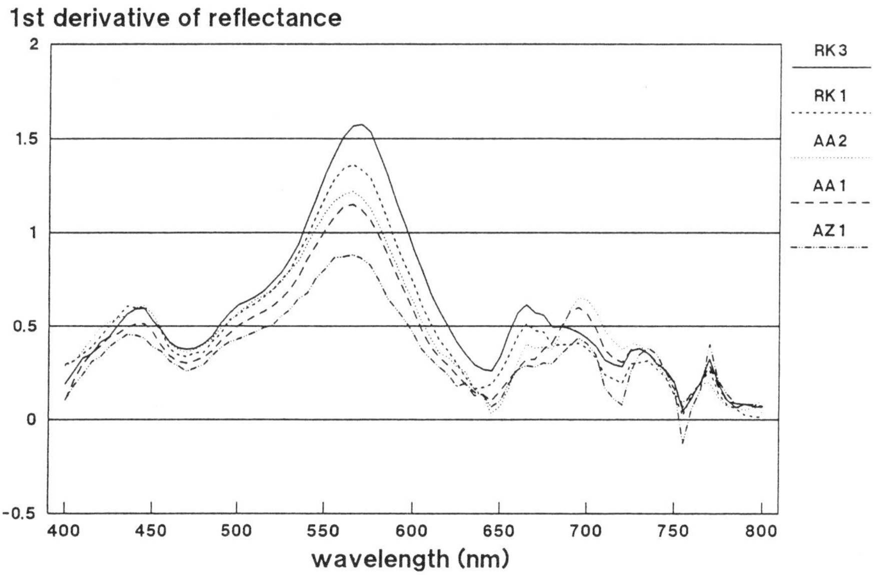

FIGURE 10. Derivative of spectra from Figure 9. The spectral shapes changes as the surface is 'desertified'.

5.3 ARIDIC SOILS SPECTRAL VARIABILITY AND REMOTE SENSING OF VEGETATION

Before concluding, the effect of soil spectral variability on remote sensing of the biomass through the computation of vegetation indices is to be reminded.

These indices rely on the concept of the 'soil line', assuming soils have a constant NIR/Red ratio. However, several studies indicate this only a rough approximation. In a recent studies we have shown that the soil induced 'vegetation signal' can be far from negligible in the case of organic or coloured aridic soils (up to 0.3 NDVI value; Escadafal and Huete, 1991). In the latter case we have suggested a preliminary correction technique using the shape in the visible part of the spectrum to model the soil 'noise' and dampen it (Escadafal and Huete, 1992). More sophisticated techniques applied to higher spectral resolution data will certainly improve the detection of low amounts of green vegetation by discriminating more efficiently the soil signal. Spectral mixture analysis is obviously a very promising one (see Hill et al., 1994; Smith et al., 1994, this volume).

6. Conclusions

A large effort is currently done to develop remote sensing techniques for retrieving parameters concerning the Biosphere and the Geosphere. Between these two main 'spheres', soils form an interface the functions of which are more and more taken into consideration. Imaging spectrometry which is already known as a powerful technique to study vegetation and minerals, is a promising technique to assess soils parameters of environmental significance. It can be used in two main ways:

- *synthetic approach*: in this case, spectrometric data will be used to spectrally discriminate different types of soil materials in a given area. Typical synthetic variable will be soil colour and dominant minerals, variables which soil scientists classically collect during field work. They will use their expertise to combine this spectra-derived information with data on the climate, on parent material and landscape morphology to infer the soil 'types'. Each of these soil types can be assigned to classes of properties relevant in environmental studies, such as organic carbon level, total water capacity, infiltration ability, sensitivity to erosion etc. Within each main soil type, subtle colour or mineralogical variations will be indicators of changes (e.g., erosion, desertification).

- *analytical approach*: techniques such as derivative spectroscopy or mixtures analysis enable to quantify the apparent content of a land surface element in known soil compounds (provided these compounds differ spectrally). Remotely sensed concentrations of carbonates, sulphates, organic carbon would be helpful as input in models of biogeochemical cycles for instance.

However, this assumes that soils are mixtures of equally sized and distributed particles. This is far from reality as soils are highly differentiated natural bodies. They present various structures from the microscopic level (ex.: clay coatings on coarser grains) to the landscape level ('catenas'). The apparent content observed at the soil surface must then be interpreted using the soil scientist's expertise. Knowing the soils of the concerned area, he will be able to infer the type of soil particle distribution and the relationships between the surface and the lower soil layers (horizons).

Whatever approach is used, it appears that only a 'soil knowledge-based' interpretation of spectral analysis can produce sound results on the global soil composition. It is also obvious that the reflectance contribution of soils, being part of most of the remotely sensed surface elements, must be taken into account even when they are not the primary focus of the study.

7. References

Arnold, R.W., I. Szabolcs, and V.O. Targulian (1990) 'Global soil change', *Report of an IIASA-ISS-UNEP Task force on the role of soil in Global Change*, International Institute for Applied System Analysis, Laxenbourg, Austria, 110 p.

Asseline, J., R. Escadafal, and A. Mtimet (1989) 'Etude expérimentale de la dynamique superficielle d'un sol aride (Bir Lahmar, Sud tunisien)', *Sols de Tunisie*, 14, 17-62.

Baumgardner, M.F., L.F. Silva, L.L. Biel, and E.R. Stoner (1985) 'Reflectance properties of soils', *Adv. in Agronomy*, 38, 1-44.

Bedidi, A., B. Cervelle, J. Madeira, and M. Pouget, (1992) 'Moisture effects on visible spectral characteristics of lateritic soils', *Soil Science*, 153(2), 129-141.

Bowers, S.A. and R.J. Hanks (1965) 'Reflectance of radiant energy from soils', *Soil Science*, 100, 130-138.

Casenave, A. and C. Valentine (1989) 'Les états de surface de la zone sahélienne. Influence sur l'infiltration', Orstom, Paris, 229 p.

Condit, H.R. (1970) 'The spectral reflectance of American soils', *Photogramm. Eng.*, 36, 955-966.

Courault, D., M.C. Girard, and R. Escadafal (1988) 'Modélisation de la couleur des sols par télédétection', *Actes du 4e Coll. int. 'Signatures spectrales d'objets en télédétection'*, Aussois, Janvier 1988, 357-362.

Curran, P.J. (1994) 'Imaging spectrometry - its present and future rôle in environmental research', in J. Hill and J. Mégier (eds.) *Imaging spectrometry - a tool for environmental observations*, Kluwer Academic Publishers, Dordrecht, (this volume).

Epema, G. (1990) 'Effect of moisture content on spectral reflectance in playa area in southern Tunisia', *Proc. Int. Symp. 'Remotes sensing of water ressources'*, Enschede, The Netherlands, August, 20-24, 1991, 301-308.

Escadafal, R. (1989) 'Caractérisation de la surface des sols arides par observations de terrain et par télédétection', Etudes et thèses, Orstom, Paris, 312 p.

Escadafal, R. (1992) 'Remote sensing of soil color: principles and applications', *Remote Sensing Reviews*, in press.

Escadafal, R. and Y. Callot (1991) 'Monitoring Saharan dust sources areas with multispectral imagery', *Proc. Eighth Thematic Conference Geol. Remote Sensing*, April 29-May 2, 1991, Denver, Colorado (USA), 1473-1483

Escadafal, R., M.C. Girard, and D. Courault (1989) 'Munsell soil color and soil reflectance in the visible spectral bands of Landsat data (MSS and TM)' *Remote Sensing of Environment*, 27, 37-46.

Escadafal, R. and A.R. Huete (1991a) 'Influence of the viewing geometry on the spectral properties (high resolution visible and NIR) of selected soils from Arizona', *5th Intern. Coll. 'Mésures physiques et signatures en télédétection'*, Courchevel, France, 14-18 Janvier 1991,

European Space Agency, SP-319, 401-404.
Escadafal, R. and A.R. Huete (1991b) 'Improvement in remote sensing of low vegetation cover in arid regions by correcting vegetation indices for soil 'noise',*C.R. Acad. Sc. Paris*, 312, Sér.II, 1385-1391.

Escadafal, R. and A.R. Huete (1992) 'Soil optical properties and environmental applications of remote sensing', *Int. Arch. Photogramm. Rem. Sens.*, vol.29(B7), 709-715.

Escadafal, R., A.R. Huete, and D. Post (1990) 'Estimating soil spectral properties (visible and NIR) from color and roughness field data', *Proc. 23d Int. Symp. Rem. Sens. Environment*, Bangkok, Thailand, April 18-25 1990.

Escadafal, R. and M. Pouget (1989) 'Comparaison des données Landsat MSS et TM pour la cartographie des formations superficielles en zone aride (Tunisie méridionale)', *Proc. Workshop 'Earthnet pilot project on Landsat Thematic Mapper applications'*, Dec. 1987, Frascati (Italie), ESA publ. SP-1102, 301-307

Fernanadez, R.N. and D.G. Schulze (1987) 'Calculation of soil color from reflectance spectra', *Soil Sci. Soc. Am. J.*, 51, 1277-1282

Hill, J. and J. Mégier (1991) 'The use of imaging spectrometry in mediterranean land degradation and soil hazard assessments', *Proc. 5th Int. Coll. 'Physical measurements and Signatures in Remote Sensing'*, Courchevel, France, 14-18 jan., ESA SP-319, 185-188.

Hill, J., W. Mehl, and M. Altherr (1994), 'Land degradation and soil erosion mapping in a Mediterranean ecosystem', *in* J.Hill and J. Mégier (eds.) *Imaging spectrometry - a tool for environmental observations*, Kluwer Academic Publishers, Dordrecht, (this volume).

Huete, A.R. and R. Escadafal (1991) 'Assessment of biophysical soil properties through spectral decomposition technique', *Remote Sensing of Environment*, 35, 149-159.

Jackson, R.D., P.M. Teillet, P.N. Slater, G. Fedosejevs, M.F. Jasinski, J.K. Aase, and M.S. Moran (1990) 'Bidirectional measurements of surface reflectance for view angle of oblique imagery', *Remote Sensing of Environment*, 32, 189-202.

Latz, K., R.A. Weismiller, G.E. van Scoyoc, and M.F. Baumgardner (1984) 'Characteristic variations in spectral reflectance of selected eroded alfisols', *Soil Sci. Soc. Am. J.*, 48, 1130-1134.

Mabbutt, J.A. and C. Floret (eds.) (1980) 'Case studies on desertification', *Natural Resources Research XVIII*, Unesco, Paris, 279 p.

Madeira, J. (1991) 'Etude quantitative des relations constituants minéralogiques-réflectance diffuse des latosols brésiliens. Application à l'utilisation pédologique des données satellitaires TM (région de Brasilia)', Thèse de doctorat, Université Pierre et Marie Curie, Paris, 232 p.

Mougenot, B. (1991) 'Caractéristiques spectrales de surfaces salées à chlorures et sulfates (Sénégal)', *in* M. Pouget (ed.) *'Caractérisation et suivi des milieux terrestres en régions arides et tropicales'*, Colloques et Séminaires, Orstom, Paris, 49-70.

Mulders, M.A. (1987) 'Remote sensing in soil science', *Developments in Soil Science*, 15, Elsevier, Amsterdam, 379 p.

Price, J.C. (1990) 'On the information content of soil reflectance spectra', *Remote Sensing of Environment*, 33, 113-121.

Schwertmann, U. and R.M. Taylor (1977) 'Iron oxides', *in* Dixon and Weed (eds.) *'Minerals in soil environment'*, Soil Sci. Soc. Am., Madison (USA), 145-180.

Smith, M.O., J.B. Adams, and D.E. Sabol (1993) 'Mapping sparse vegetation canopies', *in* J. Hill and J. Mégier (eds.) *Imaging spectrometry - a tool for environmental observations*, Kluwer Academic Publishers, Dordrecht, (this volume).

Stoner, E.R., M.F. Baumgardner, L.L. Biehl, and B.F. Robinson (1980b) 'Atlas of soil reflectance properties', L.A.R.S., Purdue University, 75 p.

Torrent, J., U. Schwertmann, H. Fechter, and F. Alferez (1983) 'Quantitative relationships between soil color and hematite content', *Soil Science*, 136, 354-358.

Wilson, M.F., and A. Henderson-Sellers (1987) 'Sensitivity of BATS to the inclusion of variable soil characteristics', *J. Clim. Appl. Meteorol.*, 26, 341-362.

Wysecki, G. and W.S. Stiles (1982) 'Color science: concept and methods, quantitative data and formulae', Wiley, New York, 2nd edition, 950 p.

DATA ANALYSIS - PROCESSING REQUIREMENTS AND AVAILABLE SOFTWARE TOOLS

WOLFGANG MEHL
Institute for Remote Sensing Applications
Commission of the European Communities
Joint Research Centre
I-21020 Ispra (Va), Italy

ABSTRACT. Methods for analysing remote sensing data are conditioned by sensor technology as much as by the particular application. However, in the past twenty years a number of techniques have proved to be widely applicable and have become standard tools available in public domain or commercially offered software packages. Many of these tools are no longer applicable to data acquired with imaging spectrometers either because of the high dimensionality of the data which increases processing time beyond practical bounds (e. g. maximum likelihood classifiers), or because of the high statistical variability of data which again is directly linked to the data space dimension (e. g. clustering algorithms), or simply because implementations have arbitrary limits for the processable number of bands. New techniques have evolved which exploit the continuity of spectral information available with imaging spectrometers (e. g. correlation and fitting algorithms) or the detail observed in the short wave infrared region. Also techniques which have not been generally accepted in the past have been reconsidered within the context of image spectrometry (e. g. spectral unmixing).

1. Introduction

This paper cannot give a beginner's course in imaging spectrometry (IS) data processing. While there are no text books yet about that specific subject, the amount of material to be presented would by far exceed the available space here. Its intention is not even to describe algorithms or methods in a detail that would allow to implement them; algorithms will be described only to such an extent that the reader will know what mathematical tools would be required to understand them. The main emphasis will be given to what makes the processing of IS data different from processing of optical data we are more familiar with, and why; to show where methods and tools developed for remote sensing data analysis can be applied to IS data; and to point to some possible usefulness of techniques commonly associated with IS data processing for remote sensing in general.

Methodology in remote sensing has always been conditioned as much by the characteristics of sensors and their vehicles as by a particular application. Therefore we need to understand what makes an IS sensor different from a multispectral scanner:

An IS has at least n bands; there is no clear definition for n but existing instruments have between 16 and 288 bands.

J. Hill and J. Mégier (eds.), Imaging Spectrometry - a Tool for Environmental Observations, 89–107.

The band width of an IS is less than x nm (or $\frac{bandwidth}{wavelength} < y$*);* again, *x* or *y* are not defined but existing instruments have band widths of usually less than 30 *nm* in the visible, near infrared and short wave infrared region.

An IS has contiguous bands in any spectral region; multispectral scanners with narrow bands exist but their bands are not contiguous. Imaging spectrometers typically have enough bands to cover a whole region between major atmospheric absorption bands.

IS should have absolute radiance calibration and therefore we are speaking of spectrometry, not spectroscopy. However, most real instruments haven't ...

Most IS are experimental sensors, meaning that they have no standard data formats, are not space borne, are not necessarily described very well nor stable in time.

Some IS cover unusual wavelength ranges. Although the Thematic Mapper sensor is covering the 1.5 μm and 2.5 μm regions (each with one band), only IS instruments permit to appreciate the richness of these spectral ranges.

2. Data Properties and Processing Requirements

The design of imaging spectrometers, as any instrument design, is governed by a number of compromises since different requirements are often incompatible with each other. Among the design parameters which have an impact on required or factual data properties are *number of bands*; *band width, data quantisation, system noise*, mechanical, electrical and thermal *system and components stability, data flow band width, spatial resolution*, to name a few but in particular those (with the possible exception of spatial resolution) which require major attention when designing IS instruments. The energy *E* collected by a sensor element while measuring a single radiance value is given by $E = I \cdot C \cdot \Delta t \cdot \Delta\lambda$, where *I* is the average radiation intensity arriving at the sensor, Δt is the sampling time interval which in turn is related to the spatial sampling interval, $\Delta\lambda$ is the band width and therefore determined by the dispersing optical element, and *C* is determined by parameters of the optical components of the instruments (aperture, optical transmission loss, focussing length etc.). Low values of *E* will result in low signal/noise (*S/N*) ratios. High values of $\Delta\lambda$, meaning lower spectral resolution, may allow large S/N ratios but require the data to be digitized with higher precision if the detectability of small spectral features is to be preserved. Assuming that a given spectral region is to be covered contiguously by the instrument, data quantity is inversely proportional to $\Delta t \cdot \Delta\lambda$ (and therefore to *E*). From those relationships we see easily the tradeoff between the various design parameter listed above.

Some of the characteristics of the probably most advanced IS instruments, the Airborne Visible/Infrared Imaging Spectrometer (AVIRIS) of JPL, are summarised as follows (Green *et al.*, 1992):

spectral range: *0.4* μm *- 2.5* μm contiguous

spectral band width: *10 nm*

nr. of bands: *224*

S/N ratio (bright natural target): *400* at $0.6\ \mu m$ and $1.0\ \mu m$, ≥ 200 in most of the range $0.5\ \mu m$ - $1.1\ \mu m$, ≥ 150 at $1.2\ \mu m$ - $1.3\ \mu m$ and ≥ 100 throughout most of the full spectral range

noise equivalent radiance $\approx 40 \frac{nW}{cm^2 nm\, sr}$ throughout most of the range, $< 20 \frac{nW}{cm^2 nm\, sr}$ at 1.5 - 2.5 μm

ground sampling (nadir): *20 m*

pixels / line:*614*

data word: *16 bit*

calibrated data radiance resolution: $5 \frac{nW}{cm^2 nm\, sr}$

An image processing system handling such data should provide:

large buffers: a full AVIRIS line contains 137536 values.

handling of at least 16 bit wide image values: the minimum for preserving the S/N ratio is 10 bit. For practical reasons data is delivered as 16 bit signed integer words.

3-D display capabilities for data visualisation: the representation of spatial properties and a spectral continuum at the same time defeat false colour image and 2-D graphical visualisation.

extended precision arithmetics: even for simple computations like the Euclidean distance between two data vectors, 24 mantissa bits of an IEEE single precision floating point value are not sufficient to prevent data loss.

Clearly, such data is demanding also more raw computational resources than conventional remote sensing imagery requires; this will be discussed in more detail later.

3. Preprocessing

Most remote sensing applications require that the raw imagery to be analysed will pass through procedures which will make it more suitable for subsequent interpretation, be it automatic, computer-aided, or visual, but are not strictly belonging to the interpretation process themselves. Those preprocessing steps deal with *reformatting*, *radiometry* (calibration or radiometric normalisation, atmospheric correction), *geometry*, *noise processing* (spatial or frequency filtering, image restoration), and *information reduction*.

3.1 DATA REFORMATTING

Remote sensing data is delivered in an infinite variety of formats. Almost all image processing software systems expext the data to be in only one of the following organisations:

Band Interleaved by Pixel (BIP): all bands of the first pixel in the first line are followed by all bands of the second pixel in the first line, etc.; each successive line is stored after all information for the previous line.

Band Interleaved by Line (BIL): the first band of all pixels in the first line are followed by the second band of all pixels in the first line, etc.; each successive line is stored after all information for the previous line.

Band Sequential (BSQ): the first band of all pixels in the first line are followed by the first band of all pixels of the second line, etc.; each successive band is stored after all information for the previous band, within the same file for direct access storage media; for this data ordering scheme the bands are always separated by file marks on sequential media.

Band Separated by File (BSF): the storage order for each single band is like BSQ, but different bands are stored in different files. For sequential storage media (magnetic tapes), the same organisation is usually denominated as BSQ.

Raw data for many sensors is often organised in ways which cannot be categorised as above since the sensors are gathering their measurements in an order which is determined by their geometrical arrangement relative to scan optics, dispersion optics, flight and scan direction, and by delays in the electronic signal processing. Moreover the data length is often not a multiple of a common storage unit. Another frequent cause for reformatting is the difference in data representation between different processing hardware. It is common experience that data sets must be reformatted various times throughout their utilisation and that this apparently simple processing step turns out to be one of the more complex, often aggravated by insufficient or inaccurate format documentation!

3.2 CALIBRATION AND RADIOMETRIC NORMALISATION

While purely statistical classification algorithms on single multispectral imagery does not require much calibration as long as the sensor system has a sufficiently linear response, for almost any application of IS data conversion of measured radiance values into some reference system is required. Most sensor systems do have a sufficiently linear radiance response; therefore a transformation of measured values d into a calibrated value r is just determined by two constants, a *slope* (or *gain* or *calibration factor*) s and an *intercept* (or *bias* or *offset*) c, such that $r = d \cdot s + c$. These constants could be determined by values r_0 and r_1 of reference measurements taken from two targets with digital counts d_0 and d_1, and $s = \frac{r_1 - r_0}{d_1 - d_0}$, $c = r_0 - sd_0$. In practice, d_i

and r_i are always averaged over a number of measurements to remove noise. Another method of determining these constants uses statistics over corresponding inhomogeneous targets. If $\mu(x_i)$ and $\sigma(x_i)$ denote mean value and standard deviation over a population $\{x_i\}$, then $s = \frac{\sigma(r_i)}{\sigma(d_j)}$, and $c = \mu(r_i) - s\,\mu(d_j)$. Note that here values in d_i and r_i do not necessarily map one to one, the populations need only to be statistically equivalent. If the sensor to be calibrated is noisy or not perfectly linear, a better estimation of s and c are computed using a linear regression, with $s = \frac{\mu(d_i r_i) - \mu(d_i)\mu(r_i)}{\sigma^2(d_i)}$ and $c = \frac{\mu(d_i^2)\,\mu(r_i) - \mu(d_i)\,\mu(d_i r_i)}{\sigma^2(d_i)}$. Regression analysis also supplies a validation of sensor linearity and can be extended to non-linear calibration.

Depending on the reference source we distinguish several types of calibration procedures:

Inter-line and/or inter-pixel calibration is the minimum required for any data but becomes more complex with the larger sensor arrays typical for the more advanced imaging instruments.

Inter-band calibration is particularly significant for IS sensors where spectral ranges are dispersed over sensor arrays. For example, the CASI instrument uses a detector array of $288 \cdot 512$ elements where the spectral range is dispersed across the 288 rows, the spatial range across the 512 columns. This means that 147456 detectors need to be cross-calibrated. When spectral bands of an IS instrument (or of a spectroradiometer) are not intercalibrated, small spectral features may not be detectable and artefacts generated.

Inter-flight calibration is needed for multi-temporal data analysis but may also provide a measure for instrument reliability, in particular if calibration coefficients change randomly.

Inter-instrument calibration is important for multi--sensor data analysis but together with inter-flight calibration it is also a way to obtain or maintain coefficients for absolute calibration.

Absolute calibration in physical radiance units can be achieved under laboratory conditions with standardised radiation sources. Quite often, the in-flight environment is sufficiently different from laboratory conditions to invalidate a laboratory calibration.

Two issues may interfere strongly with radiometric calibration and may have a significant impact on the interpretability of IS data:

Spatial registration of bands is a preprocessing step required also for more traditional imaging sensors, including the Thematic Mapper. Preprocessing of TM data is indeed rather complex since the data streams of its 100 sensors are starting at different times within a non-linear scan.

Spectral band position calibration is typical for IS instruments where due to the small band width and the band contiguity, the correspondence between wavelength and spectral band number is not easily determined and may even change because of mechanical or thermal effects during operation of the instrument.

3.3 Atmospheric Correction

The purpose of atmospheric correction is the determination of reflectance properties of the observed surface as they would appear without intervening atmospheric absorption or scattering. The simplest approach is methodologically equivalent to a calibration of sensor data against simultaneous ground measures, and the methods described in 3.2 apply. However, this approach has some shortcomings:

Not always are simultaneous measurements of ground targets available.

Atmospheric effects are non-linear. The reflection of the area surrounding a target will contribute to the signal arriving from the target direction to the sensor because of atmospheric scattering. The geometrical configuration of incident light and look direction varies within the image, as does the air mass between sensor and target.

Therefore model-based atmospheric correction procedures are still preferred for orbital platforms and are necessary for airborne sensors, hence for IS instruments. Still some model parameters can be derived from imagery data, either through simultaneous ground measurements and possibly ground based incident direct and diffuse light measurements within extended homogeneous areas, or by using extended targets with known reflectance properties, an approach which implies more uncertainty about the incoming radiation. With IS data it is also possible to derive information about the quantity of some atmospheric constituents, by observing the profile or the depth of absorption bands of these constituents in spectral ranges which are sufficiently featureless for the expected ground targets. As an example see Gao *et al.* (1993, 1990): the ratios $\frac{\rho(0.94\,\mu m)}{0.55\rho(0.865\,\mu m)+0.45\rho(1.03\,\mu m)}$ and $\frac{\rho(1.14\,\mu m)}{0.55\rho(1.23\,\mu m)+0.45\rho(1.05\,\mu m)}$ are used to determine the column water vapor densities ($\rho(\lambda)$ is the apparent target reflectance at the sensor at wave length λ).

Calculation of good radiance transfer models can be demanding in processing time if applied on a by pixel bases to imagery, but there is no significant difference in required resources between IS and other images; processing time is proportional to the number of points and the number of bands, and the computation may be done for each band separately. Often imagery will be corrected using simplified tabulated models, while full multiple scattering radiance transfer models are applied only for atmospheric parameter estimation and removal of atmospheric effects in ground target based calibration.

For high resolution image data and complex topography, a full atmospheric correction requires the use of digital elevation data, to account for the variation in path length sun $\rightarrow$ target $\rightarrow$ sensor. Digital elevation models (*DEM*) permit moreover to improve the estimation of target illumination, assuming Lambertian reflection. The latter assumption however is reasonable only for few natural surfaces and may be grossly violated for vegetated areas. Again, no particular considerations apply for IS data versus more traditional imagery.

3.4 GEOMETRIC RECTIFICATION

Methods and problems of geometric rectification, either against a master image or into a map coordinate grid, are not peculiar for IS imagery. Like for any airborne imagery, geometric correction should use all of flight attitude data, ground control points (*GCP*), and digital elevation models. Care must be taken to derive a compound geometric model from the input sources (attitude, GCP, DEM) before performing the rectification because any resampling will cause radiometric or positional errors, which would be accumulated in case of repeated geometric transformations. The rectification procedure may be done separately for each band, implying that memory resources are not different from any other image rectification, and processing time is proportional to the number of spectral bands. Clearly the better strategy is not to perform geometric rectification on the full multiband image but only on derived products which will have a much smaller number of channels, and reference any ground target to the unprocessed image by applying the inverse geometric transformation.

3.5 FILTERING AND IMAGE RESTORATION

We can classify image degradation processes essentially into the following categories:

coherent noise, i. e. periodic noise, caused generally by calibration differences between the elements of a sensor array, or by some sort of AC interference in the electronics of the instrument.

Image blur due to defocussing, motion, or image sensor memory effects.

Data loss, i. e. blocks or lines in the image which do not contain image information.

Incoherent banding

Random noise, e. g. thermal noise, detector quantum noise, digitisation noise.

The first two processes induce autocorrelation while the others diminish autocorrelation in the image.

Coherent image banding can of course be reasonably well corrected in the Fourier domain since the frequencies of the disturbance will result in isolated peaks in the Fourier transform which can be averaged out. If the banding is caused by calibration differences in a detector array, techniques described in 3.2 will provide comparable results using less computing and memory resources; the same methods are the only choice for irregular banding along image lines or columns.

Also image blur may be corrected in the frequency domain if the blurring is translation-invariant or can be made so through some transformation, and the Fourier transform of the point spread function (also called modulation transfer function, *MTF*) is invertible. For MTFs with a small footprint (i. e. small non-zero extension), applying the corresponding spatial deconvolution filter may be more cost-effective. In any case, there are two problems, one of which is the determination of the MTF. The second one is due to the fact that any blurring effect will average over some image values, with information loss due to the limited dynamic range of sensor systems. Image deconvolution may therefore significantly enhance instrument noise.

Coherent noise, banding effects, and blurring are usually processed independently for each recorded spectral band; neither their occurance nor their correction are specific for IS data.

Lost data blocks are usually interpolated from the surrounding data. Spectral band continuity permits for IS instruments to interpolate from adjacent bands as well.

Linear digital filters trade random noise against resolution and are therefore not very helpful in improving the potential for information extraction from images. The median filter will also reduce image noise since it replaces isolated values with the centre value of the radiance distribution inside the filter window. This filter tends to improve the contrast along the border between two homogeneous image areas but trades this against an uncertainty in locating the border; moreover it removes thin linear structures. A number of averaging spatial filters have been designed which adapt their coefficients according to some homogeneity or edgeness measure. One of the more successful filters of that type is the one proposed by Nagao and Matsuyama (1979) which measures the variance in subwindows of the filter window which include the target pixel, and replaces that pixel with the average value over the subwindow with the lowest variance. This filter can be improved significantly for multiband data by replacing variance with a criterion which measures multispectral homogeneity, and by adapting dynamically the size of the overall filter window to the local radiance distribution (Gerig *et al.*, 1985, Hill *et al.*, 1993). Needed computing resources for that filter are proportional to the number of spectral bands but the quality of the filter seems to improve likewise.

Another valuable approach to spatial random noise filtering is also based on multiband analysis. For a spectral band to be filtered, the new value of the central pixel of the filter window is estimated through a multiple linear regression between that band and some reference bands within the filter window (Tom 1986). This filter is particularly suited to cleaning of low S/N bands if coregistered better quality bands are available, and is applicable even if the reference bands are derived from a different sensor. Probably the band continuity of IS data will improve the quality of this filter.

In IS data, random noise will appear not only as uncorrelated spatial variation but also in the spectrum of each point; typically the noise level will be constant over an extended part of any spectral range. Since spectra are one-dimensional signals, context information to be used in filtering them is more limited; on the other hand, for high resolution data the autocorrelation to be expected for a clean spectrum is larger than the spatial autocorrelation. Clearly any spatial filtering will influence (hopefully diminish) noisiness of the spectra and viceversa. However it may be interesting to consider 3-dimensional filters, like a variant of the filter of Nagao and Matsuyama which includes an estimation over the local gradient over the spectrum, as alternative to the derived multiband filter described above.

3.6 Data Compression

If image analysis is understood to be information extraction, it means also a process of diminishing the amount of data. For IS data more than for traditional RS imagery, it is desirable to remove some of the information contents algorithmically before the image is analysed. Techniques for removing information invertibly exist and could have a larger efficiency for IS data with its high correlation between many spectral bands than for less autocorrelated images. They are not important for our discussion since data compressed with a reversable process usually is too mangled to be suitable for interpretation, and will be decompressed before use, and furthermore they are presently not used. In general we shall not consider algorithms which require an image restoration step before further processing will be possible.

The most obvious technique for data compression consists in selecting those spectral bands or ranges which contain the most information for the specific application. This is most obvious for water body applications where only spectral regions with sufficiently high transmittivity in water and possibly some infrared band for the only purpose of atmospheric correction are retained. Also within spectral regions of interest for a particular application it may be unnecessary to retain the full available resolution, and bands may be averaged along those subregions where the spectral information is expressed in the overall reflectance level rather than in small bandwidth features. Such averaging is particular useful in low signal regions where the relative noise level will mask out spectral details.

The vector space spanned by the spectral bands of an image is often called *feature space* (*FS*). A spectral signature forms a single point in that space, the whole image or parts of it form a cloud of points; spatial information is not retained in the FS. A commonly used class of data compression algorithms consist in linear transformations of this space onto a subspace. The best known algorithm of this type is the *principal component* transformation, an orthogonal tranformation onto the space spanned by the eigenvectors of the covariance matrix of the data cloud. In the transformed space, all covariances will be zero, and the variance along each transformed axes is an eigenvalue of the original (and also of the transformed) covariance matrix. The feature space will therefore be projected onto the subspace with the largest covariance matrix eigenvalues. In general more than 90 % of data variance is contained in a low dimension subspace of the FS, even for IS data, hence the shape of the full spectrum may be predicted from few components of its transformed FS with only a small error. It is common experience, however, that the deviation of the actual shape from the one predicted from the subspace projection is carrying important information for the image analysis in particular for IS data; this emphasises the need for high S/N ratio for that data.

While the principal component transformation maximises the variances along the first axes, a better strategy for minimising cost of a supervised classification purposes consists in projecting onto a subspace such that class separation will be maximised (for Gaussian distribution of signatures for each class, see Colwell *et al.*(1983a), p. 801). The cost of linear data reduction is determined by $an^2 + bnm$, where n is the dimension of the FS, m the dimension of the transformed space, a describes the complexity of calculation of the transformation and b the number of data points. Typically $b >> a$.

The FS projection techniques described so far are oriented towards the mapping can also be used to extract directly thematic information, and some pa cost reduction of subsequent data analysis, in particular of classification algorithms. Linear FS mapping can also be used to extract directly thematic information, and some particularly promising IS data analysis methods are based on that approach. Examples will be given in 4.3 .

4. Data Analysis

Every single application field in remote sensing, or even more specific in RS of earth resources, probably has its own collection of problems and techniques in data analysis. We can cover here only a small selection related to our own experience.

4.1 VISUALISATION

Traditional image display/interaction systems permit to represent single image bands in black and white or pseudo colour, three bands in false colour, change magnification or look at a subwindow in a different magnification, create interactively contrast stretching or colour look up tables, extract numerical values for a small subwindow in a single band or for few points in all bands, switch between few bands or band sets, define polygon boundaries for the extraction of image statistics, display histograms and scattergrams, overlay vector information and annotation. More recent display systems are often conceived as front ends to powerful image analysis packages and permit to spy into each processing step, allowing to interactively tune procedure parameters. This development is partly due to the advances in hardware and user interface development; display hardware which few years ago was considered specialised equipment is now integrated in the console of powerful and affordable workstations. Therefore expectations of the RS user community towards image visualisation have been easily met or surpassed. IS data however adds literally another dimension to the image data; it is indeed not trivial to imagine how spectral shape could be represented contemporaneously with two-dimensional spatial context. The display of profiles is straightforward, either as an image one axis of which carries the spatial information while the other one spans across the spectral range, or in a perspective representation, through a series of profiles or as a shaded surface. The first display technique can easily be obtained on traditional image analysis systems by reshaping the data cube; it can also be slightly generalised to a three pixels wide transsect with a false colour display which may be difficult to interpret however. The perspective display is not yet common in RS software packages but is available in various general purpose scientific data visualisation packages (UNIGRAPH, PV-WAVE, IDL, MATLAB, MATHEMATICA, MAPLE, MACSYMA, GNU-PLOT etc., see 5). A full visualisation of the data cube however stresses the limits of present display technology. Possible solutions would be:

animation or interactive displacement along a spatial axis,

animation or interactive displacement along the spectral axis,

display as translucent solid cube.

While two of the above mentioned program packages provide animation tools, the technique is demanding in data flow requirements; for passing through the spectrum of a whole AVIRIS scene, about 70 MBytes of data must be passed to the screen within few seconds; a 256×256 window would still amount to 15 MBytes. The solid cube representation is yet to be investigated; an algorithmically simplified and proven version could be derived from the ``grand tour" data representation methodology developed at the University of Washington, Seattle; the reflectance in a spectral band would be indicated by a point cloud density. The "grand tour" data viewer permits, among other display possibilities, to animate or rotate interactively perspectively drawn 3-D scattergrams, and has proven to be a powerful tool for the understanding of data properties.

4.2 Statistical Analysis and Classification

In the history of digital processing of RS imagery, automatic classification of land cover has been one of the research topics closest to operational applications. Attempts to use spatial context information derived from the imagery had success only under favourable boundary conditions and for particular applications; more generally applicable approaches are described under the topic "edge preserving filtering" in 3.5. However the inclusion of context information from other sources (thematic maps, stratification boundaries, DEM, etc.) has become part of the common strategies of automatic classification, to be complemented by multitemporal and/or multisensor data sets where it is possible and adequate. Here we shall be concerned with single pixel classification algorithms, which may be categorised into:

Parametric Supervised Classification requires statistical properties of predefined classes to be determined and supplied to the classifier.

Nonparametric Supervised Classification requires the algorithm to be trained with sample points from each predefined class.

Unsupervised Classification partitions the feature space (*FS*) without any a priori definition of classes, into ``natural" clusters of data points; the number of clusters may be predefined or determined by the algorithm.

Parametric classifiers usually derive for each class a measure for the distance or *minimising discriminant function (MinDF)* of an arbitrary point in FS to the class. For each image point the distances to all classes are computed, and the point is attributed to the class to which it is closest; it may be considered unclassifiable if the distance to the closest class is above a threshold or not sufficiently distinct from the distance to another class. Sometimes instead a similarity function or *maximising discriminant function (MaxDF)* is used, i. e. a point is attributed to the class to which the similarity value is largest; of course, any distance could be expressed as similarity function and vice versa. Some commonly used discriminant functions are:

Road block distance: the sum of all projections of the Euclidean distance between point and class centre point onto the FS axes. If x_i is the value of the point in the *i*th band and $\mu_i(n)$ the mean value for class *n* in band *i*, the distance *d* between them is $d = \sum_i |\mu_i(n) - x_i|$. The corresponding classifier is known as *parallel-epiped classifier.*

Euclidean distance: The used distance is really the square of the Euclidean distance between class centre and point in FS: $d = \sum_i (\mu_i(n) - x_i)^2$. The algorithm is known as *em minimum distance classifier.*

Variance-normalised Euclidean distance: $d = \sum_i \frac{(\mu_i(n)-x_i)^2}{\sigma_i^2(n)}$, where $\sigma_i^2(n)$ is the variance of band *i* in class *n*.

Gaussian probability density: if the class distribution can be described sufficiently well as a multivariate Gaussian distribution, the probability $P(\mathbf{x}|n)$ for $\mathbf{x}$ to be in class n is $P(\mathbf{x}|n) = P(n)(2\pi)^{-\frac{m}{2}}|\Sigma(n)|^{-\frac{1}{2}}\exp\left(-\frac{1}{2}(\mathbf{x}-\mu(n))^{\perp}\Sigma^{-1}(n)(\mathbf{x}-\mu(n))\right)$, where *P(n)* is the a priori probability for class *n*, *m* the dimension of the FS (number of bands), Σ the covariance matrix of the class distribution, $\mathbf{x} = [x_i]$ the feature vector to be classified, and ⊥ denotes the transpose operator. The distance used is $d = -2\log P(n)$. This classifier is known as *maximum likelihood classifier*. It describes reasonibly well oblique feature clouds, i. e. where for points belonging to the class, bands are correlated (which is almost always true). The Euclidean and variance-normalised distances can be interpreted as special cases of the maximum likelihood distance where the covariance matrix is unity and diagonal, respectively; therefore a variance-normalised classifier on a principal component transformed data set will often give results comparable with the maximum likelihood classifier.

Angular distance: The used MaxDF s is the cosine of the angle between feature vectors, or the normalised scalar product of the feature vectors: $s = \frac{\mu(n)\bullet\mathbf{x}}{\sqrt{\mu(n)\bullet\mu(n)\,\mathbf{x}\bullet\mathbf{x}}}$, where • denotes the scalar product: $\mathbf{x}\bullet\mathbf{y} = \mathbf{x}^{\perp}\mathbf{y} = \sum_i x_i y_i$. This similarity or its naturally corresponding MinDF $d = \arccos s$ (the angle between feature vectors) are invariant with respect to uniform scaling, i. e. $d(\mathbf{x},\mathbf{y}) = d(a\mathbf{x},\mathbf{y})$ for any nonzero positive constant a, and therefore less sensitive to illumination differences in an image.

Correlation is a measure not traditionally used for classifiers, because it is rather meaningless for few points (meaning here few bands) but is being used in IS data analysis as we will discuss in 4.3 . It is computed as $s = \frac{\overline{\mathbf{x}\cdot\mathbf{y}}-\overline{\mathbf{x}}\,\overline{\mathbf{y}}}{\sqrt{(\overline{\mathbf{x}\cdot\mathbf{x}}-\overline{\mathbf{x}}\,\overline{\mathbf{x}})(\overline{\mathbf{y}\cdot\mathbf{y}}-\overline{\mathbf{y}}\,\overline{\mathbf{y}})}}$, where $\overline{\mathbf{x}} = \frac{1}{m}\sum_i x_i$ for $\mathbf{x} = [x_i]$ and m bands, and $\mathbf{x}\cdot\mathbf{y} = [x_i\,y_i]$. It is translation invariant, $s(a+\mathbf{x},\mathbf{y}) = s(\mathbf{x},\mathbf{y})$ (with $a+\mathbf{x} = [a+x_i]$), and invariant against uniform scaling.

While the computation of discriminant functions which describe the extension of a class along the axes of the FS requires a number of elementary arithmetic operations which is a linear function of the number of bands, distance measures which describe also the orientation of the class in FS require typically a number of computations which grows with the square of the number of bands, and which become prohibitive for the large dimensions of IS feature spaces. If classification of raw data is required, it may be therefore more efficient to describe a class through a number of subclasses which cover the desired shape in feature space and to use the simpler distances.

The representation of an image in terms of discriminant function values with respect to class prototypes can be a useful image analysis tool; this has been implemented in particular for IS applications using the angular distance (Spectral Angular Mapper), or the Euclidean distance, see 5 .

Nonparametric classifiers partition the FS such that the total cost of misclassification computed for each point of predefined clusters is minimal. One strategy is the iterative determination of surfaces (linear or higher order) until for each pair of clusters the class separation is optimised. The separation surfaces are expressed by equations $s_i(\mathbf{x}) - s_j(\mathbf{x}) = 0$, where s_i are to be determined to become MaxDF for class i. The functions s_i are initialised for all classes in any convenient way, and adjusted by cycling repeatedly through all points of cluster pairs i, j such that for $\mathbf{x}$ belonging to class i, the two discriminant functions will be maintained for $s_i(\mathbf{x}) > s_j(\mathbf{x})$, modified such that $s_i(\mathbf{x})$ will increase and $s_j(\mathbf{x})$ decrease otherwise. Algorithms of this type are discussed in Nilsson (1965) The iteratively adjusted discrimination functions can be used as MaxDF for the final classification. The adjustment of discriminant functions may require a large number of iterations; for each iteration the number of elementary operations is at least in the order of cnm, where c is the number of training clusters, n is the total number of points in all training clusters combined and m the number of bands. This type of classification algorithm is seldom used since parametric classifiers deliver the same performance at lower cost.

Recently interesting results have been obtained using neural network classifiers, see Kanellopoulos *et al.* (1992). However, the methodology is currently too demanding in computing resources to be of practical interest in particular for IS data.

Unsupervised classification algorithms may either successively split from a single cluster comprising all points of the data set, or may merge clusters repetively from an initial state of each point in the data set being a separate cluster, or may start with a predefined number of arbitrarily selected cluster centres (which are not necessarily points of the data set). Splitting and merging are based on a distance measure like those discussed above, in particular Euclidean distance or some variation thereof (attractor function). Some algorithms use both splitting and merging strategies. One example of a parameter-free merging criterion is the mutual nearest neighbour criterion: two points are merged if and only if each one is closer to the other one than to any other point. The choice in cluster algorithms is too wide to be covered in any reasonable way here, only some general observations can be made.

Most cluster algorithms are iterative. Generally they either converge uniquely but require a number of elementary computing operations which grows at least with the square of the number of points in feature space, or they require less computing resources but the result depends on the order in which points are considered, or one some other arbitrary choice. One important application for cluster algorithms is the separation of thematic classes into spectral subclasses; the clustering is done only on points of training sets, permitting thus the use of more sophisticated algorithms.

4.3 Spectral Property Analysis

The extraction of category maps does not exploit the full potential of RS data and much less of IS data as each single point comprises a number of measurements, and should therefore intrinsically allow the extraction of quantitative parameters. Such an approach is also more likely to extract meaningful information about an object (resolution element) which almost always is made up of various more or less unrelated components, each one of which gives a contribution to the measured signal. The extraction of quantitative parameters which describe observed objects on a by pixel basis has indeed been applied in RS either as primary goal or for an application driven feature selection. Examples are:

Extraction of the concentrations of phytoplanction (or rather their chlorophyll contents) and sediment load in ocean water.

Bathymetry in clear shallow waters

Measurement of atmospheric parameters (see 3.3)

Determination of vegetation cover through *vegetation indices*

Expression of feature space in terms of *brightness, greenness* and *wetness* with the *tasseled cap* transformation.

Most of the measured items above are derived from simple nonlinear functions of radiances or reflectances, like ratios of linear combinations of bands. A noteworthy exception is the tasseled cap transformation, a linear projection of feature space into a subspace spanned by vectors characteristic for three features: bright soil, vegetation, water, see Crist *et al.* (1984).

In addition to measuring radiances or reflectances, IS data permits to determine location or changes in location of spectral features. Intensive research has been undertaken to extract information about vegetation stress from the shape of vegetation spectra in the range *670 nm-870 nm*, between a chlorophyll absorption band and the beginning of the plateau of high reflectivity caused by the leaf cell structure. The position and gradient of this pronounced slope in vegetation spectra is determined by fitting with a Gauss-integral or (easier but with a similar degree of confidence) with a third order polynomial, i. e. using four anchor points.

Fitting of spectral absorption features in the short wave infrared region with elementary functions shaped around corresponding bands of minerals is also employed for identifying the presence and quantity of those minerals.

Globally fitting a spectrum has also been employed for extracting or enhancing spectrally local features, by analysing the difference between spectrum and fitting curve. An example is the following classification algorithm, where the fitting function is actually a constant: for each point $\mathbf{x}$ in feature space, as also for the prototype spectra, the vector $\mathbf{d}(\mathbf{x}) = D(\mathbf{x} - \bar{\mathbf{x}})$, where $D(x_i) = 1$ for $x_i > 0$ and $D(x_i) = -1$ otherwise. The classification uses the MaxDF $s = \mathbf{d}(\mathbf{x}) \bullet \mathbf{d}(\mathbf{y})$, which can be implemented extremely fast through a Boolean *xor* operator and bit counting (or table lookup) if the vectors $\mathbf{d}(\mathbf{x})$ are generated as bit patterns (coding zero for -1). The algorithm has been improved by generating a second bit for each band which is set according to the local slope around this band. Note that the MaxDF s is essentially a correlation measure. In IS applications this algorithm has been successfully used for detection of minerals which have relatively sharp absorption features. For other thematic applications results have been less convincing; however, the algorithm may be improved by employing other fitting functions.

Rather than by fitting globally a spectrum, local spectral properties can be isolated also by generating an enveloppe to that spectrum; there exists a unique upper convex hull $\mathbf{h}(\mathbf{x})$ for any spectrum $\mathbf{x}$, and $\mathbf{h}(\mathbf{x})-\mathbf{x}$ will permit to measure local spectral properties independent of the presence of large scale features which mostly are convex over large parts of the spectrum (or can be made so by nonlinearly rescaling the wavelength axis). Also other upper enveloppe functions have been used.

A very general approach for spectral analysis, *spectral unmixing*, will be discussed in detail in subsequent lectures. The most basic implementation of this technique consists essentially in performing a multiple linear regression of the spectrum $\mathbf{x}$ (which is not necessarily the whole spectrum measured by the IS instrument) with a number of base spectra $\mathbf{b}_i$, i. e. coefficients f_i are determined such that $\mathbf{x}$ will be fitted with $\sum_i f_i \cdot \mathbf{b}_i$. More precisely, $\mathbf{f} = [f_i]$ is determined to satisfy $\mathbf{x} = \mathbf{r} + \sum_i f_i \cdot \mathbf{b}_i$ (or $\mathbf{x} = \mathbf{r} + \mathbf{f}^{\perp}\mathbf{b}$ for $\mathbf{b} = [\mathbf{b}_i^{\perp}]$) such that $\overline{\mathbf{r}}\mathbf{r}$ is minimal. The spectrum $\mathbf{r}$ is called the residual spectrum. Although this approach has found particular attention in the context of IS data analysis, it has shown to be applicable as well with traditional RS data.

5. Software Tools

Already before 1975 fairly complex and complete RS image processing packages like LARSYS (Purdue University), ORSER (Pennsylvania State University), or VICAR (Jet Propulsion Laboratory) were available. These systems supplied procedures for image formatting, spatial and frequency filtering, geometric rectification, arithmetic manipulation of images, linear transformation based feature selection, image and training statistics, supervised and unsupervised classification. These systems were implemented on mainframes and executed in batch mode, with modules which communicated through tape or disk files.

Development of RS packages proceeded mainly in the following areas:

Platforms: mainframe → minicomputer and display station → workstation, PC.

User interface: batch deck → menu control and display unit interaction → WIMP (Window-Icon-Mouse-Pointer). Much better visualisation of data.

Speed and capacity: progress was achieved through faster hardware and better storage devices.

Portability: modern packages are often available on several hardware platforms.

Commercialisation: several companies offer complete packages.

Algorithms: major advances in geometric correction packages; interfaces to data bases and GIS systems. Image editing and annotation facilities. Remarkably little evolution in other areas as far as commercial packages are concerned. Research oriented software is obviously in continuous development, in all areas described in the previous chapters.

5.1 GENERAL PURPOSE REMOTE SENSING PACKAGES

There is now a wide choice of commercial packages available although some previous suppliers are no longer in the market. Examples for available systems are

ERDAS: available for a number of workstations, and for IBM PC compatibles.Probably the most successful vendor. The ERDAS product line is now replaced with the IMAGINE product. IMAGINE is a complete rewrite of the former ERDAS package, with a well designed graphical user interface. At the writing of this article IMAGINE has not yet the full functionality of the previous ERDAS package. Imagine supports the ERDAS data file formats but uses different data formats which are not publicly documented. Both ERDAS and IMAGINE have strong links to the ARC/INFO GIS system. IMAGINE does support some major UN*X platforms, but presently not yet as consistently as packages of other vendors.

PCI EASI/PACE: available for a number of workstations, and for IBM PC compatibles. Like ERDAS, this system has a long history but it has also a very good record of backwards compatibility. Its various implementations offer the same functionality and user interface throughout all supported platforms. This system is very open for integration with user software and other packages. In particular many of its modules permit to integrate external data formats by direct access from all processing functions rather than through import/export operations (this facility is currently supported for BIP, BSQ, and BSF formats but not for BIL formats). There is no single consistent graphical user interface to all functions but current improvements to the package address also that problem. The system does not cover much GIS functionality and future developments are rather oriented towards providing links with GIS packages of other vendors.

ERM ER-MAPPER: powerful system with a very well designed graphical user interface. The system is currently available for a number of UN*X workstations. Like the PCI product, it coexists well with other packages, uses publicly described formats, and permits to access directly image files in any external BIL format.

IDIMS: System with a long tradition but not much evolution except for user interface. Available for some platforms. Future development is uncertain (change of mother company).

The list can be continued. It should be clear, however, that there is no ideal all purpose system for everybody, and each application scenario needs its own research. There are also some low cost or public domain systems, like PC-IMEGA, CHIPS (both are low cost systems for IBM PC compatibles), GRASS (mainly a Geographical Information System), KHOROS (general image processing not oriented particularly towards remote sensing, but having an advanced object-oriented user interface and process control paradigm).

It should be noted that most if not all of these packages are not well suited to handle IS data; some of the above software is not able to handle consistently a large number of spectral bands, nor does it provide sufficient possibilities of displaying spectral data with its annotation: operations like showing the superposition of different spectra or annotation with a wavelength scale do not exist, spectra cannot be filtered, band sets cannot be excluded except by subsetting, etc. Moreover most of the algorithms provided by those codes are not of particular interest for IS data analysis.

Those packages however evolve as a consequence of user requests; since the processing of large band sets is not uniquely domain of IS data processing but has become a requirement also for time series processing of data from operational sensors, or for merging image data with a large number of GIS layers, restrictions on the number of usable bands are being relaxed and tools for processing of large band sets integrated in the more recent versions of these products. For example, the processing of several hundred spectral bands is supported in the ERM and partially in the PCI products, and an optional PCI module implements linear spectral unmixing.

5.2 Imaging Spectrometry Packages

Packages developed specifically for IS data processing are existing by now, and at least three are available to the public:

SPAM of JPL is the first widely available package, and is distributed free of charge, as source code. It contains image and spectra display functions, spectral library handling, the binary spectral matching/classification algorithm described in 4.3 including a clustering algorithm based on the same discriminant function, and a spectral unmixing algorithm. Although most of the code is portable, the display part is hardware dependant. Development of this system has been discontinued.

SIPS of CSES/CIRES, University of Colorado, is written partially in the scientific visualisation language IDL, and is relatively portable although it is distributed ready to run only for SUN and DEC RISC workstations. It is distributed free of charge as source code but requires the commercial software IDL. It contains image and spectra display functions, spectra library handling, sensor calibration procedures, the atmospheric correction procedure described in Gao *et al.* (1993), Spectral Angular Mapper algorithm (see 4.2), and spectral unmixing algorithms. This system is suited also for handling of traditional RS data. The package is technically open to user extensions but those should be made available to CSES/CSIRO to be integrated into the distributed package. The future status of this package is uncertain as financial support for further development is lacking; however CSIRO attempts to continue support (bug fixes, maintaining compatibility with future versions of IDL). The commercial package ENVI (described below) is based on this package.

GenIsis of WTJ Software Services is a commercial product implemented on IBM PC oompatibles. It contains image and spectra display functions, spectra library handling, and an Euclidean distance mapper. It is designed for very fast simple processing of large quantities of image data (IS data, image time series) on PCs.

Algorithm Development Tools

The approach demonstrated with the SIPS package, i. e. the use of a data visualisation language, is particularly interesting for algorithm development for IS data analysis since such languages (IDL, PV-WAVE, MATLAB) provide primitives not only for advanced visualisation (including animation) but also tools for the simplified handling of vectors, matrices, and higher dimensioned arrays, and powerful built-in functions. Such data visualisation languages provide a viable environment for prototyping of image processing algorithms which can be formulated in concise syntax and the intermediate results of which can be visualised in 2-D or projected and sectioned 3-D, or even adding another dimension through animation operators (which are integrated in some of these languages). It should be noted however that these languages are not directly translatable in machine code but rather interpreted or compiled into interpreted pseudo code. Therefore procedures formulated in such a language tends to use considerably more resources than a similar program well written in a classical compiled programming language, and will achieve similar performance only if the algorithm can make strong use of intrinsic array operators or functions, and does not require conditional or control operations to be repeated many times.

A combination of RS image processing package and complex data visualisation environment will be represented by the commercial ENVI system, based on SIPS but written completely in IDL and to be marketed by RSI. The system is expected to be generally available in Spring 1994. It will include the full functionality of SIPS and cover a range of functions comparable with the major RS packages mentioned above. Being completely embedded in the IDL environment (unlike SIPS), it will be particularly strong in complex mathematical and display operations and will be offered for a variety of platforms including personal computers. It will support direct access of external data formats similar to EASI/PACE and ER-MAPPER, for BIP, BIL, and BSQ formats. Similarly to EASI/PACE, it does not provide GIS functions by itself but rather through data links to external packages.

Clearly this package will have the same potential limitations for operational work as its underlying language system; it could however become a viable alternative or complement to a RS software of the kind described in 5.1, where the data analysis is oriented more towards an exploration of sensor characteristics or of the data space and its relation to the observed reality rather than towards a routine application of proven algorithms to large quantities of data.

6. References

Colwell, R.N., D.S. Simonett and F.W. Ulaby (eds.) (1983a), '*Manual of Remote Sensing*', *Volume 1: Theory, Instruments and Techniques*, American Society of Photogrammetry, Falls Church.

Colwell, R.N., D.J.E. Estes, and G.A. Thorley (eds.) (1983b), '*Manual of Remote Sensing*', *Volume 2: Interpretation and Applications, American Society of Photogrammetry*, Falls Church.

Conel, J.E., R.O. Green, R.E. Alley, C.J. Bruegge, V. Carrere, J.S. Margoslis, G. Vane, T.G. Chrien, P.N. Slater, S.F. Biggar, P.M. Teillet, R.D. Jackson and M.S. Moran (1988) 'In-flight radiometric calibration of the Airborne Visible/Infrared Imaging Spectrometr (AVIRIS)', *in* P.N. Slater (ed.) *Recent advantages in sensors, radiometry, and data processing for remote sensing*, Proc. SPIE, vol. 924, 179-195.

Crist, E.P., and R.C. Cicone (1984), 'A physically-based transformation of Thematic Mapper data - the TM tasseled cap', *IEEE Trans. Geoscience & Remote Sensing*,vol. GE-22, no. 3, 256-263.

Gao, B.-C., and A.F.H. Goetz (1990), 'Column atmospheric water vapor and vegetation liquid water retrievals from airborne imaging spectrometer data', *J. Geophys. Res.*, 95, 3549-3564.

Gao, B.-C., K. Heidebrecht, and A.F.H. Goetz (1993), 'Derivation of scaled surface reflectances from AVIRIS data', *Remote Sensing of Environment*, vol. 44, nos. 2/3, 165-178.

Gerig, G., and K. Seidel (1985), 'Structural description of a LANDSAT TM scene for improved region-based classification', *Proceedings of IGARSS' 84 Symposium*, Straßburg, Aug 27-30, ESA Publication SP-215, 101-105.

Green, R.O., J.E. Conel, C.J. Bruegge, J.S. Margolis, V. Carrere, G. Vane, G. Hoover (1992), 'In-flight calibration of the spectral and radiometric characteristics of AVIRIS', *in* R.O. Green (ed.) *Summaries of the Third Annual JPL Airborne Geoscience Workshop June 1-5, 1992,* JPL Publication 92-14, vol. 1, 1-4.

Kanellopoulos, I., A.Varfis, G.G. Wilkinson, and J. Mégier (1992), 'Land-cover discrimination in SPOT HRV imagery using an artificial neural network - a 20-class experiment', *Int. J. Remote Sensing*, vol. 13, no. 5, 917-924.

Hill, J. (1990), 'Radiometric comparison and calibration of remotely sensed data from polar-orbiting earth observation satellites', *Proc. 5th Australasian Remote Sensing Conference, Perth, Oct 8-12, 1990*, vol. 1, 42-53.

Hill, J., W. Mehl, and J. Mégier (1993), 'Analysis of GER imaging spectrometry data for the identification of soil and vegetation parameters in land degradation studies', Proceedings of the final EISAC Workshop, Ispra 1991, *ESA Publication SP-360*, 27-33.

Nagao, M., and T. Matsuyama (1979), *'A structural analysis of complex aerial photographs'*, Plenum Press, New York.

Nilsson N. J. (1965), *'Learning Machines'*, McGraw-Hill, New York.

Press, W.H., B.P. Flannery, S.A. Teukolsky, and W.T. Vetterling (1986), *'Numerical Recipes: The Art of Scientific Computing (FORTRAN and PASCAL Edition)'*, Cambridge University Press, Cambridge, New York, Melbourne.

Press, W.H., B.P. Flannery, S.A. Teukolsky, and W.T. Vetterling (1988), *'Numerical Recipes in C: The Art of Scientific Computing'*, Cambridge University Press, Cambridge, New York, Melbourne.

Tom, V.T. (1986), 'A synergistic approach for multispectral image restoration using reference imagery', Proc. of the IGARSS' 86 Symposium, Zürich, Sep 8-11, *ESA Publication SP-254,* 559-564.

RETRIEVING CANOPY PROPERTIES FROM REMOTE SENSING MEASUREMENTS

MICHEL M. VERSTRAETE
Institute for Remote Sensing Applications
Commission of the European Communities
Joint Research Centre
I-21020 Ispra (Va), Italy

ABSTRACT. A wide variety of tools have been designed to extract information on the nature and structure of terrestrial ecosystems from remote sensing data. Two approaches are described here in some detail: the use of spectral indices, and the design and inversion of physically-based models of the interaction between the radiation field and the surface. The advantages and drawbacks of these approaches are discussed, and the importance of combined field and laboratory measurement campaigns is stressed.

1. Introduction

The conceptual basis of remote sensing was described in some detail in Verstraete (1993) earlier in this volume. It was seen that some applications of remote sensing are possible without requiring heavy-duty tools such as advanced modeling. For example, mapping urbanization or land use may be relatively easy to implement to the extent that the desired information is entirely contained in the spatial variability or distribution of the properties of the environment. The main requirement for the applications is that the various surface types should exhibit sufficiently different reflectances, in at least one observed spectral band. Similarly, the detection of temporal changes in the spatial distribution of surface properties can be achieved by comparing two or more scenes of the same location taken at different times.

To support environmental monitoring, resource management, pollution control, and many other activities, we need reliable, accurate quantitative information on the state and evolution of the terrestrial ecosystems. In many cases, it is feasible to dispatch observers to take in-situ measurements, but in many other cases this is too costly or not practical. Remote sensing, especially from space platforms, provides a convenient source of data, globally and repetitively. The problem is that the data collected by these instruments do not correspond to the information we need. The claim of remote sensing as a discipline, and the hope of end-users, is that the radiation data collected can in fact be effectively interpreted in terms of the desired information.

The retrieval of useful information on the state of terrestrial surfaces from remote sensing data is a complex process, and will be the subject of this paper. The question is not so much to identify a particular surface type (e.g. a forest), but to characterize it (e.g. to measure its biomass). A range of methods have been proposed, but two extreme approaches will be presented. First, an empirical approach based on spectral indices will be outlined and discussed. Later on, the advantages and drawbacks of fully explicit models will be presented, paying particular attention to their numerical

J. Hill and J. Mégier (eds.), Imaging Spectrometry - a Tool for Environmental Observations, 109–123.

inversion. It will be seen that the selection of one approach over another depends on the nature and accuracy of the desired information, and on the availability of resources to obtain this information.

2. Spectral indices and empirical models

The simplest approach to derive useful information from satellite data is to assume that these data are directly related to the desired information. For example, let's assume that we would like to estimate the yield of agricultural activities, systematically and repetitively over a given region. We can send agricultural inspectors on the field to provide 'ground truth'. If we simultaneously observe the same small area from space, we can, in principle, correlate the radiative measurements made on-board the satellite with the field observations, and use this relation over the entire region and over time to obtain the desired information rapidly and at a reduced cost. In a way, remote sensing data is used to interpolate the ground measurements, taking advantage of the high spatial resolution of the instrument and its repetitive coverage of the area. Since space measurements are usually made in specific spectral bands, the simplest approach consists in correlating the ground data with one channel, or with a simple combination of channels: this is a main use of vegetation indices.

2.1 CLASSICAL VEGETATION INDICES

A number of such vegetation indices have been proposed in the literature, either for a specific instrument or mission, or as a generic tool (Curran, 1981; Perry and Lautenschlager, 1984). Green vegetation exhibits a strong increase in reflectance between the visible range ($0.3 < \lambda < 0.7$ micrometer) and the near-infrared region ($0.7 < \lambda < 1.3$ micrometer) (see Figure 2 of Verstraete, 1994). Vegetation indices attempt to exploit this spectral contrast to identify the presence of vegetation and evaluate its characteristics. The most widely used indices are the Simple Ratio (SR) and the Normalized Difference Vegetation Index (NDVI), defined as follows:

$$SR = \frac{\rho_{NIR}}{\rho_{RED}}$$

$$NDVI = \frac{\rho_{NIR} - \rho_{RED}}{\rho_{NIR} + \rho_{RED}} = \frac{SR - 1}{SR + 1} \qquad (1)$$

where ρ_{RED} and ρ_{NIR} are the measured reflectances in the red and in the near-infrared spectral regions, respectively. Figures 1 and 2 show the values taken by these indices for all possible values of the spectral measurements. It is seen that the two surfaces are mathematically equivalent: they could both be generated by a moving straight line rotating about the z axis. The SR surface has been truncated because the value of this index increases without bound as ρ_{RED} approaches 0. These indices have been used mainly with the AVHRR data from NOAA satellites, or with the high resolution instruments, such as SPOT and LANDSAT.

The main advantage of SR and NDVI is that they can be computed easily. Also, since they are constructed as ratios, they implicitly 'normalize' the data for the effect of any extraneous processes that affect both individual channel reflectances equally. Relative success in exploratory or specific

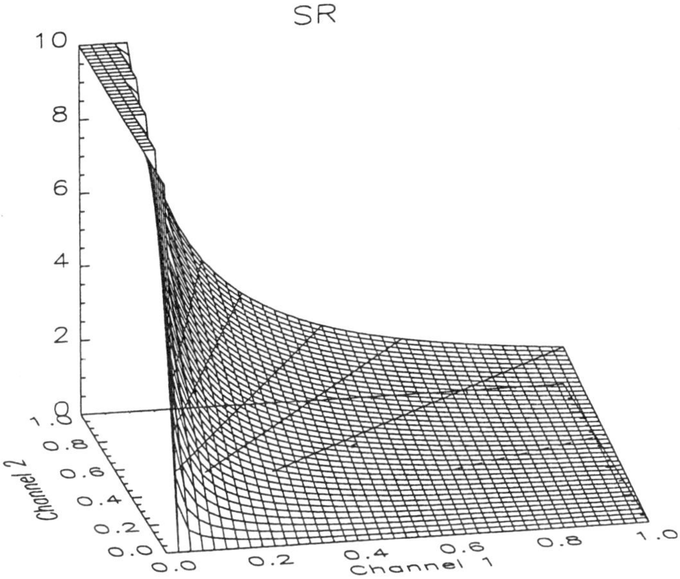

FIGURE 1. Values taken by the Simple Ratio (SR) index for all possible values of the spectral measurements (being truncated at a value of 10). Channel 1 refers to a spectral band in the visible red (0.5-0.6 μm), Channel 2 to a spectral band in the near-infrared (0.7-0.8 μm).

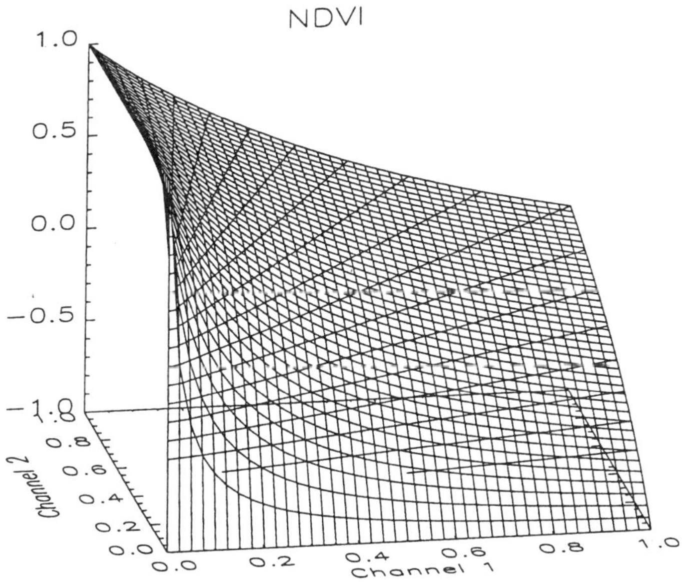

FIGURE 2. Values taken by the Normalized Difference Vegetation Index (NDVI) index for all possible values of the spectral measurements. Channel 1 refers to a spectral band in the visible red (0.5-0.6 μm), Channel 2 to a spectral band in the near-infrared (0.7-0.8 μm).

applications has encouraged their use as a general tool to interpret satellite data (e.g. Goward et al., 1985; Justice et al., 1985; Malingreau, 1986).

These vegetation indices do, however, have their drawbacks. On a very practical level, they do remain sensitive to processes not related to the state of the vegetation. For example, the NDVI may be strongly affected by varying atmospheric conditions (Kaufman, 1989; Pinty and Verstraete, 1992), or by changes in soil colour due to a modification of their water content (Colwell, 1974; Huete, 1988). Apparent changes in NDVI may also result from differences in illumination or viewing geometry, because the anisotropy of the surface is not the same in all spectral bands (Pinty et al., 1993).

2.2 NEW INDICES

The dependency of the measured signals on a multiplicity of surface and atmospheric factors is best addressed with physically-based models, as explained later in this paper, but it is nevertheless possible to define other vegetation indices, designed specifically to reduce these side effects. For example, Huete (1988) introduced a modified NDVI called the Soil Adjusted Vegetation Index (SAVI) to reduce sensitivity to soil colour changes:

$$SAVI = \frac{\rho_{NIR} - \rho_{RED}}{\rho_{NIR} + \rho_{RED} + L}(1 + L) \tag{2}$$

where L is a soil-dependent parameter. Empirical studies have shown that the value L=0.5 applies in most cases. This new index has been successfully used in field studies but preliminary investigations tend to show that it may be more sensitive than NDVI to atmospheric conditions. The SAVI should therefore be computed from satellite data only after atmospheric corrections have been applied to the data.

Another recently proposed index is the Global Environment Monitoring Index (GEMI) (Pinty and Verstraete, 1992). This non-linear vegetation index was specifically designed to be less sensitive than NDVI to atmospheric conditions and to soil colour changes. This goal is achieved by incorporating an average atmospheric correction into the index. The GEMI is specific to the AVHRR instrument, since other instruments have different spectral bands and therefore depend differently on the state of atmosphere. GEMI turns out to be rather insensitive to soil colour changes when the latter are dark or medium, i.e. for most agricultural soils, but quite sensitive to changes in soil brightness over light soils.

$$GEMI = \eta(1 - 0.25\,\eta) - \frac{\rho_{RED} - 0.125}{1 - \rho_{RED}} \tag{3}$$

where

$$\eta = \frac{2\left(\rho^2{}_{NIR} - \rho^2{}_{RED}\right) + 1.5\,\rho_{NIR} + 0.5\,\rho_{RED}}{\rho_{NIR} + \rho_{RED} + 0.5} \tag{4}$$

The overall shape of this function is shown in Figure 3.

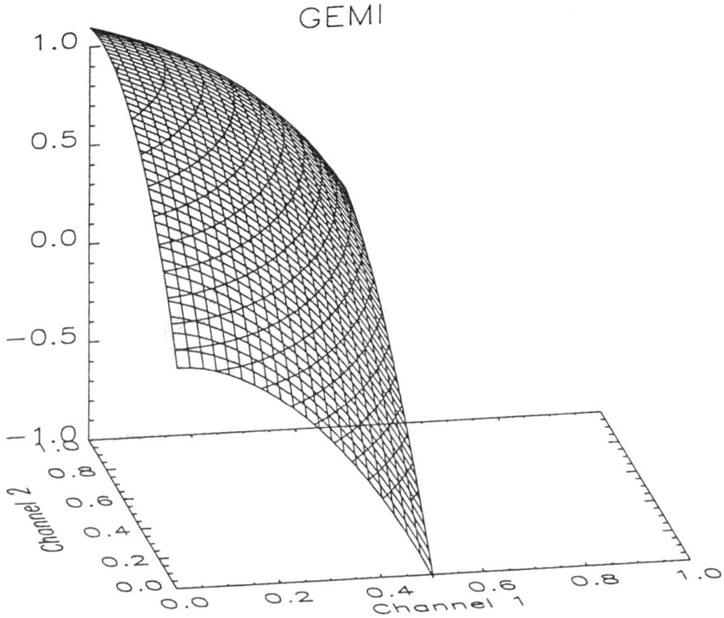

FIGURE 3. Values taken by the Global Environment Monitoring Index (GEMI) index for all possible values of the spectral measurements. Channel 1 refers to a spectral band in the visible red (0.5-0.6 μm), Channel 2 to a spectral band in the near-infrared (0.7-0.8 μm).

Future remote sensing instruments, with more advanced characteristics, were introduced in Verstraete (1994). It should be clear that the availability of additional spectral channels offers new opportunities to develop better indices, since more information may potentially be gathered. Kaufman and Tanré (1992), for example, have proposed a three-band Atmospherically Resistant Vegetation Index (ARVI) for use with Landsat TM and the forthcoming MODIS instrument. This index is also a modified version of the NDVI, where an additional channel in the blue region is used to correct the value in the red channel for the effect of aerosols:

$$ARVI = \frac{\rho_{NIR} - \rho_{RB}}{\rho_{NIR} + \rho_{RB}} \tag{5}$$

where

$$\rho_{RB} = \rho_{RED} - \gamma\left(\rho_{BLU} - \rho_{RED}\right) \tag{6}$$

where γ can take different values but $\gamma = 1$ has been found to be adequate for most surfaces and aerosol types. ARVI has been shown to be less sensitive than NDVI to atmospheric aerosols (especially for small and moderate particle sizes), and standard corrections for molecular and ozone absorption are still recommended. It may be pointed out that the developers of new indices designed for use with future high spectral resolution instruments do not need to be overly concerned about water vapor absorption effects, provided they are based on narrow spectral bands located outside of the molecular absorption features of the major atmospheric constituents.

2.3 ADVANTAGES AND DRAWBACKS OF SPECTRAL INDICES

Being simple combinations of individual channels, spectral indices can be computed easily, and do not require any additional ancillary data. Another significant advantage is that they are routinely available for very large areas and over long time periods. Vegetation indices often allow a quick qualitative estimation of the presence of vegetation, and that information may be sufficient for specific applications, such as early warning systems. Their use in more quantitative applications, however, must be carefully evaluated because these indices suffer from a number of significant limitations. Some of these are discussed here.

Despite adjustments to their computations or the derivation of improved indices, it is generally not possible to produce a vegetation index which is totally insensitive to all the possible meteorological situations and to any change in soil colour. Similarly, it is not easy to avoid changes in index values due to differences of surface anisotropy in various channels.

Some authors have suggested compositing techniques (where data from different days are merged according to some criterion), to try to reduce some of these undesirable atmospheric effects. The best known is the Maximum Value Composite (MVC) technique (Holben, 1986), which selects the channel values corresponding to the largest NDVI for each pixel of a scene over a given period of observation. This procedure is intended to also reduce the cloud contamination of the data.

This MVC technique provides a partial solution to the problem and may be adequate for some applications; but it is not ideal because it preferentially selects certain ranges of illumination and view angles, depending on the type of ecosystem and atmospheric conditions (e.g. Goward et al., 1991; Gutman, 1991; Meyer et al., 1993). As a result, even composited images exhibit linear features at the boundaries of the satellite tracks. The final interpretation of the spatial variability in the final product may also be more difficult, since adjacent pixels now come from different days, when the angular geometry and the atmosphere may have been quite different.

A more fundamental limitation of vegetation indices is that they exploit only part of the information contained in the original data. Defining the state of the surface requires, in principle, the specification of a large number of parameters to describe the soil and vegetation. To the extent that only two independent measurements are made, at most two such parameters can be specified. All terrestrial surfaces are therefore projected into a 2-dimensional space of reflectance values, typically measurements in the red and near-infrared spectral regions, and the state of the system is entirely determined, as far as remote sensing is concerned, by the 2 coordinates ρ_{RED} and ρ_{NIR} of the surface in this spectral space.

The choice of a different coordinate system is of course completely arbitrary: For example, one could rotate the coordinate axes by 45° with respect to the old ones, and define $\rho_{NIR} - \rho_{RED}$ and $\rho_{NIR} + \rho_{RED}$ as the new system of reference[1]. The difference $\rho_{NIR} - \rho_{RED}$ is obviously an un-normalized vegetation index, while the sum $\rho_{NIR} + \rho_{RED}$ is closely connected to the albedo, a very important geophysical property of the surface. In other words, to characterize the surface, we could work either with the individual channels, or with this index and the albedo.

Alternatively, we could also specify the 'position' of our target in spectral space with the help of a distance to the origin and an angle from the ρ_{RED} axis. In fact, by inspecting the shape and position of the NDVI isolines (Figure 2), it can be seen that the latter correspond to a bundle of straight

---

[1] These two systems are equivalent in the sense that the set of two individual numbers x and y contains exactly the same information as the set of the sum $x + y$ and the difference $x - y$ of these numbers. In particular, from any set of two values, we can re-create the other set, and conversely. Furthermore, the removal or hiding of any more member of these sets prevent the computing of all members of the other set.

lines intersecting at the origin, and that the value of NDVI acts much like an angular coordinate. It should be obvious, however, that the target cannot be properly characterized with the value of the NDVI alone: an infinity of different target surfaces actually have exactly the same NDVI. This implies that there is a fundamental indeterminacy (and probably a crucial deficiency) in any algorithm that attempts to assess the state of the surface on the basis of a single vegetation index.

Mathematically, the computing of a vegetation index therefore appears like a change of coordinates, where the index is a new coordinate. The characterization of the state of the observed system, on the basis of the available measurements, requires the definition of another coordinate, preferably orthogonal to the chosen index.

Now, it may be that a particular application can exploit one of these transformed coordinates better than the other, and does not even require the computation of the other. But it remains that it uses only 'half' of the information initially present in the data, and that very few definite statements can be made about the surface on the basis of this single coordinate only. These considerations can be extended to any index based on an arbitrary number $n > 1$ of channels.

This inescapable mathematical fact has another important consequence: that at most n different and independent pieces of information can be retrieved from an analysis of n independent channel measurements, and that at most one piece of information could be retrieved from any single vegetation index. It is surprising, in this regard, to list all the variables that have been 'successfully' correlated with one vegetation index or another: not only those mentioned above, but also parameters such as precipitation, evapotranspiration, primary productivity, carbon fluxes into or out of the biosphere, and many others. For sure, no single author has proposed to use a vegetation index to derive all these pieces of information for the same site and at the same time, but it must be realized that if more than one variable of interest at the surface can be correlated with a vegetation index such as the NDVI, this means that they are themselves correlated and therefore not independent.

Last but not least, it must be emphasized that even though the definitions of vegetation indices do follow some logic (enhancing the spectral contrast of vegetation while minimizing the influence of other processes), their usefulness derives exclusively from the empirical relations that may be found between them and other variables of interest. They lack a fundamental physical basis, and this is reflected in the difficulty inherent in their interpretation: from an observation of maps of vegetation indices, one gains the feeling that the index is at least sensitive to the presence of vegetation, in the sense that higher index values are more likely to be associated with vegetated surfaces than lower index values. But the lack of precise universal relations with leaf area index, biomass, productivity or any other variable hinders their application in quantitative evaluations of surface properties on large scale or long time projects.

In summary, vegetation indices appear to provide a simple tool to explore or quickly evaluate the state of the vegetation, over relatively large areas. This may be useful in a number of applications for which fast access to a general but not necessarily detailed or very accurate information may be sufficient. Classical vegetation indices tend to be significantly affected by numerous processes not related to the state of the surface (e.g. instrumental issues, atmospheric effects, anisotropy of the surface, etc.). New indices have been introduced recently to compensate for some of these effects, and may be more appropriate for these applications. Nevertheless, anyone of these indices is only telling part of the story, and cannot suffice, alone, to fully describe the surface.

3. Physically-based models and their inversion

At the other end of the spectrum of tools available to extract information about the surface from remote sensing data are the physically-based models. Because these models derive directly from the laws of physics and describe in more or less detail the transfer of radiation through the atmosphere and its scattering at the surface, they offer a very solid basis for the interpretation of these data. This approach is not without difficulties, however, as will be seen below.

There is a rather large literature on surface reflectance models (e.g. Goel, 1988; Pinty and Verstraete, 1992), and the reader is referred to these reviews for further information and pointers to other publications. In this paper, we will only outline the rationale and design of these models, and the technique of their inversion, focusing instead on the advantages and drawbacks of this approach.

3.1 MODELING THE MEASURED SIGNALS

Remote sensing instruments in the optical spectral range measure the light reflected by a target. In most practical situations, this light originates from the Sun, and is therefore incident onto the target at a specific angle. It is well-known that the reflectance of most objects depends on the position of the source of illumination. By the same token, measurements made by an instrument with a narrow field of view are also directional measurements, hence the expression 'bidirectional reflectance'. The physical interpretation of these measurements requires the development of theories and models to describe how the various media interact with the radiation to produce the observed sensor response.

Repeated observations have shown that the reflectance of a natural surface depends on at least four types of independent variables. Formally, this is written

$$\rho = \rho(\mathbf{x}, t, \lambda, \boldsymbol{\Theta}) \tag{7}$$

where $\mathbf{x}$ refers to the spatial and t to the temporal variations, λ indicates that the reflectance is spectrally dependent, and Θ represents the set of angles that specify the geometric configuration of the source of light, of the target, and of the observer. Bold symbols represent vector quantities.

In principle, the variability of ρ in any one of the independent variables can be used to extract information on the medium under observation. In practice, however, spectral signatures are used to distinguish different objects, and the spatial contrast of the signal is interpreted in terms of the geographical extent and distribution of these objects. Similarly, any time variation of the observed reflectance is used to describe the dynamic evolution of the system, and that leaves the directional variance of the signal as the principal source of information on the physical processes that must have produced the observed radiation. This is the main reason why physicists who attempt to extract quantitative information from remote sensing data focus on the angular variations of the measurements.

As explained earlier (Verstraete, 1994), the reflectance of a natural surface is not an explicit function of these independent variables. Rather, it is a function of the intrinsic optical and structural properties of the surface - we will call these the physical parameters - and these, in turn, are functions of the independent variables. A physically-based bidirectional reflectance model is therefore essentially a mathematical equation which predicts the reflectance ρ of the surface in terms of its properties:

$$\rho = f(y_1, y_2, \ldots, y_m) \quad (8)$$

where $y_j = y_j(x, t, \lambda, \Theta)$. To make the overall functional dependency of ρ more apparent, however, we will rewrite the basic equation as follows,

$$\rho = f(x_1, x_2, \ldots, x_n; y_1, y_2, \ldots, y_m) \quad (9)$$

where the variables x represent the independent variables and the parameters y stand for each of the relevant surface properties. The bulk of the modelling effort consists in specifying the nature of this functional dependency, but this topic will not be covered further here.

Having derived this formal analytical expression, we are now able to compute the reflectance of a surface on the basis of its properties, given the value of the independent variables. What would be really useful, however, is to do the reverse: given the reflectance observations and the values of the independent variables, retrieve the surface properties. This is known as an inverse problem and constitutes the fundamental issue of remote sensing.

3.2 MODEL INVERSION

If the reflectance of the surface could be computed on the basis of only one parameter y, then the inversion could be performed, analytically or numerically (depending on the complexity of f), to yield a unique solution: to each value ρ corresponds a unique value of the parameter y of interest. If, however, there are two or more parameters y_j, the problem is ill-posed, as we have more unknowns than equations: assessing the values of the parameters y_j from a knowledge of ρ and of the independent variables x_i cannot be done as a matter of principle on the basis of a single equation.

This situation occurs frequently in science and engineering, and solutions can still be found, albeit from a slightly different perspective. Denoting a particular measurement of the predicted variable ρ by the symbol $\hat{\rho}$, we first note that measurements may usually be repeated, say M times. This does not solve our problem, however, because multiple measurements may take time, refer to slightly different locations, involve different instruments, etc. There is *a priori* no guarantee that the same equation applies to describe each measurement, so that we get, in principle, a system of M equations in m x M unknowns:

$$\hat{\rho}_1 = f_1(x_{11}, x_{12}, \ldots, x_{1n}; y_{11}, y_{12}, \ldots, y_{1m}) + \varepsilon_1$$

$$\hat{\rho}_2 = f_2(x_{21}, x_{22}, \ldots, x_{2n}; y_{21}, y_{22}, \ldots, y_{2m}) + \varepsilon_2$$

$$\vdots$$

$$\hat{\rho}_M = f_M(x_{M1}, x_{M2}, \ldots, x_{Mn}; y_{M1}, y_{M2}, \ldots, y_{Mm}) + \varepsilon_M \quad (10)$$

where ε_k represents any discrepancy (due to errors of measurement or to the inadequacy of the model) between the observations $\hat{\rho}_k$ and the model-predicted values.
To successfully compensate for the lack of equations by taking multiple measurements, we must introduce additional hypotheses or constraints. Specifically:

- We will assume that the state of the system does not change appreciably while the measurements are being taken: $y_{kj} \equiv y_{lj} \equiv y_j$, for all values of $1 \leq k,l \leq M$.
- We will also assume that the nature of the relationship between the observable ρ and the various model variables x_i and parameters y_j remains identically the same: formally, $f_k \equiv f_l \equiv f$, for all values of $1 \leq k,l \leq M$.

The system of equations above then reduces to

$$\hat{\rho}_1 = f\left(x_{11}, x_{12}, \ldots, x_{1n}; y_1, y_2, \ldots, y_m\right) + \varepsilon_1$$

$$\hat{\rho}_2 = f\left(x_{21}, x_{22}, \ldots, x_{2n}; y_1, y_2, \ldots, y_m\right) + \varepsilon_2$$

$$\vdots$$

$$\hat{\rho}_M = f\left(x_{M1}, x_{M2}, \ldots, x_{Mn}; y_1, y_2, \ldots, y_m\right) + \varepsilon_M \qquad (11)$$

Our problem then becomes one of finding which values of the m parameters y_i best account for the observed variability in $\hat{\rho}_k$, i.e. minimize the error terms ε_k.

The next question to be addressed is to identify how the measurements should be taken. Clearly, if the observations were just repeated exactly under the same circumstances, the values of the independent variables x_i would be the same in each case, and the only possible variation in the values of $\hat{\rho}_k$ would be the result of errors of measurement. Since we do not want to try to make sense out of experimental errors, we should design the experimental setup in such a way that the variance in the observable ρ is not only large but much larger than the expected variance due to instrumental errors. Since the values of the parameters y_i have been assumed constant, this implies that the independent variables x_i will have to change. In fact, we will usually try to measure the observable ρ under as wide a range of x_i values as possible, subject to the condition that the analytical formulation of the dependency f cannot change either. In other words, we try to observe the same system under a wide variety of conditions, in order to derive its main characteristics from the observed variability. We have just shown that

- There must be at least one independent variable x, capable of varying over a range of values for which the corresponding measurements $\hat{\rho}_k$ will yield appreciable variability, i.e. a variance greater than the expected variability due to instrumental and observation errors.

Finally, in order to end up with a well-posed problem, we must insist on having enough equations in the system. We are thus lead to the further condition that

- We must make enough observations for the problem to be adequately constrained.

How many is enough? On purely mathematical grounds, we reckon that M must be at least equal to m. In other words, we must make at least as many observations as there are model parameters y_j to be determined for the problem to be well-posed. However, since observational errors and instrumental inaccuracies are inevitable, and since the system itself may in fact not be absolutely constant, we do not expect that each observation $\hat{\rho}$ will be exactly predicted by the model. If we have precisely the same number of equations as there are unknowns, there may be no numerical solution at all, in the sense that no particular set of values y_j can be found to predict exactly the observed values of $\hat{\rho}$ ($\varepsilon_k = 0$ for all $1 \leq k \leq M$), given the values of the independent variables x_i.

Our only way out is to have more equations than there are unknowns ($M > m$). Such a system of equation does not accept a unique well-defined solution, but there may be a 'best' solution, where the meaning of 'best' must be clearly specified. The mathematical nature of our problem has therefore changed from inverting a single equation in multiple (m) variables to solving a set of simultaneous ($M = m$) equations and now to finding the optimal (but not perfect) solution to an over-constrained system of ($M > m$) equations. Since we are no longer looking for *the* solution, but for some optimal solution,

- We assume further that the optimal solution of an over-specified problem is also the most probable true solution, i.e. the correct solution that would be found under ideal conditions.

There are many ways to define what would be a best solution. The most common approach attempts to minimize the errors of prediction, i.e. the sum of the squares of the differences between the actual measurements and the values predicted by the model. Mathematically, this is equivalent to finding the minimum of the multi-variate single-valued function

$$\delta^2 = \sum_{k=1}^{M} \left[\hat{\rho}_k - f\left(x_{k1}, x_{k2}, \ldots, x_{kn}; y_1, y_2, \ldots, y_m \right) \right]^2 \tag{12}$$

The minimum of this function δ^2 is found iteratively: starting from arbitrary initial values of the parameters y_j and the known values of the independent variables x_i, one predicts the corresponding theoretical value ρ_k. This is repeated for each of the available measurements, and the sum of the squares of the differences $\varepsilon_k = \rho - \hat{\rho}$ is computed. The values of the parameters y_j are then changed slightly, and the computation is repeated, until a small enough value of δ^2 is found. The values of the parameters y_j at that point are taken to represent the optimal solution of the problem, because they best account for the observed variability.

The strategy to locate this minimum, and the decision to stop the search for even better solutions are two fundamental mathematical issues which have been extensively covered in the literature, and will not be dealt with here. It will be sufficient to say that some algorithms are better than others, in the sense that they are better able to locate the minimum quickly, or with a minimum of computation, and that the inversion algorithms must be tested for sensitivity to initial conditions and for accuracy in the presence of noise in the observational data.

3.3 ADVANTAGES AND DRAWBACKS OF PHYSICALLY-BASED MODELS

There are three main advantages to the design and use of physically-based models rather than empirical methods to derive quantitative information about the surface from remote sensing data. First, because these models are derived explicitly from a body of proven theories, they offer a formal representation of our understanding of the processes at work. In other words, a large part of the knowledge we actually have about the system under study is encapsulated in the model. As a result, such a model can be used to explain what is happening, to predict new situations, or to study its sensitivity to changes in the initial or boundary conditions, or to variations in model parameters. This is not to say that any physically-based model is necessarily perfect: in fact, discrepancies between the model results and observations are the major driving force to upgrade the model, but at least the hypotheses are explicit and the algorithms can be tested and verified.

The second advantage is that the model results can be interpreted without ambiguity: since the variables and parameters are physical quantities, there is no doubt about their meaning or implication. The third and perhaps most important advantage is that physically-based models with a small number of free parameters can be truly validated, because they can be inverted. Indeed, the fact that a model is able to account for the variability of a data set does not provide any information about the validity of the model: any polynomial (or other function) with enough adjustable parameters can be made to fit any data set. However, if the values of the parameters retrieved by inversion happen to match the values of these parameters as measured or observed, independently from the remote sensing data, and for a variety of targets, then the model can be considered validated. This is a very stringent definition of what constitutes validation; it implies that empirical models cannot be validated *a priori*. These issues have been discussed more extensively in Verstraete and Pinty (1992) and Pinty and Verstraete (1992).

Designing and implementing a physically-based model of the surface-atmosphere system so that it can be inverted against remote sensing data is a major endeavor, however (Rahman et al., 1993a, 1993b). The complexity of this task constitutes a first drawback to this approach. A second difficulty is related to the data requirements for model validation. For instance, the validation by inversion of a bidirectional reflectance model for use with remote sensing data requires the collection not only of satellite data but also of detailed physical (optical) and structural data on the properties of the scatterers making up the system under study, at the same time and for the same place as the remote sensing data is collected. Very few such combined data sets currently exist, and this seriously hinders the development of new models or the improvement of existing ones. In summary, physically-based models can be truly validated, but a significant additional effort must be expended to collect the appropriate data: model development and data collection must be coordinated into a coherent integrated scientific strategy.

The most constraining requirement for the inversion of bidirectional reflectance models against satellite remote sensing data, at this point, is the need for multiple views of the same location in as short as possible a time period but from different angles. This condition may not be frequently realized with current data sets, but the limitation will be partially alleviated by the end of the decade, when new instruments such as NASA's MISR (with 9 cameras at different viewing angles) will become available. In the mean time, the operational use of inversion techniques may be restricted to high latitude regions, because of the frequent overpass of sun-synchronous satellites, or to limited field campaigns, where data from a number of satellites can be acquired over a short period of time.

The fourth and perhaps less critical drawback is that the design and inversion of physically-based models may require larger computer resources than simple empirical methods. No such model is currently in operational use, although it is probable that this point will be reached by the end of the decade, when better instruments and faster computers will be available.

4. Conclusion

Two approaches have been described to extract quantitative information on the state of the surface from remote sensing data: one based on empirical correlations between vegetation indices and variables of interest, and the other based on fully physical models and their inversion. It should be clear that there may be a number of other approaches. For instance, it may well be that a 'semi-empirical' approach combining the advantages of both vegetation indices and models, or based on partly empirical models, may provide a useful alternative.

For the time being, vegetation indices will continue to be used for a number of operational applications, but it is hoped that the discussion in the first part of this paper will have exposed some of problems that can be associated with the uncritical use of these indices. In parallel, the development and operational use of physically-based models will progressively improve our understanding of the environmental processes at work and our capability to manage the planet on a sound basis. These models should, in time, replace the more empirical approach, especially in scientific studies which require the quantitative estimation of physical parameters with high accuracy.

5. Acknowledgments

The comments of Stéphane Flasse and Jürgen Vogt on earlier drafts of this document are greatly appreciated. The financial support of the Joint Research Centre Exploratory Research programme for our investigations of the inversion procedure is acknowledged.

6. References

Colwell, J. E. (1974) 'Vegetation canopy reflectance', *Remote Sensing of Environment*, 3, 175-183.

Curran, P. J. (1981) "Multispectral remote sensing for estimating biomass and productivity', *in* Smith, H. (ed.) *'Plants and the Daylight Spectrum'*, Academic Press, London, 65-99.

Goel, N. S. (1988) 'Models of vegetation canopy reflectance and their use in estimation of biophysical parameters from reflectance data', *Remote Sensing Review*, 4, 1-212.

Goward, S. N., D. G. Dye, and C. J. Tucker (1985) 'North American vegetation patterns observed by NOAA-7 AVHRR', *Vegetatio*, 64, 3-14.

Goward, S. N., B. Markham, D. G. Dye, W. Dulaney, and J. Yang (1991) 'Normalized difference vegetation index measurements from the Advanced Very High Resolution Radiometer', *Remote Sensing of Environment*, 35, 257-277.

Gutman, G. (1991) 'Vegetation indices from AVHRR: An update and future prospects', *Remote Sensing of Environment*, 35, 121-136.

Holben, B. (1986) 'Characteristics of maximum-value composite images from temporal AVHRR data', *International Journal of Remote Sensing,* 7, 1417-1434.

Huete, A. R. (1988) 'A soil-adjusted vegetation index (SAVI)', *Remote Sensing of Environment*, 25, 295-309.

Justice, C. O., J. R. Townshend, B. N. Holben, and C. J. Tucker (1985) 'Analysis of the phenology of global vegetation using meteorological satellite data', *International Journal of Remote Sensing,*, 6, 1271-1318.

Kaufman, Y. J. (1989) 'The atmospheric effect on remote sensing and its correction', *in* G. Asrar (ed.) '*Theory and Applications of Optical Remote Sensing*', Wiley, New York, 336-428.

Kaufman, Y. J. and D. Tanré (1992) 'Atmospherically resistant vegetation index (ARVI) for EOS-MODIS', *IEEE Transactions on Geosciences and Remote Sensing* , 30, 261-270.

Malingreau, J. (1986) 'Global vegetation dynamics: Satellite observations over Asia', *International Journal of Remote Sensing,* 7, 1121-1146.

Meyer, D., M.M. Verstraete, and B. Pinty (1993) 'The effect of surface anisotropy and viewing geometry on the estimation of NDVI from AVHRR', *Remote Sensing Reviews*, in print.

Perry, C. and L. F. Lautenschlager (1984) 'Functional equivalence of spectral vegetation indices', *Remote Sensing of Environment*, 14, 169-182.

Pinty, B. and M. M. Verstraete (1992) 'On the design and validation of bidirectional reflectance and albedo models', *Remote Sensing of Environment*, 41, 155-167.

Pinty, B., C. Leprieur, and M. M. Verstraete (1993) 'Towards a quantitative interpretation of vegetation indices. Part 1: Biophysical canopy properties and classical indices', *Remote Sensing Reviews*, in print.

Rahman, H., M.M. Verstraete, and B. Pinty (1993a) 'A coupled surface-atmosphere reflectance (CSAR) model. Part 1: Model description and inversion on synthetic data', *Journal of Geophysical Research*, in print.

Rahman, H., M.M. Verstraete, and B. Pinty (1993b) 'A coupled surface-atmosphere reflectance (CSAR) model. Part 2: A semi-empirical surface model usable with NOAA/AVHRR data', *Journal of Geophysical Research*, in print.

Verstraete, M. M. (1994) 'Scientific issues and instrumental opportunities in remote sensing and high resolution sprectrometry', *in* J. Hill and J. Mégier (eds.) *'Imaging Spectrometry - a tool for environmental observations'*, Kluwer Academic Publishers, Dordrecht (this volume).

SPECTRAL MIXTURE ANALYSIS - NEW STRATEGIES FOR THE ANALYSIS OF MULTISPECTRAL DATA

MILTON O. SMITH, JOHN B. ADAMS, and DON E. SABOL
Department of Geological Sciences AJ-20
University of Washington
Seattle, Washington, USA 98195

ABSTRACT. Instrument noise, spectral contrast among scene components and variability of spectral scene components are not explicitly evaluated as part of classification and mapping efforts using multispectral images. Yet changes in these factors directly affect mapping accuracy. An analytical framework is proposed such that these factors can be quantified within the context of spectral mixture analysis (SMA). In applying these analyses to an AVIRIS image of Owens Valley, California, U.S.A., we find that the greatest uncertainty in abundance estimates arises from spectral variability in endmembers. Spectral variability in any endmember results in abundance uncertainty of all endmembers. We propose an analytical strategy that subsets an image into regions of lowest spectral dimensionality to minimize uncertainties and to maximize detection of new materials.

1. Introduction

A major objective in remote sensing has been to map and inventory the land surface. Over the past twenty years the number of applications using multispectral images has expanded several fold. Spectral data sets from satellite and aircraft are becoming an essential part to monitor global change. Mapping and the detection of change requires an analytical tool that can be applied to consistently identify surface materials and their abundances over a diversity of atmospheric and lighting conditions. Of significance in spectral mixture analysis (SMA) is the alternative strategies that arise out of considering uncertainty in both the measurement and the analytical framework.

The focus of SMA is on the dominant factors affecting spectral variation (e.g. mixtures of surface materials, lighting geometry, atmosphere, and instrumental calibration). SMA converts encoded radiance from all bands of any multispectral camera to fractions of spectral endmembers which correspond to materials in the scene such as green vegetation, non-photosynthetic vegetation, soil, etc. Shade, an endmember common to most multispectral images, is used to separate the pixel to pixel variation in shading and shadow. The spectral endmembers have been shown to be temporally and spatially consistent for large areas and for a variety of instruments (e.g., Landsat MSS, Landsat TM, Airborne Imaging Spectrometer AIS and Airborne Visible/Infrared Imaging Spectrometer AVIRIS) (Adams et al., 1990; Smith et al., 1990a; 1990b). Common, however, in SMA studies is a limited number of spectral endmembers (e.g. usually from 3 to 5).

J. Hill and J. Mégier (eds.), Imaging Spectrometry - a Tool for Environmental Observations, 125–143.

The advantage of using a few endmembers for an entire scene is the relative simplicity of the analysis. Disadvantages of this approach are 1) that the image spectra of some materials will not fit mixtures of the few selected endmembers, and therefore, will show up as "residuals", and 2) that some image spectra will model as mixtures of the endmembers when, in fact, they are associated with different materials than the endmembers. The "errors" that result by applying simple models to complex surfaces are expressed either as misidentified materials and/or as incorrect fractions of endmembers. A means of assessing the magnitude of these errors is essential, although the acceptable "error" tolerance will be dependent on the objectives of a given study.

The fit of the image data to linear mixtures of endmembers is the primary criteria used to select and add new endmembers. High residuals in the fit of data to mixtures of endmembers indicate that either the set of endmembers is inappropriate or new endmembers are required. The presence of significant residuals indicate compositional information in the image measurements which are not present in the reference endmembers. Roberts et al. (1992) have demonstrated the ability to separate many more than the 3 to 5 endmembers by applying simple two endmember models to subsets of the image data such that the magnitude of the residuals for each subset is near the level of instrumental noise. This strategy of subsetting the image data into multiple mixture models of two endmembers has led to an increase in the compositional information than was possible using a single simple model containing from 3 to 5 endmembers. In this paper we explore the compositional and abundance uncertainties that result with such an analytical strategy.

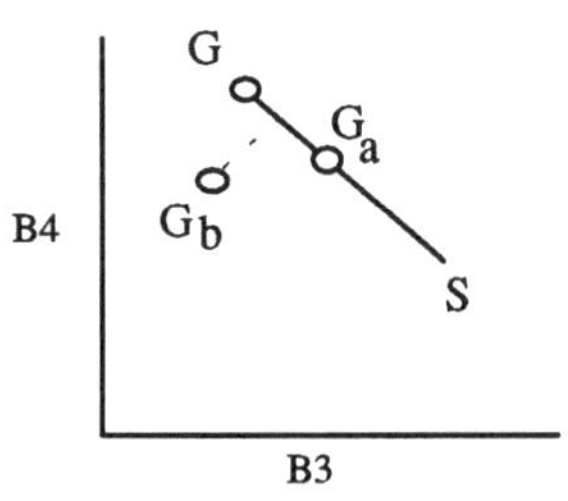

FIGURE 1. For two bands such as TM band 4 and 3 we indicate a soil (*S*) and three green vegetation spectra (*G*) to illustrate the definition of the three types of abundance uncertainty in SMA. The line between *G* and *S* is the potential position of mixed spectra of green vegetation and soil. The size of the vegetation circles is representative of the measurement uncertainty σ_i, which we assume to be gaussian. If vegetation type G_a is present in the scene we will not be able to separate it from a mixture of soil and vegetation. Thus, G_a is likely to be interpreted as a mixture of *G* and *S* when in fact it is not. This uncertainty in interpretation we refer to as σ_m, i.e., the degree to which G_a mimics a mixture of *G* and *S*. Similarly G_b is also partly a mixture of *G* and *S*, but in addition has a residual component that does not fit as a linear mixture of *G* and *S*. The effect of the residuals from reference spectra labeled vegetation on each endmember (e.g., *G* and *S*) abundance using *G* and *S* as endmembers is defined as σ_v. The abundance uncertainty for an endmember is the sum of each of the three uncertainty components.

The assumption that a single set of endmembers is applicable for an entire image data set creates a framework in SMA such that compositional uncertainty cannot be assessed. Similarly, by always using all bandpasses, it is assumed that compositional uncertainty is not expressed in the endmembers. The relevancy of these assumptions will be addressed in the context of abundance and compositional uncertainties. Incorporating uncertainty into the SMA framework provides for alternative strategies to be defined that is dependent on the mapping objectives.

Sabol et al. (1992a) showed the importance of spectral contrast and instrumental noise for SMA applications, in addition to previous constraints (e.g. goodness of fit and abundance estimates between 0 and 1). In this paper we extend the SMA framework by focusing on endmember variability with the objective of being able to minimize uncertainties in endmember abundances. We adapt procedures introduced by Sabol et al. (1992a) to evaluate the effect of instrumental noise, atmosphere, instrument sensitivity, and spectral contrast on detectability. The concepts are developed as an extension of previous analyses performed in Owens Valley, California, U.S.A. using an AVIRIS data set acquired in September 1989 (Gillespie et al., 1990; Smith et al., 1990b).

We separate endmember abundance uncertainties into three parts (figure 1), namely, 1) an instrumental noise component (σ_i), 2) the component of endmember variability which does not fit as mixtures of the endmember set (σ_v), and 3) the degree to which endmember variability mimics mixtures of the endmember set (σ_m). While each of the estimates of abundance uncertainty may be affected by non-linear mixing, the examples presented examine these uncertainties in the context of linear mixtures. The strategies which evolve depend on the objectives of the study. The objectives are: 1) to minimize the uncertainty in quantifying known endmembers; and 2) to maximize the detectability of new or unanticipated materials. While we focus on the first objective, the results provide insight into strategies for attaining the second objective.

2. Methods

Generally, spectral contrast is the relative difference in color and albedo of two objects. A quantitative definition could be approximated by the Euclidean distance between spectral measurements of two materials. The larger the Euclidean distance, the higher the spectral contrast. Using this definition of spectral contrast, it is possible to intuitively determine the relative contrast of objects by visual inspection over the visible wavelengths. For SMA, however, the contrast is between a given material and mixtures of a set of background materials. Using this definition for spectral contrast there is no longer an intuitive translation that can be made by looking at a collection of spectra or materials. For example, if we have a continuous range of spectral mixtures between a bright soil spectrum and varying amounts of shade/shadow (e.g. we defined shade/shadow as a flat spectrum with zero reflectance at all wavelengths), then surface components such as dark water and basalt flows will have a low contrast in relation to the shade-soil background. Both lakes and basalt will not look exactly like a mixture of soil and photometric shade, but the departure from the soil-shade mixture will be less than the Euclidean distance between each of the spectra measured separately.

We desire to develop a direct measure of uncertainty which incorporates the revised definition of spectral contrast and incorporates the effect of endmember variability. We begin with the linear spectral mixture rule where calibration is an integral part of the equation:

$$G_b\, DN_{em,b} = \sum_{r=1}^{Ne} F_r\, R_{r,b} + E_b \quad and \quad \sum_{r=1}^{Ne} F_r = 1 \tag{1}$$

where $R_{r,b}$ is the laboratory or field reflectance in band b for reference endmember r, and $DN_{em,b}$ is the uncalibrated radiance in band b for image endmember em. Typically the calibration gains G_b and offsets O_b for each band b are treated as constants for a given image. The number of reference endmembers Ne is also the number of image endmembers (e.g., em = 1...Ne). As previously described (e.g. Gillespie et al., 1990; Smith et al., 1990b) the F_r's, G_b's and O_b's are solved iteratively so as to minimize the E_b term for each band of Equation 1. The application of Equation 1 is restricted to determining the calibration coefficients G_b and O_b and to finding an optimal set of reference endmembers $R_{r,b}$ which linearly model the image endmembers $DN_{em,b}$.

To analyze the effects of instrumental noise, endmember variability and spectral contrast we require a solution of Equation 1 analogous to the eigenvectors in principal components analysis (PCA), but with nonorthogonal vectors in the transformation matrix; the angle between the vectors is defined by the contrast of the spectral endmembers. The uncertainty characterization is performed in uncalibrated DN's so that we incorporate the effects of the instrument sensitivity into the uncertainty estimates. After solving for calibration coefficients, we can simplify the determination of fractions from Equation 1 in matrix notation to:

$$[F]=[P]\left([DN]-[DN_{lem}]\right) \tag{2}$$

where $[F]$ is the vector of fractions containing all F_r's of reference endmembers (e.g. endmember abundances); $[P]$ is the pseudo inverse matrix (size = number of bands, number of endmembers -1) which transforms the spectral measurement vector $[DN]$ into abundances; and $[DN_{lem}]$ is a constant consisting of the last endmember in the list of endmembers (e.g. typically this vector is shade which is often the zero vector in reflectance or the vector formed by the O_b's from equation 1). By subtracting $[DN_{lem}]$ in equation 2 we incorporate the constraint that the endmember fractions sum to one. Using a modified Gram Schmidt reduction algorithm described by Golub and Van Loan (1989), we determine the pseudo inverse matrix $[P]$ from the reference endmembers. Thus, the solution of endmember abundances for an image pixel reduces to a vector subtraction and a matrix multiplication. The $[P]$ matrix can be interpreted similar to the eigenvectors of PCA in that the absolute magnitudes of the vectors indicate the relative importance of specific wavelengths in determining endmember abundances.

We desire to determine the uncertainties in $[F]$ arising from uncertainty of the spectral bandpass measurements (λ's), e.g.,

$$\sigma_i = f\left(\sigma_{\lambda 1}, \sigma_{\lambda 2}, \sigma_{\lambda 3}, \ldots\right) \tag{3}$$

where σ_i is the standard deviation of the fraction estimate determined from equation 2, and the σ_{λ_i}'s are the instrumental variances at each wavelength λ. Although there are many ways to measure the instrumental variance we will employ a method used by Sabol et al. (1992a). This method uses the E_b's from equation 1 of an area in the image to estimate the σ_{λ_i}'s after the endmembers have been selected and the image calibrated using equation 1.

Using derivations performed by Bevington (1969) to compute the propagation of errors, we express the deviations of a single fraction F in equation 1 to first order as a Taylor series expansion, e.g.,

$$F_i - \bar{F} \cong \left(\lambda_{1i} - \bar{\lambda}_1 \right) \left(\frac{\partial F}{\partial \lambda_1} \right) + \left(\lambda_{2i} - \bar{\lambda}_2 \right) \left(\frac{\partial F}{\partial \lambda_2} \right) + \ldots \quad (4)$$

By substituting Equation 4 into the definition of variance (e.g. $\sigma_x{}^2 = \frac{1}{N} \sum (x_i - \bar{x}))$ we obtain:

$$\sigma_i{}^2 = \sigma_{\lambda 1}{}^2 \left(\frac{\partial F}{\partial \lambda_1} \right)^2 + \sigma_{\lambda 2}{}^2 \left(\frac{\partial F}{\partial \lambda_2} \right)^2 + 2\,\sigma_{\lambda 1, \lambda 2}{}^2 \left(\frac{\partial F}{\partial \lambda_1} \right) \left(\frac{\partial F}{\partial \lambda_2} \right) + \ldots \quad (5)$$

We make the assumption that the measurement error of each bandpass is independent and thus the covariances of the fluctuations (e.g. the third term in equation 5) between wavelengths is near zero. Because the F's are the weighted sums of the spectral measurements in equation 2 the partial derivatives are simply the weighting constants in the $[P]$ matrix. We find that the translation of measurement error from wavelengths to fractions simplifies to the sum of the transformation vector squared and multiplied by the instrumental variance at each wavelength, e.g., to determine the variance of the first fraction F_1:

$$\sigma_i{}^2 = \sum_{\lambda=1}^{nb} \sigma_{\lambda i}{}^2 \, P_{1\lambda}{}^2 \quad (6)$$

where nb is the number of bandpasses in the image system, $[\sigma_{\lambda i}]$ the standard deviation vector defining instrumental noise, and $P_{1\lambda}$ the vector extracted from the P array corresponding to the first endmember. The results obtained from equation 6 is computationally equivalent to that presented by Sabol et al. (1992a) except that we now express the fraction variance directly as a function of instrumental variances instead of using a signal to noise ratio. The reason for this specific formalization is that it provides a direct means of calculating abundance uncertainties. Using equation 6 we can begin to formulate strategies that optimize abundance estimates under conditions of instrument noise, spectral contrast and endmember variability.

We postulate two effects of endmember variability on the estimate of abundance uncertainty. First, is that component of endmember variation that does not fit as linear mixtures of the reference endmember set (e.g., σ_v), and thus, has many similarities to instrumental noise σ_i. To calculate σ_v we use a $\sigma_{\lambda v}$ vector in equation 6 instead of $\sigma_{\lambda i}$. The difference between $\sigma_{\lambda i}$ and $\sigma_{\lambda v}$ is that $\sigma_{\lambda i}$ is computed using image measurements and $\sigma_{\lambda v}$ uses the reference spectra of figure 2. In both cases, these estimates reflect that component of the input data that does not fit as mixtures of the set of endmembers illustrated in figure 3. The reference spectra of figure 2 include reflectances acquired from different vegetation types to include foliage and stem components to determine the vector $[\sigma_{\lambda v}]$.

A second effect of endmember variability on abundance uncertainty arises from mixtures of endmembers which mimic spectra of uniquely different materials (Smith et al., 1990b; Sabol et al., 1992a). The standard deviation of the abundance estimates derived from several reference spectra (figure 2) that sample the variability of the reference vegetation endmember (figure 3) is used to define the degree of mimicry (σ_m) for a given set of endmembers. Specifically, to determine σ_m, an abundance estimate F_r is made on several reference spectra all with the same label using a single set of reference endmembers. For example, σ_m, for vegetation would be the standard deviation of the vegetation F_r's determined for several vegetation reference spectra (figure 2) by a fixed set of reference endmembers (figure 3).

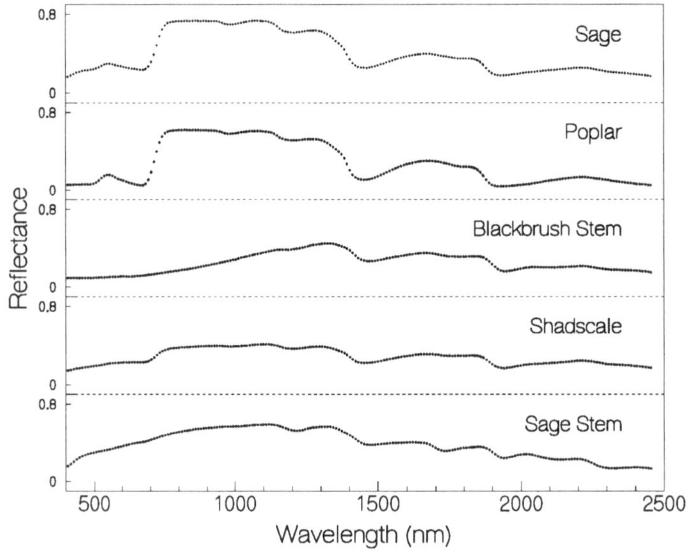

FIGURE 2. Laboratory reflectance spectra of other vegetative components on the bajada's of Owens Valley. This subset of reflectance spectra is used to access the effect of spectral variability in vegetation on SMA using a single vegetation endmember.

3. Results

3.1 Spectral Contrast

Figure 3 illustrates the four spectral reference endmembers that fit the spectral variation in an AVIRIS image of Owens Valley (Gillespie et al., 1990; Smith et al., 1990b). We desire to quantify the uncertainties (e.g., σ_i, σ_v, and σ_m) in mapping vegetation for this AVIRIS image. Although the focus is on the vegetation abundance, the concepts are readily extendible to the two soil endmembers. While a laboratory spectrum of Blackbrush (*Coleogyne ramosissima*) foliage is used as the endmember for all vegetation types we do not imply that all vegetation is Blackbrush. However, we are interested in the resulting abundances that the Blackbrush reference endmember provides in the context of representing all vegetation types given the spectral variability of vegetation. In the context of this paper references to vegetation abundances are abundances obtained from the spectrum of Blackbrush which was dominantly green foliage. The foliage was attached to the stems so naturally there is a stem component that is part of the Blackbrush spectrum.

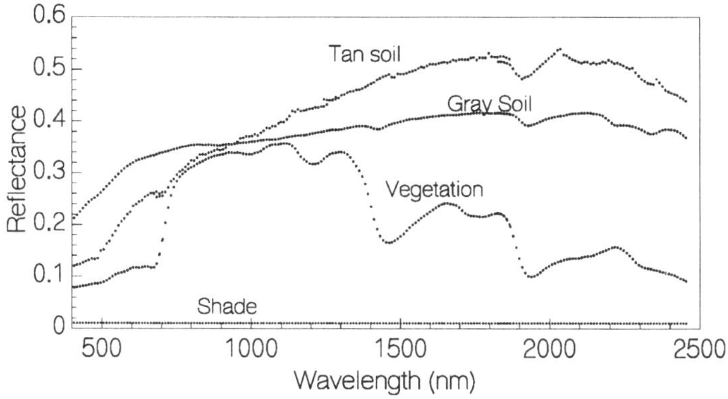

FIGURE 3. The reference endmembers for a 1989 AVIRIS image of Owens Valley, California, U.S.A. The endmember reflectances are displayed for all 224 AVIRIS channels. The endmembers consist of two soils, a vegetation endmember and photometric shade (e.g., zero reflectance in all wavelengths).

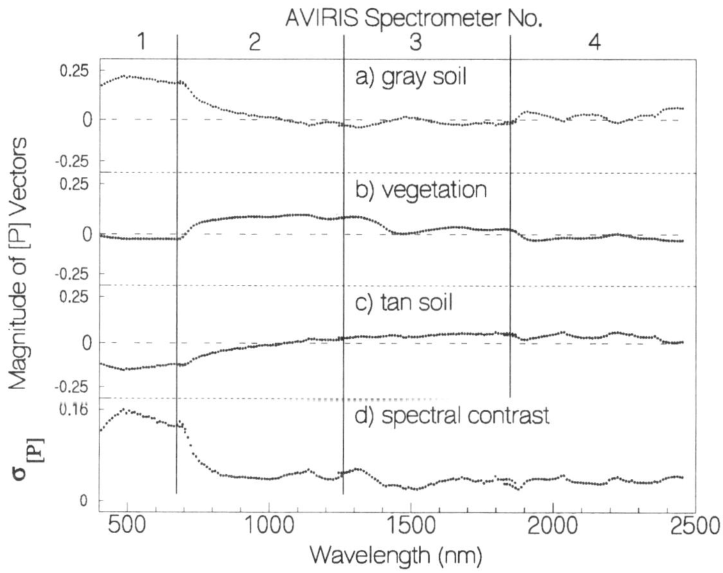

FIGURE 4. The signed magnitudes of the vectors in [*P*] as a function of wavelength can be used to access the relative importance of wavelengths contributing to abundances and uncertainties of the different endmembers. The [*P*] vectors here have been determined using reflectances of the endmembers from figure 3. The spectral contrast of the endmembers is expressed as the standard deviation of the vectors in [*P*]. The greatest spectral contrast in these endmembers occurs in the visible and near infrared regions of the spectrum.

Figure 4a-c illustrate the vectors associated with each endmember of the [P] matrix determined from the laboratory reflectance spectra in figure 3 of all 224 AVIRIS wavelengths. These vectors are used to determine the endmember abundances (equation 2) and also the uncertainties of endmember abundances (equation 6). Positive values indicate wavelengths which are positively correlated to the endmember while negative values are indicative of wavelengths inversely proportional to endmember abundance. The magnitude of the values of [P] is proportional to the weight a given wavelength contributes to an abundance estimate. The values of [P] are dependent on the endmember set and not any particular endmember so that a change to any endmember will affect all the vectors in the [P] matrix.

The standard deviation of the vectors in [P] (figure 4d) is used as an estimate of the spectral contrast at each wavelength with respect to linear mixtures of the endmember set. For this case, we find that the visible and near infrared wavelengths have the highest contrast as they correspond to the highest variance of the vectors in [P]. The wavelength region includes the red edge typical of vegetation, but not the infrared plateau as the spectral contrast of the Blackbrush with respect to the soils diminishes in this region. The variance in figure 4d is valid only in the context of all wavelengths used to compute the [P] matrix. Removing or adding wavelengths from those used in figure 4 changes the spectral contrast and thus all elements of the [P] matrix.

Application of equation 2 to the image data requires that some wavelengths be removed from analysis due to instrumental noise and atmospheric absorption. For the Owens Valley 1989 AVIRIS image we removed 73 bands leaving a total of 171 bands for analysis. In figure 5 the vectors in [P] are plotted again but include the effect of the atmosphere and instrument sensitivity for the 171 bands. The differences in spectral contrast between figures 4d and 5d are due to changes in instrumental sensitivity and atmospheric absorption as a function of wavelength. The magnitudes of [P] (figures 4a-c) corresponding to spectrometer 4 have near zero weights for all endmembers indicating the lack of influence of these wavelengths in affecting abundance determinations. The near zero values of [P] is due to the lack of instrumental sensitivity over this wavelength region. Thus, independent of the spectral contrast in the endmember set for wavelengths covered by the fourth spectrometer, these wavelengths will have negligible influence on endmember abundances. The drop of contrast at the start of the visible region (spectrometer 1) is due to the high atmospheric backscatter at these wavelengths in comparison to the ground reflectance. The differences between figures 4d and 5d indicate the necessity to incorporate the effect of instrument and atmosphere to evaluate uncertainties in endmembers and spectral contrast in endmember sets.

3.2 Instrumental Uncertainty

The effect of instrument noise $\sigma_{\lambda i}$ on abundance uncertainty requires determining the instrumental noise as a function of wavelength as described by Sabol et al., 1992a. The $\sigma_{\lambda i}$'s (figure 6) for the 171 AVIRIS bands show a relatively flat noise response over all wavelengths. This noise characterization of the AVIRIS instrument is also typical of the instruments recent performance (Sabol et al., 1992a). Upgrades made to AVIRIS effect the sensitivity, and thus, are noticed as features in [P] as opposed to increases in instrument channel noise $\sigma_{\lambda i}$. In contrast to instrumental noise is the endmember variability $\sigma_{\lambda v}$ which is significantly greater than $\sigma_{\lambda i}$ as illustrated by figure 6.

The variance as calculated in figure 5d is used to determine the magnitude of σ_i as a function of the number of AVIRIS bands used in the analysis. For each set of bands the [P] matrix is computed and the Blackbrush endmember band with the lowest variance in [*P*] is removed. The effect of using different bandpasses cannot be intuited from figure 5d (e.g., a single [*P*] matrix determined from 171 bandpasses). As each band is removed a new [*P*] must be calculated because the spectral contrast has changed.

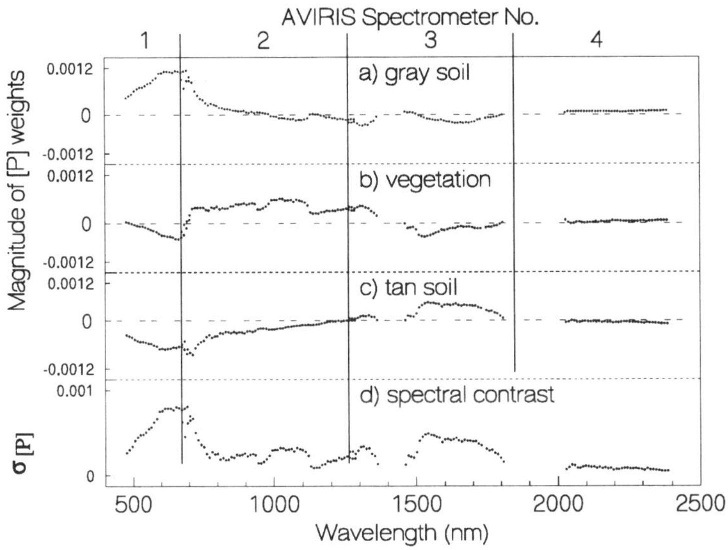

FIGURE 5. The vectors in [*P*] plotted for the set of laboratory endmembers in figure 3 which include the effect of instrument sensitivity and atmosphere. From the original 224 AVIRIS bands, only 171 bands were sensitive to spectral variability of the surface. Instrument sensitivity and atmosphere significantly alter the spectral contrast normally found in the reflectance spectra of the endmembers as observed by comparing this figure to figure 4.

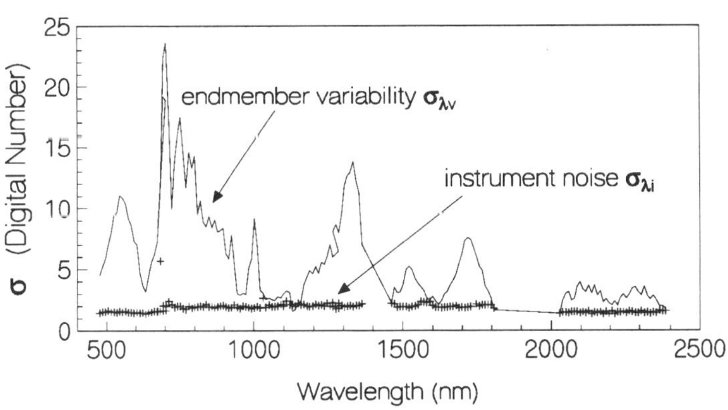

FIGURE 6. The uncalibrated instrument variability (σ_i) is compared to that introduced by laboratory endmember variability (σ_v). The effect of endmember variability is determined by computing the standard deviation of E_b's in Equation 1 using the laboratory spectra from figure 2.

The effect of the number of bandpasses is minimal on all σ_i's until approximately 60 bandpasses (figure 7). Because the noise response is relatively flat with wavelength the dominant factor determining uncertainties is the unsigned magnitudes of $[P]$ (e.g. spectral contrast). The procedure for determining figure 7 is only approximately optimal as indicated by the dips and peaks in σ_i's at relatively low numbers of bandpasses. Removing the band of lowest contrast in a set of n bands is only an approximation of the best contrast in a set of bands of n-1. The approximation works best for large numbers of bandpasses. As the number of bands used in the analysis goes below ~50 bands it is necessary to reconsider bands that have been eliminated in previous steps. A strategy indicated by figure 7 alone is that 6 bands provide σ_v's near 0.01 for all three endmembers, thus more than 6 bands provide little additional benefit for the case of these four endmembers. This strategy considers only the effect of instrumental noise and does not include effects of endmember variability.

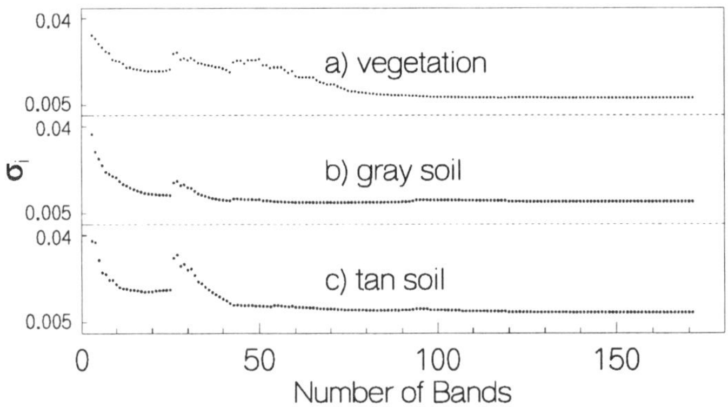

FIGURE 7. The effect of the number of bandpasses on σ_i. The magnitude of error is relatively constant down to approximately 50 bands. For each point plotted, a new $[P]$ matrix is computed after removing the band with the lowest spectral contrast as defined by the standard deviation of the vectors in $[P]$.

The effect of endmember variability on the reference vegetation endmember is illustrated in figure 8. In this figure, σ_v for the vegetation endmember has a much higher value than observed in figure 7 where only the instrumental variability was considered. The effect of variability in the vegetative endmember (figure 6) gives more significance to the larger number of bandpasses in maintaining lower uncertainties in vegetation abundances than uncertainties determined using only instrumental noise. Regardless of the subset of reference spectra used to determine vegetation variability, the effect of $\sigma_{\lambda v}$'s will dominate over those determined from instrumental noise.

3.3 Number of Endmembers

An important consideration in the application of SMA is the number of endmembers chosen for analysis. Why not include the reference spectra displayed in figure 1 as new endmembers so that abundances include a larger range of vegetation types? In figure 9 the effect of a variable number of endmembers on σ_i of the vegetation endmember is illustrated. Independent of the vegetation endmember added or the number of bands used there is an increase in the Blackbrush σ_i's associated with increasing the number of endmembers. Only a small increase in σ_i occurs between 3 and 4 endmembers because the endmember added was a soil which changes the contrast only slightly for the Blackbrush endmember. The magnitude of the Blackbrush σ_i (0.04) for the 6 endmember case is sufficiently low to indicate that reasonable abundance estimates could be obtained for this set of endmembers. The σ_i for a local area is as predicted by figure 9 using all 171 bands. However, the distribution of the vegetation endmember abundances is not related to the distribution of vegetation types. The vegetation endmembers begin to map subtle soil differences.

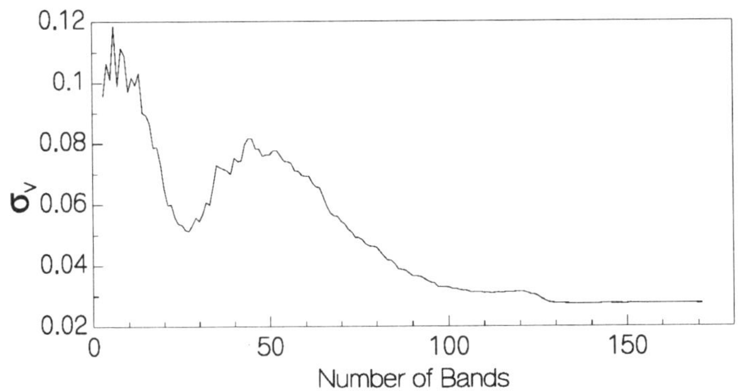

FIGURE 8. The standard deviation of the vegetation endmember, σ_v, is determined from the reference spectra of figure 2. The magnitude of σ_v is significantly greater than that determined from the instrument as indicated by comparison to figure 7a. The importance of many spectral measurements or bandpasses is much more apparent in the context of endmember variability.

3.4 NEW ENDMEMBERS

Adding an endmember cannot be determined by the magnitude of σ_i alone. It is necessary to also determine if the added endmember significantly reduces the E_b terms of Equation 1. However, when the mean root mean square (rms) E_b term is small (e.g. around 0.01 to 0.02 reflectance) for the 171 AVIRIS bands adding almost any new reference spectrum as a new endmember has a small noticeable improvement, even if the added endmember has no association with the spectral attributes of the surface. This condition arises due to the degenerate nature of spectra for natural surfaces. When E_b's are large the unfit portion is typically unique in terms of spectral identity. However, a small rms in the E_b's is characterized often by small wavelength regions which are not fit by mixtures of endmembers. Thus, while adding new endmembers may reduce E_b, the criteria for adding new endmembers to obtain a better fit is no longer valid for small E_b's.

3.5 MIMICKING

Only part of the uncertainty in abundance estimates can be considered in the context of instrumental noise σ_i or unfit endmember variability σ_v. The variability in σ_m's (table 1), which result from fitting each spectrum of figure 2 to the Owens Valley endmembers, is significantly greater than the σ_v contribution (figure 8). The magnitude of σ_m is significantly reduced by removing both soil endmembers (e.g., the endmembers with high contrast) from the endmember group (table 1).

The σ_m's in table 1 apply if the abundances determined from the spectral measurements in the image range from 0 to 1 for each endmember. The σ_m's overestimate the magnitude of uncertainty because on the bajada of Owens Valley there is a limited range of abundances for each endmember. From earlier studies (Smith et al., 1990) spectral abundances of vegetation vary from 0 to 0.3. The σ_m's scale linearly with the range of abundance so that the effect of endmember variability is more likely one third the magnitude calculated in table 1.

Given the pronounced effect of σ_m dominating the uncertainty in vegetation abundance we focus on reducing this influence. By plotting the variability of the vegetation spectra (figure 1 and Blackbrush) we can determine wavelengths most susceptible to endmember variability (figure 10). To achieve more accurate fractions when applying a simple model of 2-5 *em*'s the objective is to identify the bands that minimize σ_v and σ_m. When all bands are included, endmember variability directly affecting the vegetation σ_m dominates. By removing wavelengths with maximum variability in vegetation, the component σ_v increases to a point where it is equivalent to σ_m. However, the same bands provide information on other spectral components in the scene and would be used in a multi-endmember analysis.

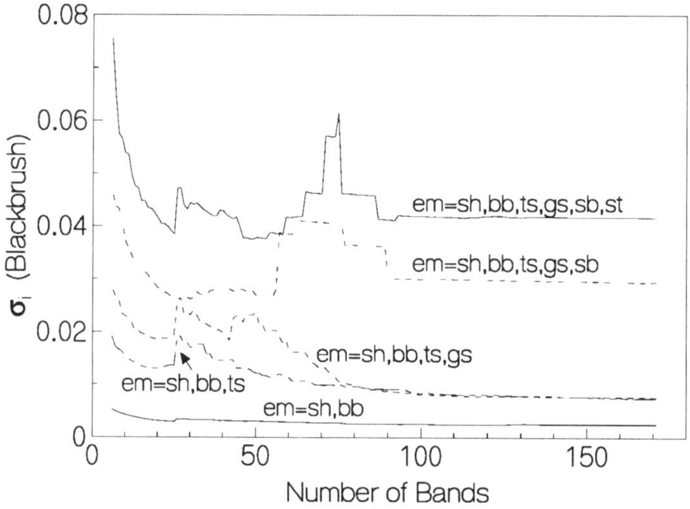

FIGURE 9. The effect of number of endmembers on the Blackbrush endmember uncertainty, using only the noise induced by the AVIRIS instrument. Abbreviations are defined as follows: sh - shade, bb - Blackbrush, ts - tan soil, gs - gray soil, sb - sagebrush, st - sagebrush stems. For any new endmember added the uncertainty in the Blackbrush abundance increases. The magnitude of increase in σ_i is dependent on the spectral contrast to this endmember. For example, adding a second soil (gs) for a total of 4 endmembers increases the Blackbrush σ_i less than adding the sage or sage stem endmembers.

Figure 11 illustrates the effect of removing bands most sensitive to endmember variability. As with earlier results the unfit component of endmember variability, σ_v, increases independent of the order in which bands are removed. The initial reduction of σ_m as bands are removed is due to endmember variability caused by vegetation. As the number of bands is reduced σ_m decreases to a minimum at about 110 bands, and with removal of additional bands σ_m increases again, because many of the deleted bands provide optimal spectral contrast to separate vegetation from soils.

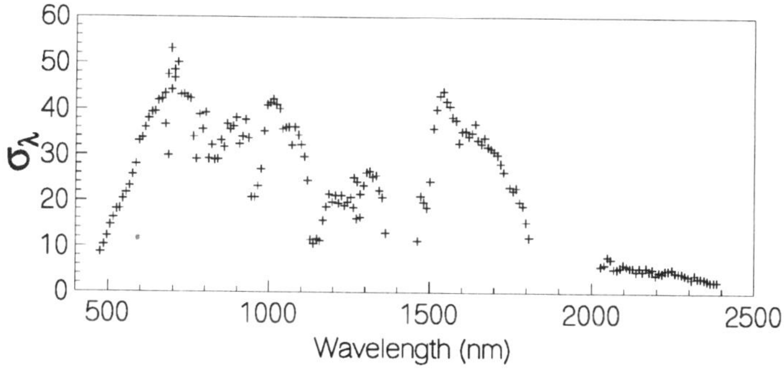

FIGURE 10. The uncalibrated variability of vegetation spectra shown in figure 2 for each of the 171 AVIRIS wavelengths. Wavelengths of highest variation in the vegetation endmember are also associated with maximum contrast between vegetation and soil endmembers.

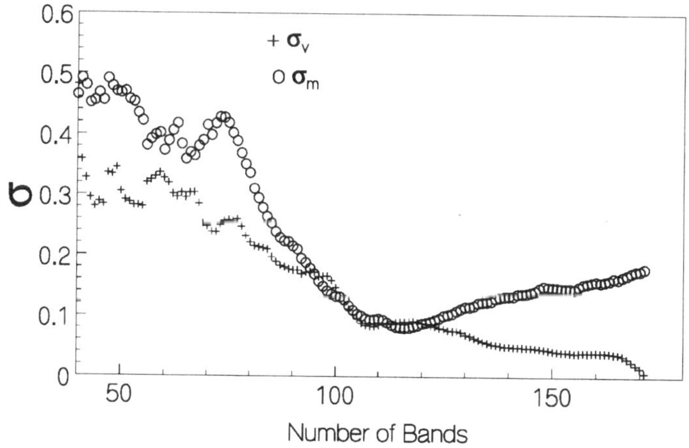

FIGURE 11. Wavelengths were selected to minimize the uncertainties in vegetation abundance by minimizing both σ_m and σ_v. Bands were sequentially removed starting from the maximum variability of the vegetation as defined in figure 10. The increase of both σ_m and σ_v at lower number of bands is due to the decrease in spectral contrast that occurred in removing wavelengths of highest variability.

σ_m	Endmember	Endmember Group
0.753	Blackbrush	Blackbrush, Gray Soil, Tan Soil, Shade
0.668	Gray Soil	"
0.573	Tan Soil	"
0.753	Blackbrush	Blackbrush, Tan Soil, Shade
0.549	Tan Soil	"
0.311	Blackbrush	Blackbrush, Shade

TABLE 1. The effect of the laboratory spectra in figure 2 on σ_m determined from different endmembers of the four Owens Valley spectral endmembers consisting of two soils, a photometric shade and vegetation. Deleting endmembers with high spectral contrast to vegetation (e.g. the soils) reduces the uncertainty in the vegetation σ_m caused by spectral differences among types of vegetation.

3.6 Application to Owens Valley

In applying a new band list to estimate vegetation abundance we find the results to be in accordance with field observations. First the analysis of Smith et al., 1990b indicated a much wider range of vegetation variability on the bajada than expected for the September AVIRIS image acquisition which is consistent with a high σ_m. The range in vegetation abundance was more similar to the May 1985 TM vegetation estimates than to the December 1982 TM estimates (Smith et al., 1990a). The burned area (figure 12) has 0.09 more vegetation than we obtain from the reduced band list. This difference is consistent with spectral differences in vegetation on and off the burned area. Burns typically are invaded by sagebrush with a higher reflectance in the VIS-NIR wavelengths. The sagebrush will yield a higher abundance than the Blackbrush for the same projected aerial coverage. By using bandpasses that minimize the difference between Blackbrush and sagebrush we obtain a more uniform mapping of the vegetation over the bajada. The effect of σ_m is observed to occur over low frequency spatial scales. This is the case applicable to earlier analysis of AVIRIS data in Owens Valley (Smith et al., 1990b).

A comparison of the RMS images (e.g., root mean square of the E_b's in equation 1) obtained from the reduced band list (figure 13a) and the original 171 bands (figure 13b) are consistent with a change in σ_v resulting from vegetation variability. Remaining features still exist in the reduced band list rms image (figure 13a) depicting riparian communities and agricultural fields. These areas could contain spectral variability of vegetation not taken into account by the small subset of reference spectra (figure 2). The highest RMS values in E_b's are obtained by performing SMA on bands with the highest spectral variability of vegetation (not shown). This change in fit of the image to mixtures of the endmembers with selective removal of bandpasses is consistent with spectral variability in the vegetation endmember.

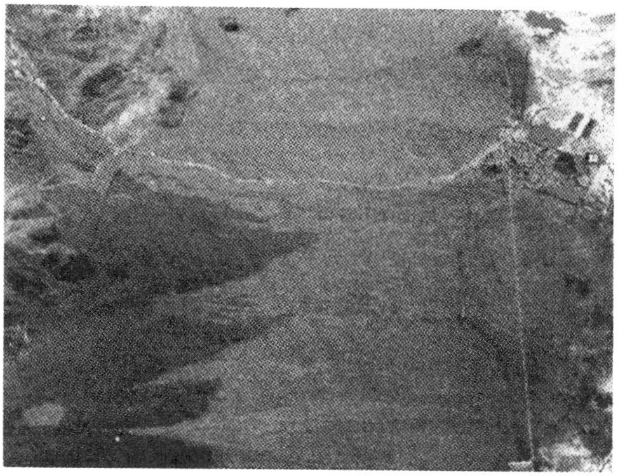

FIGURE 12. A difference image made by subtracting the Blackbrush abundance computed using all 171 bands from the Blackbrush abundance determined using wavelengths which minimize the variability of the Blackbrush endmember. The image computed from the subset of wavelengths has a reduced range of vegetation abundances in comparison to the analysis using 171 bands.

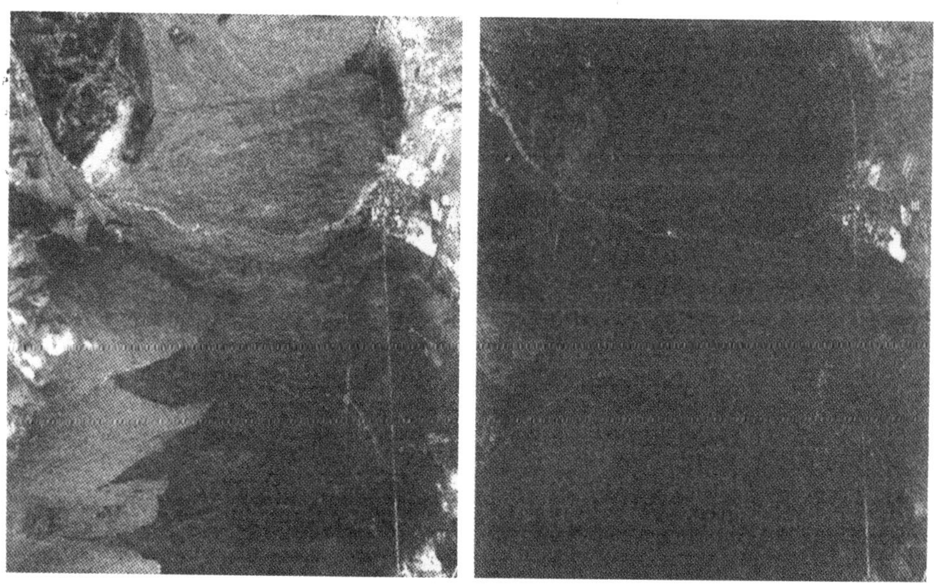

FIGURE 13. A comparison of the rms errors determined in the fit of equation 1 for two cases: a) all 171 bands and b) a reduced set of wavelengths to minimize the effect of endmember variability. Spatial patterns are less apparent in b) than a) while speckle noise is higher in a) than b) indicating the effect of the vegetative endmember variability.

4. Discussion

Obtaining variable abundances of endmembers from a single multispectral data set is presented as evidence that endmember variability is a significant factor in mapping endmembers using SMA. Although variable abundances could also be produced by nonlinear mixing among endmembers the correlation of abundance uncertainties with endmember variability as a function of wavelength strongly suggests the significance of spectral variation in endmembers. Since we cannot eliminate spectral variability in an endmember, our objective is to understand the magnitude and propagation of uncertainty as a function of variability and to determine possible strategies which may be incorporated in mapping analysis to minimize abundance uncertainty.

There is an implicit inverse relation between σ_m and σ_v in the application of SMA. To illustrate this relation we provide a qualitative description of the mathematical coupling between F_r and E_b's from equation 1 to the multispectral measurement. For a given multispectral system there is a finite n-dimensional measurement volume associated with the input dimensionality of the system. For AVIRIS this volume can be computed as a series of products of the signal to noise in each spectral band. SMA divides the spectral measurement volume into three separate volumes. The first is the mixture volume which is often referred to as the convex hull. Within this volume all fractions of the endmembers are between 0 and 1, e.g., $0 \leq F_r \leq 1$. The dimensionality of this volume is equal to Ne - 1. Outside the convex hull is the volume where the fractions of at least one endmember is less than 0 or greater than 1. The remaining measurement dimensionality which is not fit by linear mixtures of the endmembers comprises the remainder of the spectral measurement volume. Uncertainties in σ_v will be incorporated in that volume not fit by linear mixtures of the endmembers, while σ_m will be incorporated within the first two volumes which includes the measurement dimensionality defined by the convex hull.

The size of the mixture volume is increased by adding new endmembers. Because the measurement volume is finite, an increase in the mixture volume will cause a corresponding decrease in the volume not fit by mixtures of the endmembers. By increasing the number of endmembers that we use to model an image it follows that we increase the uncertainty in abundance estimates arising from σ_m and decrease σ_v. For example, if the mixture volume is the same dimensionality as the measurement volume then σ_v is zero and all uncertainty due to endmember variability will be incorporated in σ_m. We find for the Owens Valley AVIRIS image (table 1) such an increase in σ_m with more endmembers. The magnitude of the change is dependent on the variability in the added endmember and its spectral contrast to the other endmembers. There is an insignificant change for large numbers of bands in σ_v when the gray soil endmember was added (table 1). The gray soil endmember changed the mixture volume by only 2 percent in comparison to the previous three endmember model.

The incremental volume added to the convex hull by each endmember also affects σ_m and σ_v. The change in mixture volume which results from adding an endmember is proportional to the spectral contrast of the added endmember to the existing endmembers. If we add endmembers which result in little or no change in the mixture volume then we increase both σ_m and σ_v for not only the added endmember, but also for those endmembers near it. The increase in the Blackbrush σ_i with an increase in endmembers observed in figure 9 is also indicative of the change of σ_v with increasing endmembers. σ_i increases in figure 9 because the endmembers being added have little spectral contrast (e.g., < 1 percent increase to the mixture volume) to the Blackbrush.

In the application of SMA, σ_m is the most difficult to quantify because it is embedded in the abundance estimates of the endmembers. Thus, it is likely that changes in abundance of an endmember are interpreted as real rather than due to endmember spectral variability. Comparative analysis of multitemporal data sets using SMA may show abundance differences which are difficult to interpret (e.g., are differences the result of spectral variability or a real change in endmember abundance?). To minimize σ_m for an endmember it is necessary to estimate the variability of that endmember as a function of spectral bandpasses and then to remove bandpasses with maximum variability as illustrated in figure 11. In removing bands, one must consider the potential reduction in spectral contrast among endmembers and the resulting increase in σ_v. It may be the case that the resulting decrease in spectral contrast between endmembers in multispectral images such as Landsat TM may not warrant removing any bands from analysis.

Reduction in the number of bands, however, is not the only strategy available to minimize σ_m. By using the minimal set of endmembers applicable for an area within an image, we maximize the volume not fit by mixtures of endmembers in addition to the contrast of endmembers. The impact of this strategy will result in higher σ_v. Uncertainties of σ_v are more tolerable because we can quantify them as the unfit component of mixture analysis. However, this strategy also insures that the endmembers will have the best spectral contrast thus minimizing σ_v . In past applications, a single set of endmembers has been applied to an image when in fact there are many areas which could be fit by a smaller set of endmembers. However, by separating an image into regions of similar spectral variability of reduced dimensionality we minimize uncertainty in abundances caused by endmember variability. The procedure is more complex in that different endmembers are applied for each region.

Endmember variability for the AVIRIS instrument is likely to dominate abundance uncertainties for most cases involving natural surfaces in comparison to instrumental noise. The effect of instrumental noise has a relatively minor effect on abundance estimates compared to σ_m and σ_v . The magnitude of σ_m will typically be significantly larger than σ_v. Evidence for this is illustrated using spectral libraries which may contain 1000 rock, soil and vegetation spectra, but all 1000 spectra can be modeled by linear mixtures of less than 5 or 6 of these spectra with fits of ~ 0.02 reflectance (Gillespie et al., 1990).

These results suggest a totally different strategy than is typically applied in mapping vegetation with band ratios or other indices such as NDVI. Indices attempt to utilize spectral data only in regions with the maximum spectral contrast. This strategy maximizes effects of endmember spectral variability because wavelengths of greatest contrast also are most variable among different vegetative components. Using wavelengths of maximum spectral contrast also maximizes detection of a endmember. A primary goal in using images derived from hyperspectral systems is to begin understanding the relative importance related to the mapping objectives. As illustrated, there are at least two contrasting strategies depending on whether one desires to maximize detectability or uniformity of abundance estimates.

These results indicate two guiding constraints to optimize vegetation mapping of endmembers. First, if one adds more spectral endmembers than are present in the image data then the uncertainty of abundance estimates increase (Sabol et al., 1992a). If, however, not all endmembers are taken into account then uncertainty in abundance estimates also increase. It is necessary to include precisely the valid endmembers to minimize abundance uncertainties. Evaluation of σ_m, σ_v and residuals E_b's alone are not sufficient to guarantee that the endmember set is meaningful. These estimators of uncertainty, however, do help constrain and guide the analysis.

Roberts et al. (1993) have attributed different vegetation abundances obtained by using wavelengths in the visible and near infrared regions to nonlinear mixing caused by increased scattering of light over wavelengths characterized by the infrared plateau of vegetation spectra. In this study we observe similar results, but attribute these results to endmember variability of vegetation spectra. The Great Basin communities of Owens Valley are sparse (< 30% cover) and the foliage relatively opaque (except perhaps riparian poplar stands) so that it is difficult to believe that differential scattering is a significant factor. The results are sufficiently similar to warrant further investigation into the effect of leaf transmission and scattering versus endmember variation on abundances. Intuitively, the scattering processes that occur within the canopy are similar to that inside the leaf except for scale, so that proof of the cause may be difficult. The introduction of a vegetative shade endmember by Roberts et al. (1993) would have a similar effect on the current data set in that errors would be reduced.

The bands removed from the Owens Valley AVIRIS image to maximize uniformity of vegetation abundances are also the bands which provide compositional information regarding vegetation variability. Maximizing the number of endmembers by subsetting the image into smaller areas similar to that described by Roberts et al. (1992) requires bandpasses that maximize contrast between endmembers. A research focus into utilizing the vegetative bandpasses of highest contrast is needed to quantify subtle differences in vegetation. It is likely that these efforts will aid in defining vegetative endmembers which correspond to the spatial scale of AVIRIS pixels rather than the spatial scale typical of laboratory and field spectral measurements.

5. Conclusions

We find that

- all bandpasses are not necessarily optimal for determination of endmember abundances. The spectral contrast of endmembers and endmember variability are required to select wavelengths to minimize uncertainty in abundance estimates. For the Owens Valley AVIRIS image the minimum uncertainties in vegetation abundances were obtained using wavelengths that did not maximize contrast between the soils and the vegetation endmember.

- minimizing σ_m's is achieved by using the minimum number of endmembers, thus implying a strategy that attempts to model abundances in images as regional subsets of two endmembers rather than a fixed set of a much larger number of endmembers applied over the entire image.

- optimizing abundance estimates of each endmember involves invoking different strategies dependent on endmember variability and spectral contrast.

- the accuracy of mapping is not single valued as is assumed using classification and fitting algorithms, but dependent on the spectral contrast of the background and endmember variability.

- high spectral resolution measurements are necessary to minimize effects of endmember variability. Simple ratios and indices which depend on only a few bands with maximum contrast to perform mapping are susceptible to significant σ_m uncertainties which are not easy to quantify.

6. Acknowledgments

We gratefully acknowledge the W.M. Keck Foundation for computer equipment and support and also the Joint Research Centre, Ispra, Italy for their financial support in developing these concepts.

7. References

Bevington, P.R. (1969) *'Data Reduction and Error Analysis for the Physical Sciences'*, McGraw-Hill Book Company, New York, N.Y.

Gillespie, A.R., M.O. Smith, J.B. Adams, S.C. Willis, A.F. Fischer, and D.E. Sabol (1990) 'Interpretation of residual images: spectral mixture analysis of AVIRIS images, Owens Valley, California', *Proceedings of the Airborne Science Workshop, JPL Publ.,90-54*, 243-270.

Golub, G.H., and C.F. Van Loan (1989) *'Matrix Computations'*, John's Hopkins University Press, Baltimore, MD.

Roberts, D.A., M.O. Smith, D.E. Sabol, J.B. Adams, and S.L. Ustin (1992) 'Mapping the spectral variability in photosynthetic and non-photosynthetic vegetation, soils and shade using AVIRIS', *in* R.O. Green (ed.), *'Summaries of the Third Annual JPL Airborne Geoscience Workshop '*, vol. 1, AVIRIS Workshop, June 1-2, 1992. pp. 38-40.

Roberts, D.A., M.O. Smith, and J.B. Adams (1993) 'Green vegetation, non-photosynthetic vegetation and soils in AVIRIS data', *Remote Sensing of. Enivronment,* 44, 255-269.

Sabol, D.E., J.B. Adams, and M.O. Smith (1992a) 'Quantitative subpixel spectral detection of targets in multispectral images', *J. Geophys. Res.*, 97, 2659-2672.

Sabol, D.E., D.A. Roberts, M.O. Smith, and J.B. Adams (1992b) 'Temporal variation in spectral detection thresholds of substrate and vegetation in AVIRIS images', *in* R.O. Green (ed.), *'Summaries of the Third Annual JPL Airborne Geoscience Workshop '*, vol. 1, AVIRIS Workshop, June 1-2, 1992. pp. 132-134.

Smith, M.O., S.L. Ustin, J.B. Adams, and A.R. Gilliespie (1990a) 'Vegetation in deserts I: A regional measure of abundance from multispectral images', *Remote Sensing of. Enivronment,*31, 1-26.

Smith, M.O., J.B. Adams, and A.R. Gillespie (1990b) 'Reference endmembers for spectral mixture analysis', *Proc. 5th Australian Remote Sensing Conference, Perth, Western Australia,8-12 October,* vol. 1, 331-340.

MODELING CANOPY SPECTRAL PROPERTIES TO RETRIEVE BIOPHYSICAL AND BIOCHEMICAL CHARACTERISTICS.

FRÉDERIC BARET and STÉPHANE JACQUEMOUD
INRA Bioclimatologie
BP 91
84143 Montfavet Cedex
France

ABSTRACT. Retrieval of canopy biophysical and biochemical characteristics from high spectral resolution data is investigated using model simulation. In the first part, we describe leaf, soil and canopy reflectance models that will be coupled to give the SPECAN model. It allows to compute canopy reflectance spectra as a function of canopy biophysical and biochemical characteristics. Then two approaches for canopy characteristics retrieval are considered.

The first one is based on wavelength shifts observed in the red edge of canopy reflectance. This spectral index characterized by the wavelength position of the inflexion point of the red edge minimizes the effects of soil optical properties, specular component and of the atmosphere. It is sensitive to leaf area index, chlorophyll concentration and leaf inclination. However, this approach is rather empirical and the link to individual canopy characteristics is not explicit.

The second approach is based on model inversion. Preliminary results indicate that it can provide good estimates of both leaf chlorophyll concentration and water equivalent thickness. However, the inversion process has to be stabilized to get reasonable values of canopy structure parameters. Further, in the inversion process, soil background optical properties are supposed known.

Finally, the model is used to investigate the sensitivity of canopy reflectance to leaf optical properties. These results initiate a discussion about the capability of high spectral resolution to remotely sense leaf biochemical composition.

1. Introduction

Present development of spectro-imaging systems requires a concurrent research effort focusing on the interpretation of this new information. Canopy biophysical or biochemical characteristics that are retrieved from remote sensing data may then be incorporated into ecosystem models to monitor and predict human or climatic impacts on the Earth's regulating capability. Authors proposed three main approaches to translate the spectral information remotely sensed into canopy characteristics:

- *Spectral mixture analysis*. In this approach, the reflectance spectrum of a given target is considered as composed of a small number of elementary objects called end-members. End-members are chosen in spectral libraries, or can be retrieved from the image itself, either by statistical methods such as principal component analysis (Huete, 1986) or from well identified pixels (Adams et al., 1991; Ustin et al., 1992). Spectral mixture analysis amounts

J. Hill and J. Mégier (eds.), Imaging Spectrometry - a Tool for Environmental Observations, 145–167.

to determine the linear combination of end-members that describes the best the observed spectrum. However, this approach obviously assumes that radiative transfer processes involved are additive. This assumption is not valid in most cases, particularly for dense canopies. A more detailed description of this approach is provided in this issue by Smith et al. (1994).

- *Spectral derivatives and shifts.* This approach focuses mainly on the red edge, the spectral domain between red and infrared where chlorophyll absorption decreases strongly with wavelength (Horler et al., 1983). One objective of this approach is to enhance spectral features that reveal canopy characteristics while minimizing confounding factors such as soil optical properties or atmospheric effects (Demetriades-Shah and Steven, 1990, Hall et al., 1990, Baret et al., 1992). This approach is rather empirical although some theoretical considerations may be used to understand the properties of derivatives or wavelength shifts. In this paper, we will present results obtained on the red edge shift.

- *Model inversion.* This approach assumes that an analytical model describes spectral variations of canopy reflectance as a function of canopy, leaf and soil background characteristics. These characteristics are both structural and optical. Structural characteristics are leaf area index, leaf angle distribution, size and clumpiness of leaves, leaf surface aspect, leaf mesophyll structure, concentration and spatial distribution of leaf biochemical constituents. Optical characteristics correspond to the complex refraction index of each biochemical constituent, the real part describing the scattering (refraction index) and the imaginary part the absorption (specific absorption coefficient). Generally, these optical characteristics are considered almost constant for a given biochemical species. Once the model is developed, it may be inverted to retrieve canopy and leaf structural properties from observed reflectance spectra.

In this chapter, we will first develop a canopy reflectance model that consists in coupling a leaf reflectance and transmittance model and a soil reflectance model to a canopy reflectance model. Then, we will use this model to investigate the two last former approaches used to retrieve canopy characteristics. We will finally address one main issue: the capability of high spectral resolution data to assess canopy biochemical composition.

2. Modeling canopy spectral reflectance

Classical canopy reflectance models allow computation of canopy bidirectional reflectance in a given wavelength using leaf reflectance and transmittance and soil reflectance as inputs. To complete a reflectance spectrum over the optical domain (400-2500 nm), a canopy reflectance model is run for each wavelength. Leaf and soil optical properties are the only input parameters that vary from one wavelength to another. Therefore, for inversion purposes, the number of unknown parameters to be estimated increases proportionally with the number of wavelengths considered. For a reflectance value observed in an additional wavelength, three additional unknowns have to be considered. They are leaf reflectance and transmittance and soil reflectance. To break down this process, a model that describes spectral variation of leaf reflectance and transmittance as a function of leaf structural and biochemical characteristics is required. A similar model is also required to describe spectral variation of soil reflectance with input parameters such as soil type, surface moisture and aspect. We will first develop a leaf model.

2.1. Modeling Leaf Optical Properties

Various approaches were used to describe leaf reflectance and transmittance spectral variation. Allen et al. (1973) and Kumar and Silva (1973) proposed a ray tracing technique to compute the 2D transfer of photons within leaf mesophyll. This technique was very powerful, but required high degree of details for the description of leaf mesophyll structure. It resulted also in large computation time. Tucker and Garrat (1977) decomposed leaf into well identified layers. They proposed a Markov chain model to describe photons transfer from one layer to an other. This approach required also a very good description of leaf mesophyll structure. Allen et al. (1968) proposed an alternative approach to model radiative transfer within a leaf. They use the Kubelka and Munk (1931) theory to compute leaf reflectance and transmittance. They were the firsts to derive scattering and absorption coefficients of fresh leaves from spectrophotometric measurements. Conversely to Allen et al. (1968), Allen et al. (1970) considered a leaf as a pile of N identical layers separated by thin slices of air (figure 1). Each layer was characterized by its diffuse reflectance and transmittance. Radiative transfer within one elementary layer was described by the plate model (Allen et al., 1969): Scattering by interfaces between elementary layer and air was computed using Fresnel equations. Absorption by biochemical constituents such as water, chlorophyll, as well as lignin, nitrogen, starch, sugar and other minor constituents may be described by the classical Beer's law. Finally, the model requires 3 parameters to compute leaf reflectance and transmittance for a given wavelength λ: (i) the number N of layers (N is not necessarily an integer), (ii) the refraction index $n(\lambda)$, and (iii) the absorption coefficient $K(\lambda)$. Starting from this theoretical background, Jacquemoud and Baret (1990) developed the PROSPECT model in which the absorption coefficient is given by

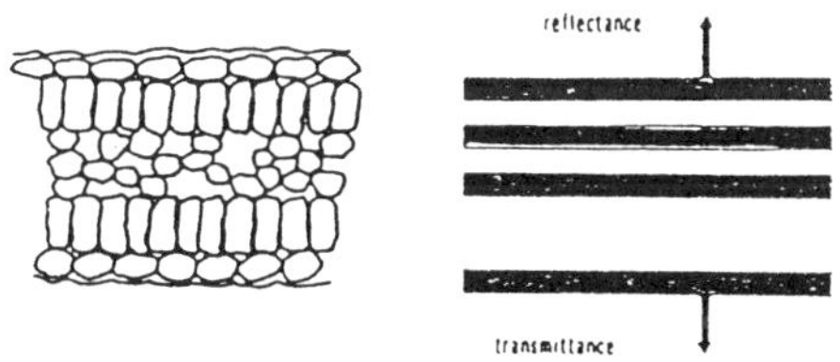

FIGURE 1. Actual and Allen's (1969, 1970) schematic representation of leaf mesophyll structure

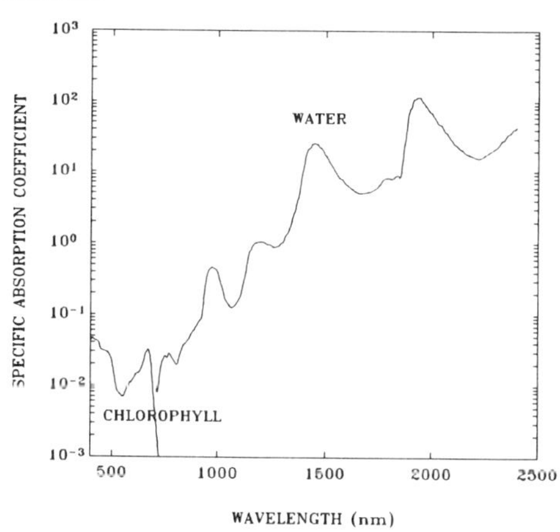

FIGURE 2. Specific absorption coefficients (derived by Jacquemoud and Baret (1992) for chlorophyll ($\mu g^{-1}cm^2$) and water (cm^{-1}).

$$K(\lambda) = \sum_i k_i(\lambda) C_i \qquad (1)$$

where $k_i(\lambda)$ is the specific absorption coefficient and C_i is the concentration per unit leaf area of individual biochemical constituent i. Applying this model to a large range of leaf types and status, they derived in vivo specific absorption coefficient spectra of chlorophyll a and b and water as a function of wavelength (figure 2). Further, they also derived leaf material refraction index spectra that were almost flat with an average value close to 1.4.

The PROSPECT model requires at least three input parameters to compute reflectance and transmittance spectra of leaves from 450 to 2500 nm:

- N, called the structure index. The compact mesophyll of young leaves or monocotyledons corresponds to N values close to 1.0. The spongy mesophyll of dead leaves or dicotyledons corresponds to N values in the range 1.5-2.5.
- C_{ab}, the chlorophyll concentration expressed per unit leaf area.
- C_w, the equivalent water thickness.

This simple model does not take into account explicitly other biochemical constituents such as lignin, cellulose, starch, nitrogen, etc. It does not either take into account leaf phase function, nor the spatial distribution of absorbing materials and the mesophyll fine structure. However, model simulations show a very good agreement with observed reflectance spectra measured over a wide range of leaves.

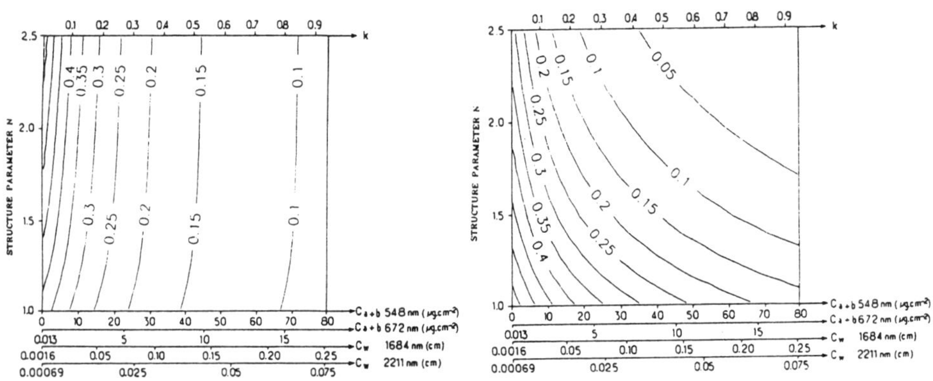

FIGURE 3. Isocontour plots of leaf reflectance (left) and transmittance (right) as a function of absorption coefficient K (top x axis) and leaf mesophyll structure index (N). Absorption coefficient equivalency to Chlorophyll concentration at 548 nm and 672 nm and water equivalent thickness at 1684 nm and 2211 nm is also given at the bottom x axis.

A sensitivity analysis was performed both on leaf reflectance and transmittance as a function of mesophyll structure index N and absorption coefficient K (figure 3). The absorption coefficient may be also expressed for a given wavelength as a function of chlorophyll concentration or water equivalent thickness according to equation (1). It shows that leaf reflectance is not very sensitive to N mesophyll structure index as compared to transmittance. The sensitivity of both reflectance and transmittance to biochemical concentration depends obviously on the absorption coefficient value. Therefore, it depends on wavelength and on the biochemical species considered. Sensitivity to variation of chlorophyll concentration is maximum at 672 nm for low values of chlorophyll concentration. Conversely, sensitivity is higher at 548 nm for higher values of chlorophyll concentration. The same results apply to water absorption domain: for low values of leaf equivalent water thickness, reflectance and transmittance sensitivity to C_w is higher around 1450 nm and 1950 nm where water absorbs strongly. For medium to high values of absorption coefficient K, leaf transmittance appears more sensitive than leaf reflectance.

This sensitivity analysis highlights one of the main specificity of high spectral resolution, the possibility to get the full range of sensitivity of leaf optical properties to changes of leaf biochemical composition and mesophyll structure through wavelength variation. This property was used to retrieve the three parameters N, C_{ab}, and C_w by inversion of the PROSPECT model (Jacquemoud and Baret, 1990). This could be used also at canopy level as we will see later.

2.2. MODELING SOIL SPECTRAL REFLECTANCE

A version of Hapke's model (1981) was successfully tested to describe spectral and bidirectional variations of soil reflectance (Pinty et al., 1989; Jacquemoud et al., 1992). This model assumes that bidirectional reflectance ρ is the sum of a single scattering ρ_s and a multiple scattering ρ_m components:

$$\rho = \rho_s + \rho_m \tag{2}$$

For simplifications, only a directional punctual source is considered. Single scattering is described by the single scattering albedo, ω, a phase function, $P(g,g')$, and a function, $B(g,h)$ representing the backscattering as a function of roughness parameter h:

$$\rho_s = \frac{\omega}{4(\mu_i + \mu_0)}(1 + B(g,h))\,P(g,g') \tag{3}$$

g and g' are the phase and antiphase (angle between specular and viewing directions) angles, μ_i and μ_o are respectively the cosine of the incident and observation zenith angles. The backscattering function is an empirical function that mimics the hot spot effect. It follows that the roughness parameter has no direct clear physical signification. The multiple scattering component is evaluated using the Chandrasekhar (1960) function, $H(\omega,\mu)$, assuming isotropic scattering:

$$\rho_m = \frac{\omega}{4(\mu_i + \mu_0)}\left(H(\omega,\mu_i)\,H(\omega,\mu_0) - 1\right) \tag{4}$$

For each wavelength, model input parameters are incidence and observation angles, roughness parameter h, single scattering albedo ω, and the 4 coefficients of the phase function $P(g,g')$ described by a modified Legendre polynomial:

$$P(g,g') = 1 + r_i \cos g + r_2 \frac{3\cos^2 g - 1}{2} + s_1 \cos g' + s_2 \frac{3\cos^2 g' - 1}{2} \tag{5}$$

A complete description of the model is provided by Jacquemoud et al. (1992). The model was inverted on a data set gathered over a wide range of soil types and status by fitting the 6 parameters (ω, h, r_1, r_2, s_1, s_2) for each soil sample and each wavelength (Jacquemoud et al., 1992). The 0.015 RMSE value (R^2=0.995, n=5460) computed over the whole data set has the same magnitude as the accuracy of the measurements indicating very good descriptive performances of the model. As noticed by Pinty et al. (1989) and Baret et al. (1992), the roughness parameter and the phase function coefficients are very little wavelength dependent. The small spectral variations observed in the real part of refractive indices of soil materials justify this result. When the model is similarly applied to the same data set assuming that the single scattering albedo is the only parameter wavelength dependent, it still describes the data set with the same accuracy (RMSE=0.015, R^2=0.995, n=5460). This important result indicates that the phase function and the roughness parameter may be assumed spectrally independent. Therefore, spectral variation of soil reflectance results only from the single scattering albedo spectral variation that should depend on soil type and surface moisture. Figure 4 shows single scattering albedo spectra for few contrasted soils. Soil moisture and soil type effects are important. Modeling effort should be directed to describe single scattering albedo spectral variation as a function of soil mineralogical composition, pedogenesis, and surface moisture to improve the applicability of this approach. Nevertheless, this simple model will be integrated to the canopy spectral reflectance model developed hereafter

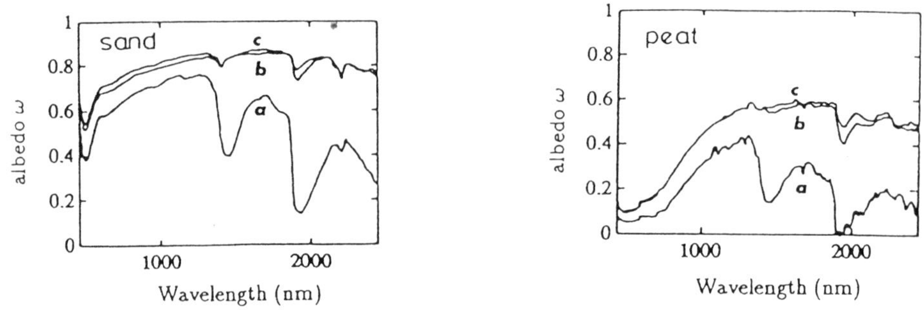

FIGURE 4. Single scattering albedo spectra obtained through model inversion by Jacquemoud et al. (1992) for sand and peat with various moisture levels: (a) very wet, (b) moderately dry, (c) very dry.

2.3. Modeling Canopy Spectral Reflectance: The 'SPECAN' Model

2.3.1. The 'SPECAN' model. Several reflectance models were developed in the last ten years. They may be sorted according to two main criteria: (i) the degree of detail used to describe canopy architecture and bidirectional properties of the elements, and (ii) the way radiative transfer processes are conducted. To get a model that might be simple enough to be manipulated and even inverted, the number of input parameters must be small. The SAIL model (Verhoef 1984, 1985) is a good compromise between simplicity and predictive performances. Leaves are supposed to be lambertian, with no finite dimensions and a random spatial distribution. Leaf inclination distribution function is here approximated by an ellipsoidal distribution (Campbell, 1986), characterized by the average leaf inclination (θ_l). Canopy reflectance is expressed as:

$$\rho_c(\lambda) = \rho_c\left(\rho_l(\lambda), \tau_l(\lambda), \rho_s(\lambda), l, \theta_l, \theta_0, \phi_0, \theta_s, f\right) \tag{6}$$

Where input variables required to compute canopy reflectance spectra are:

$\rho_l(\lambda), \tau_l(\lambda)$	respectively leaf reflectance and transmittance.
$\rho_s(\lambda)$	soil reflectance
l	leaf area index
θ_l	average leaf inclination
θ_0, ϕ_0	view geometry (respectively zenith and azimuth angles)
θ_s	sun zenith angle
f	fraction of diffuse incoming irradiance (assumed isotropic).

Variables that depend on wavelength are thus leaf optical properties and soil reflectance. Their spectral variation is described using previous models of leaf and soil optical properties. Coupling SAIL canopy reflectance model to leaf and soil optical properties models (figure 5) results in the 'SPECAN' model that gives:

$$\rho_c(\lambda) = \rho_c\left(N, C_{ab}, C_w, h, r_1, r_2, s_1, s_2, \omega(\lambda), l, \theta_l, \theta_0, \phi_0, \theta_s, f\right) \tag{7}$$

The single scattering albedo of soil, $\omega(\lambda)$, is the only parameter that remains wavelength dependent. As stated earlier, this parameter should be expressed as a function of soil intrinsic characteristics. Following Price (1990), it could be approximated by a linear combination of 4 elementary end-member spectra. However, in the following, we will assume single scattering albedo of soil as known. This assumption is justified considering that in many cases, soil types and their corresponding single scattering albedo spectra, could be estimated from external information.

The SPECAN model was validated successfully over sugar beet crops (Jacquemoud, 1992). It provides a convenient tool for sensitivity analysis.

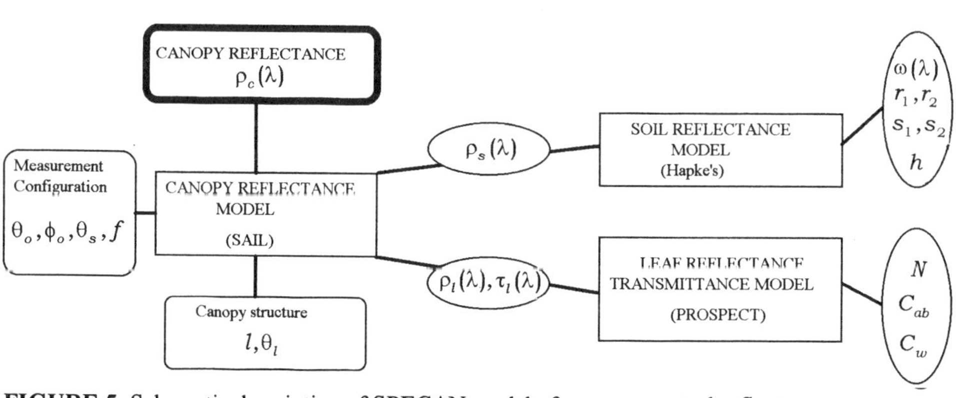

FIGURE 5. Schematic description of SPECAN model of canopy spectral reflectance.

2.3.2. Sensitivity analysis. In the following, we will vary canopy parameters to illustrate canopy spectral reflectance sensitivity to various canopy characteristics. Simulations will be carried out with the SPECAN model for a standard configuration: nadir viewing and 45° sun zenith angle. Diffuse incoming radiation is not taken into account here, and incoming radiation is considered as only directional (f = 0). For simplification, soil reflectance was set to 0.20 independently of wavelength. Computations were performed from 400 nm to 2450 nm with 2 nm steps.

- *Leaf area index* is obviously one of the main variables affecting canopy reflectance over the whole spectrum (Figure 6a). Generally, canopy reflectance increases when leaf reflectance is higher than soil reflectance. This is particularly true in the near infrared domain. Conversely, in spectral domains characterized by strong absorption features, canopy reflectance decreases when leaf area index increases.
- *Leaf angle inclination* (Figure 6b) acts very similarly to leaf area index. Confusion between these two canopy structure variables is expected when retrieving canopy characteristics from canopy reflectance spectral variations.
- *leaf mesophyll structure index*. Changes in leaf mesophyll structure index (N) induce small variations, mostly concentrated where leaves absorb very little such as in near infrared (Figure 6c). An increase in N produces a leaf reflectance increase and a decrease in leaf transmittance as seen previously (Figure 3). Since canopy reflectance was evaluated mostly from leaf reflectance and transmittance combined into leaf single scattering albedo ($\rho_l + \tau_l$), canopy reflectance is almost independent from leaf structure index as observed in figure 6c.
- *Chlorophyll concentration and water equivalent thickness* changes result in large variations of canopy reflectance. As noticed earlier, water and chlorophyll absorption spectral domains are distinct. Around 800 nm, chlorophyll and water specific absorption coefficients are so little (figure 2) that canopy reflectance is insensitive to variations in chlorophyll concentration or equivalent water thickness (Figure 6d).

Along this sensitivity analysis, a large range of canopy reflectance sensitivity to canopy structure and leaf biochemical composition (chlorophyll and water here) is observed throughout wavelengths. This was earlier noticed for leaf optical properties. For example, for low leaf area indices, spectral domains with high contrasts between leaf and soil optical properties are the most sensitive to changes in vegetation amount. Conversely, for dense canopies, spectral domains where leaves transmit and reflect light efficiently are the most sensitive to canopy changes. Similar results are observed for leaf inclination or chlorophyll and water concentrations. This differential sensitivity with wavelength is one of the most interesting properties provided by high spectral resolution systems. In the following, we will use indirectly this property to retrieve canopy biophysical or biochemical characteristics.

3. Strategy for canopy biophysical/biochemical characteristics retrieval

3.1. SPECTRAL INDICES

One way to retrieve canopy characteristics is to develop empirical or semi-empirical relationships. However, as expressed by equation (7), even for the simple theoretical canopy described through the SPECAN model, at least 10 to 15 parameters or variables act. Fortunately, some are perfectly known such as those characterizing the measurement configuration ($\theta_0, \phi_0, \theta_s$). Some others are not known and not very pertinent for vegetation studies. This is the case of soil reflectance that influences strongly canopy reflectance (Baret and Guyot, 1991). Atmospheric effects for space-borne observations also belong to this type of variables termed confounding variables (noted Θ in

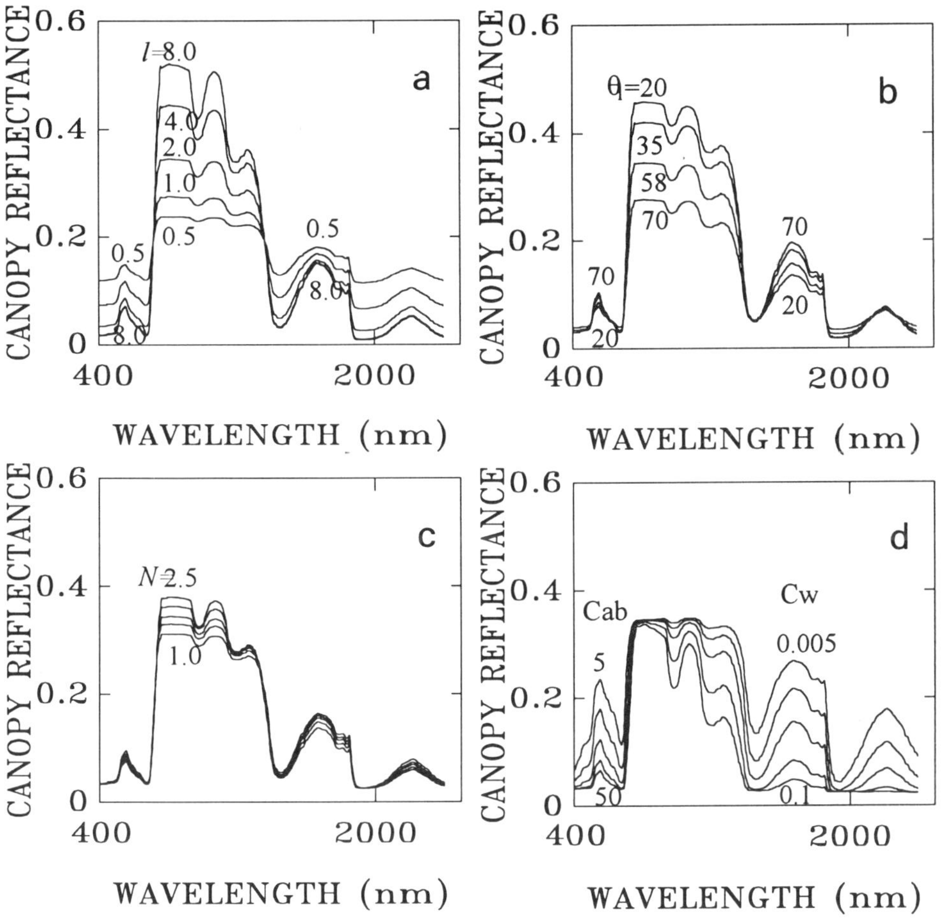

FIGURE 6. Sensitivity of canopy spectral reflectance to canopy characteristics. Variations around a 'standard' canopy are considered. This standard canopy is characterized by: l = 2.0, θ_l = 58°, C_{ab} = 35 µg·cm^{-2}, C_w = 0.02 cm, θ_o = 0°, ϕ_o = 0°, θ_s = 45°. The soil is assumed lambertian with a constant reflectance value of 0.2. The first graph (a) presents variations with leaf area index (l = 0.5 1.0 2.0 4.0 8.0). The second one (c) presents variations with leaf inclination (θ_l = 20, 35, 58, 70). The third one (c) presents variations with leaf mesophyll structure index (N = 1.0, 1.25, 1.5, 2.0, 2.5). The last one (d) presents simultaneous variations of leaf chlorophyll concentration that affects the visible part of the spectrum (C_{ab}= 5, 10, 20, 35, 50 µg·cm^{-2}) and leaf water equivalent thickness that affects the near and middle infrared parts of the spectrum (C_w= 0.005, 0.01, 0.02, 0.05, 0.10 cm).

the following). On the other hand, leaf area index, leaf inclination, chlorophyll and water concentrations are relevant biophysical/biochemical variables for canopy functioning models. Classical spectroscopic methods were indeed used as a first approach to retain particular features from complex spectra. Many of these techniques are based on spectral derivatives (Demetriades Shah and Steven, 1990). For leaves or canopies, most authors focused on the red edge that represents the sharpest variation with wavelength of leaf and canopy optical properties. Because the red edge is outside water absorption domain, only chlorophyll concentration will be explicitly investigated. In this restricted spectral domain, spectral shifts may be characterized by the wavelength position λ_i of the inflexion point of canopy reflectance spectrum. Therefore, by definition, the second derivative of canopy reflectance, $g = d^2\rho(\lambda)/d^2(\lambda)$, zeroed for $\lambda = \lambda_i$. The g function depends both on wavelength λ and on confounding variables Θ that may or may not depend on λ. We thus define the implicit function $\lambda_i(\Theta)$ such as $g(\lambda_i, \Theta) = 0$. Basic properties of implicit functions lead to:

$$\frac{d\lambda_i}{d\Theta} = -\frac{dg}{d\Theta}\bigg/\frac{dg}{d\lambda} \tag{8}$$

This equation demonstrates the equivalence between λ_i and the second derivative g ($d\lambda_i/d\Theta = 0 \Leftrightarrow dg/d\Theta = 0$) for the independence to the confounding factors Θ. Consider now that some confounding variables depend on wavelength: $\Theta = \Theta(\lambda)$. This corresponds for example to soil reflectance or atmospheric effects. Canopy spectral reflectance can be written as:

$\rho_c = \rho_c(\Omega(\lambda), \Theta(\lambda))$, where $\Omega(\lambda)$ corresponds to input variables distinct from the confounding ones. Input variables $\Omega(\lambda)$ include canopy relevant variables to be retrieved such as leaf area index or chlorophyll concentration. Using these definitions, evaluation of the second derivative with respect to the wavelength gives:

$$g = \frac{d^2\rho_c}{d\Theta^2}\left(\frac{d\Theta}{d\lambda}\right)^2 + \frac{d\rho_c}{d\lambda}\frac{d^2\Theta}{d\lambda^2} + \frac{d^2\rho_c}{d\Omega^2}\left(\frac{d\Omega}{d\lambda}\right)^2 + \frac{d\rho_c}{d\lambda}\frac{d^2\Omega}{d\lambda^2} \tag{9}$$

To get the sensitivity of g with respect to Θ, we compute $dg/d\Theta$ considering that by definition, Ω is independent from Θ ($d\Omega/d\Theta = 0$):

$$\frac{dg}{d\Theta} = \frac{d^3\rho_c}{d\Theta^3}\left(\frac{d\Theta}{d\lambda}\right)^2 + 3\frac{d^2\rho_c}{d\Theta^2}\frac{d^2\Theta}{d\lambda^2} + \frac{d\rho_c}{d\Theta}\frac{d\lambda}{d\Theta}\frac{d^3\Theta}{d\lambda^3} \tag{10}$$

Equation (10) simplifies and gives the condition under which g and λ_i do not depend on Θ:

$$\frac{d^3\rho_c}{d\Theta^3}\left(\frac{d\Theta}{d\lambda}\right)^2 + 3\frac{d^2\rho_c}{d\Theta^2}\frac{d^2\Theta}{d\lambda^2} + \frac{d\rho_c}{d\Theta}\frac{d\lambda}{d\Theta}\frac{d^3\Theta}{d\lambda^3} = 0 \tag{11}$$

This condition is verified when: $\frac{d^2 \rho_c}{d\Theta^2} = 0$ and $\frac{d^2 \Theta}{d\lambda^2} = 0$ (12)

Conditions (12) are obtained if canopy reflectance is a linear function of the input variable Θ, and when Θ varies linearly with wavelength within the spectral domain considered. We will now list some of the confounding variables that do not affect spectral shifts of the red edge.

- *Soil reflectance.* Baret (1988) shows that canopy reflectance could be approximated as a linear function of soil reflectance. Further, soil reflectance spectra are quasi linear in the red edge. These two basic properties verify condition (12) and demonstrate that λ_i is independent from soil optical properties. This result was confirmed using SPECAN model simulations (Baret et al., 1992)
- *Specular component.* The λ_i spectral index should also remove the effect of specular light reflected by canopies. As noticed by Rondeaux and Vanderbilt (1992), canopy reflectance can be considered as the sum of specular and non specular components. The specular component may represent a large fraction of light reflected by canopies that does not penetrate leaves and thus does not carry information about leaf biochemical composition. Since light specularly reflected by leaves, thus canopies, is almost spectrally flat, spectral shift λ_i should be independent from canopy specular features.
- *Atmospheric effects.* The same principles applied to atmospheric effects: apparent satellite level reflectance, ρ^*_c can be linearly related to actual ground level canopy reflectance ρ_c if the surrounding effects are neglected (Conel et al., 1988):

$$\rho^*_c = \alpha(\lambda)\,\rho_c + \beta(\lambda) \tag{13}$$

In the red edge domain and outside gaseous absorption bands, parameters $\alpha(\lambda)$ and $\beta(\lambda)$ vary linearly with wavelength (Baret et al., 1992). As these properties satisfy condition (12), thus λ_i is independent from atmospheric conditions. These results were verified by Baret et al. (1992) using the SPECAN model and the 5S model (Tanré et al., 1990) to simulate atmospheric effects.

We showed that spectral shifts observed in the red edge are independent from soil optical properties, canopy specular component and atmospheric effects. We will now investigate λ_i sensitivity to canopy biophysical and biochemical characteristics. The λ_i spectral index was simulated using analytical derivatives of the SPECAN model (Baret et al., 1992). Since leaf area index is one of the main variables governing canopy processes, we will therefore always analyze λ_i sensitivity to each variable in interaction with leaf area index variations. Figure 7a shows that spectral shifts are primarily governed by leaf area index and chlorophyll concentration. An increase of chlorophyll concentration or leaf area index shifts λ_i towards longer wavelengths. This is in good agreement with experimental results observed by Rock et al. (1988) or Demetriades-Shah and Steven (1990). Between extreme situations, a 35 nm shift of the inflexion point is observed. However, for usual chlorophyll concentrations (35 μg·cm^{-2}) the range of λ_i variation due to leaf area index variations reduces to about 15 nm. Leaf angle inclination effects are also significant (figure 7b). This was experimentally confirmed by Vanderbilt et al. (1988) who observed red edge shifts associated to canopy structure changes due to the wind. This effect is more pronounced for vertical structures. Changes in mesophyll structure index (figure 7c) induce moderate shifts. a 0.25 increase of N value results in about 1 nm shift towards longer wavelengths.

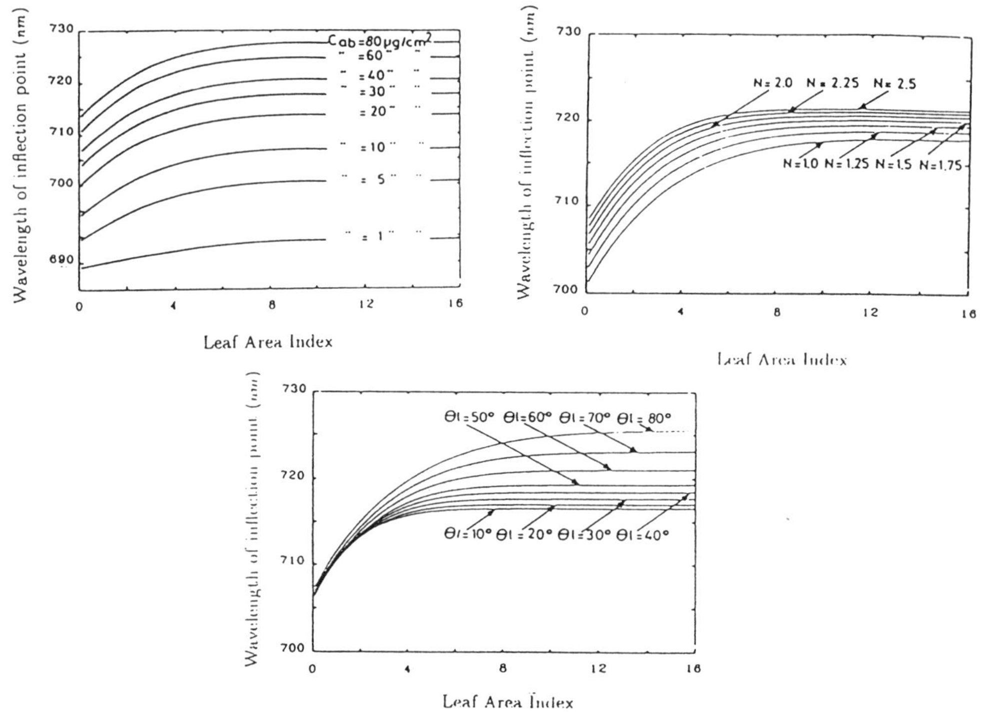

FIGURE 7. Sensitivity of spectral shift of the red edge (λ_i) to leaf area index.(l), chlorophyll concentration (C_{ab}) (7a), mesophyll structure index (N) (7b) and leaf inclination (Θ_l) (7c). Results obtained from SPECAN model simulations.

This brief simulation study shows that spectral shifts of the red edge present interesting performances: they efficiently minimize the influence of confounding factors such as soil reflectance, specular component or atmospheric effects. They are sensitive to canopy characteristics such as leaf area index, chlorophyll concentration, and leaf inclination. However, this spectral index, like most indices, did not provide explicit information about canopy biophysical/biochemical characteristics. It should be related to more global variables such as canopy photosynthetic capacity. The dynamic range is not very important and will require a good radiometric and wavelength resolution to be properly interpreted. In the next section, we will investigate an alternative approach that will potentially provide detailed information on canopy biophysical/biochemical characteristics.

3.2. Model Inversion

3.2.1. Theoretical considerations. The SPECAN model simulates canopy reflectance spectra in the direct mode when canopy characteristics are given. Inverting the SPECAN model consists in retrieving the best combination of input variables that produces a canopy reflectance spectrum as close as possible to the measured one. As seen earlier, input variables are grouped into 3 categories:

- *Canopy biophysical/biochemical characteristics* such as leaf chlorophyll concentration, leaf equivalent water thickness, mesophyll structure index, leaf area index and leaf inclination angle. These are the variables to be retrieved from measured canopy reflectance spectra.
- *Soil reflectance* is not a variable that can be easily incorporated into canopy functioning models. For this reason, we will not try to infer soil optical property from reflectance spectra through SPECAN inversion process. Further, retrieval of soil characteristics such as input parameters of the soil reflectance model described earlier, requires the single scattering albedo spectra to be parameterized. In this study, we will simply consider soil background optical properties as known. We will discuss later consequences of this assumption.
- *Measurement configuration variables* such as view and source geometry, and the fraction of diffuse incoming radiation. These parameters are generally known.

Jacquemoud (1992) showed that the SPECAN model was numerically invertible: from a simulated reflectance spectrum, canopy characteristics may be retrieved through SPECAN model inversion. However, simulations showed that various sets of canopy characteristics could result in canopy reflectance spectra that are very close together. This could lead to serious problems when inverting actual spectra contaminated by instrumental noise or residual atmospheric effects. Further, actual canopies do not necessarily verify all the approximations introduced through leaf and canopy models used. Complementary analysis carried out by Jacquemoud (1992) showed that when a 5% level relative noise was added to simulated canopy reflectance spectra, canopy characteristics retrieved from model inversion were still very close to the original ones. These theoretical results are thus to be evaluated on actual canopies.

3.2.2. Application on actual canopies. High spectral resolution reflectance measurements were performed over sugar beet canopies at Broom's Barn experimental station (U.K.) in July 1989 (Malthus et al., 1989).The IRIS spectroradiometer used recorded spectra from 450 nm to 2400 nm in 975 narrow spectral bands. From this data set, reflectance values in the 224 spectral bands of AVIRIS were computed to simulate the capabilities of space-borne systems. Soil reflectance was supposed known. The SPECAN model was inverted from reflectance spectra acquired over plots having various leaf area indices, with all canopy characteristics to be retrieved at the same time (N, C_{ab}, C_w, l, Θ_l). Spectra simulated with the retrieved values were very close to the original spectra with a RMSE value of 0.05. Retrieved values of leaf biochemical composition are also close to the measured values. However, retrieved values of canopy and leaf structure characteristics such as leaf area index, leaf inclination and leaf mesophyll structure index are often very far from the measured values. That suggests unstable inversion processes.

Previous sensitivity analysis indicated that leaf structure parameter N affects very little canopy reflectance. Further, both leaf area index and leaf inclination angle influence in similar ways canopy reflectance. We thus decided to set N and Θ_l to their average values for sugar beet crops. Inverting SPECAN model on the same data set using these constraints resulted in a slight increase of the RMSE characterizing the distance between measured and simulated spectra (RMSE = 0.07). However, leaf area index was now well estimated (Figure 8c). Chlorophyll concentration retrieved values were also very close to the measured values (Figure 8a). The average leaf equivalent water thickness was also quite well estimated (Figure 8b). However, the small range of variation of the measured values prevents more detailed evaluation on this variable.

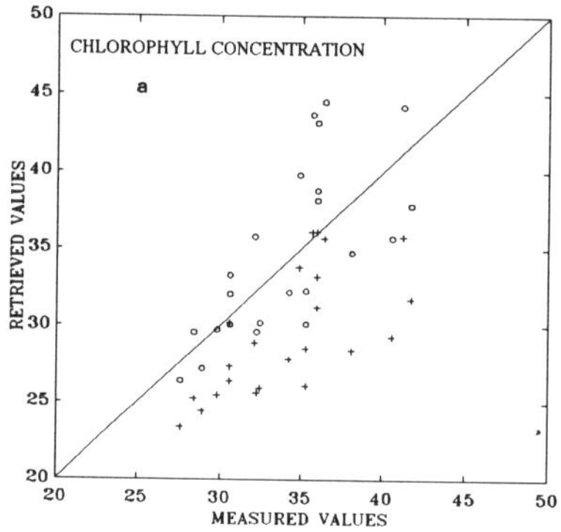

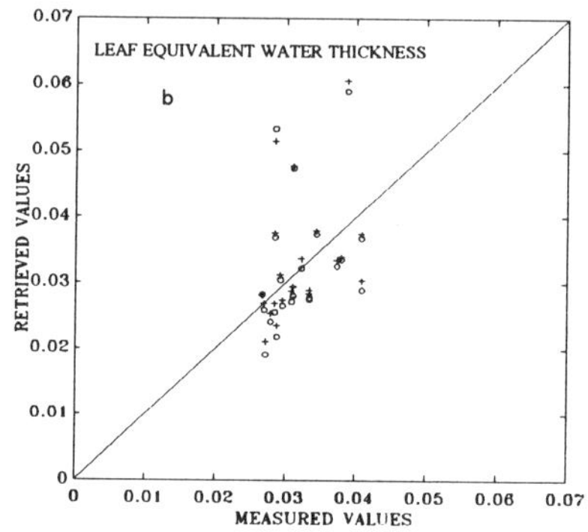

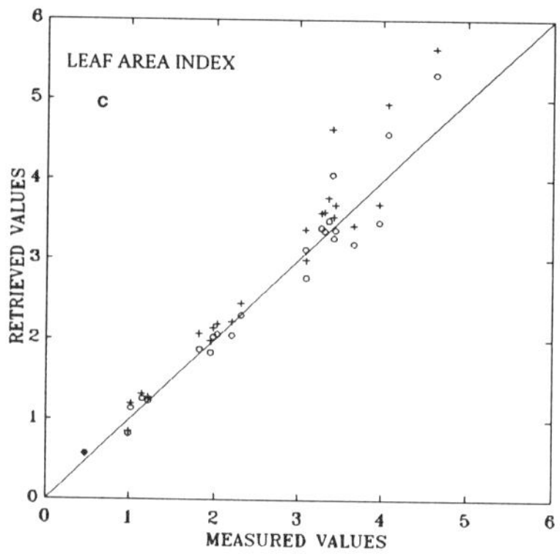

FIGURE 8. Comparison between measured canopy characteristics and values retrieved from SPECAN model inversion. Soil optical properties, leaf structure mesophyll and leaf inclination are assumed to be known. Circles and plus correspond to the inversion process applied respectively to AVIRIS and Landsat TM bands. Chlorophyll concentration is expressed in $\mu g.cm^{-2}$, and leaf equivalent water thickness in cm.

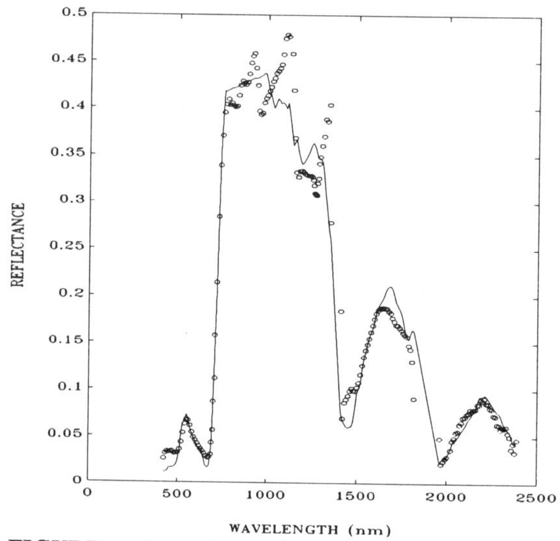

FIGURE 9. Comparison between AVIRIS atmospherically corrected measured Alder forest canopy spectrum (circles) and spectrum simulated with SPECAN model using input parameters retrieved through the inversion process (solid line).

An example of SPECAN model inversion is provided by Johnson et al. (1992). In this study, forest canopy reflectance spectra were recorded with AVIRIS during OTTER project (Johnson and Peterson, 1991). AVIRIS data were converted into ground level reflectance using LOWTRAN-7 atmospheric correction code (Kneizys et al., 1989). The inversion process was conducted considering soil reflectance and leaf mesophyll structure index as known.

Canopy reflectance spectrum simulated with the retrieved values of canopy and leaf characteristics is in good agreement with the observed spectrum (Figure 9). However, disagreement appears between 1100 and 1400 nm, presumably due to problems in water vapor atmospheric corrections. Retrieved value (C_w = 0.010 cm) of leaf equivalent water thickness is very close to the measured one (C_w = 0.011 cm). However, chlorophyll concentration is underestimated (C_{ab} = 20.4 μg·cm^{-2}, as compared to the measured value of 32.0 μg·cm^{-2}). Conversely leaf area index is overestimated (l = 7.52 as compared to l = 4.6). This discrepancy could be due to the understory vegetation that had a significant leaf area index but was not considered in this experiment. It could be also attributed to stability problems when estimating concurrently both leaf area index and leaf inclination. Retrieved values of leaf inclination was 65.4°, although no discussion of this value was possible because of the lack of detailed measurements.

3.2.3. Comparison between high spectral resolution and broad band inversion performances. It is interesting to compare the performances of the inversion process when applied to high spectral resolution or to broad band data. For this purpose, the 6 Landsat TM broad bands were simulated from the sugar beet IRIS spectra recorded at Broom's Barn in 1989. SPECAN model was inverted over the same set of data and with the same constraints as the ones used previously with AVIRIS data. Retrieved values were globally in good agreement with values derived from high spectral resolution AVIRIS data (Figure 8). However, chlorophyll concentration was underestimated.

These results indicate that the inversion approach could provide an efficient way to estimate canopy chlorophyll or water equivalent thickness. However, the inversion process presented here is incomplete because we assumed that characteristics such as soil optical properties were known. Jacquemoud (1992) showed that this was not very important for dense canopies. Conversely, for sparse canopies, the problem is important and requires further research effort oriented towards the improvement of both soil spectral reflectance model and the inversion techniques used.

On the other hand, inversion using only the spectral variation of reflectance resulted in instability on canopy structure parameter such as leaf area index and leaf inclination. Research effort should be oriented towards an other more global parametrization of canopy reflectance models and on the concurrent use of directional and polarization variability of canopy reflectance.

Comparison between high spectral resolution and broad band data suggests that spectro-imaging systems should improve slightly the capabilities of retrieving canopy characteristics such as leaf area index or concentration of the main absorbers (chlorophyll and water) through model inversion process. This is due to the differential sensitivity of reflectance to canopy characteristics observed along wavelength. However, we did not address leaf biochemical constituents inference capability that might be one specificity of high spectral resolution data.

3.3. APPLICATION TO LEAF BIOCHEMICAL COMPOSITION ASSESSMENT.

Biochemicals playing a key role in ecological processes have specific absorption features in the 1000-2500 nm spectral domain. Absorption mechanisms in this spectral region result from the fundamental stretching vibrations of organic bonds between light atoms (C-H, O-H, N-H, C-O, C-C).(Curran, 1989). This property is widely used by people studying forage quality. They developed near infrared spectroscopic techniques (NIRS) that provide a convenient way to estimate biochemical contents, avoiding the tedious use of classical wet chemistry analyses (Marten et al., 1989). However, these NIRS techniques are applied on dried optically thick ground materials. It is obviously more complex when applying these techniques directly to canopies that have often a significant background contribution, and always a structure that modifies largely the biochemical absorption signal. Except for chlorophyll and water, the biochemical signature is much more tiny as compared to the well known strong effect of canopy structure. It is thus difficult but important to isolate the biochemical signal from confounding factors such as canopy structure, view and source configuration, and soil background. We will investigate one possible way to evaluate canopy reflectance sensitivity to leaf biochemical composition.

3.3.1. Theoretical study. As seen earlier, canopy reflectance (ρ_C) is a function of several variables including the wavelength, λ, and leaf biochemical composition C

$$\rho_c = \rho_c\left(S, \theta, \rho_s(\lambda), L(\lambda, C), B(\lambda)\right) \tag{14}$$

where S represents variables governing canopy structure such as leaf area index or leaf inclination, θ represents view and source configuration, ρ_S is the soil reflectance, L represents leaf optical properties such as reflectance and transmittance, and B the optical properties of other vegetation elements such as bark. Some of these variables depend on wavelength (ρ_S, L, B) and some others are obviously not wavelength dependent (S, θ). Canopy reflectance sensitivity to leaf biochemical content C for a given chemical constituent such as lignin, cellulose or nitrogen is expressed by the derivative:

$$\frac{d\rho_c}{dC} = \frac{d\rho_c}{dL}\frac{dL}{dC} \tag{15}$$

This means that canopy reflectance sensitivity to leaf biochemical composition is partitioned into (i) canopy reflectance sensitivity to leaf optical properties and (ii) leaf optical properties sensitivity to leaf biochemical composition. The second term of equation (15), $dL\ /\ dC$, may be derived directly from a leaf optical property model. It can also be approximated by observed variations of leaf optical properties when biochemical composition changes while leaf structure governing the scattering remains unchanged ($\Delta l\ /\ \Delta C$). The first term, $d\rho_C\ /\ dL$, may be derived directly from canopy reflectance models. However, canopy modeling is a very hard task that requires intensive effort to account for the complexity of vegetation structure. Another approach based on experimental observations may also answer the question.

Because canopy reflectance is a function of wavelength as stated by equation (14), its derivative as a function of wavelength is:

$$\frac{d\rho_c}{d\lambda} = \frac{d\rho_c}{d\rho_s}\frac{d\rho_s}{d\lambda} + \frac{d\rho_c}{dL}\frac{dL}{d\lambda} + \frac{d\rho_c}{dB}\frac{dB}{d\lambda} \tag{16}$$

Canopy reflectance sensitivity to leaf optical properties $d\rho_c\ /\ dL$ is then extracted from equation (16):

$$\frac{d\rho_c}{dL} = \left(\frac{dL}{d\lambda}\right)^{-1}\left(\frac{d\rho_c}{d\lambda} - \frac{d\rho_c}{d\rho_s}\frac{d\rho_s}{d\lambda} - \frac{d\rho_c}{dB}\frac{dB}{d\lambda}\right) \tag{17}$$

Equation (17) shows that canopy reflectance sensitivity to leaf optical properties may be evaluated through spectral variations observed concurrently at canopy and leaf levels. When contribution of bark or soil background is negligible as for very dense canopies, very dark soil or bark ($d\rho_C\ /\ d\rho_S \approx 0$ or $d\rho_C\ /\ dB \approx 0$) or when either bark and soil optical properties are considered spectrally flat in a given spectral domain ($d\rho_s / d\lambda \approx 0$ or $dB / d\lambda \approx 0$), equation (17) reverts to the very simple form:

$$\frac{d\rho_c}{dL} = \frac{d\rho_c}{d\lambda}\left(\frac{dL}{d\lambda}\right)^{-1} \tag{18}$$

replacing $d\rho_c / dL$ in equation (15) by its expression in equation (18) leads to:

$$\frac{d\rho_c}{dC} = \frac{d\rho_c}{d\lambda}\left(\frac{dL}{d\lambda}\right)^{-1} \frac{dL}{dC} \tag{19}$$

This demonstrates that, for a restricted spectral domain where soil background and other vegetation materials have a neglected contribution to canopy response or are spectrally flat, canopy reflectance sensitivity to leaf biochemical concentration may be derived directly from concurrent measurements of spectral variations of leaf and canopy optical properties. However, it is assumed that a leaf model including explicitly biochemical composition is developed. Unfortunately, such a leaf model does not exist yet. In the following, we will restrict the analysis to canopy reflectance sensitivity to leaf optical properties.

3.3.2. Canopy reflectance sensitivity to leaf optical property. We simulated with the SPECAN model variations of canopy reflectance induced by changes in leaf optical properties. Leaf absorptance ($1 - \rho_l - \tau_l$) was chosen to represent variations of leaf optical properties due to changes in biochemical constituents. The range of leaf absorptance was obtained using a range of absorption coefficient that covers usual values of chlorophyll or water concentrations over the whole spectrum (figure 2). For simplification, the refraction index was assigned to a constant average value (n = 1.4). The spectrally flat pattern of n and the low sensitivity of leaf optical properties to small changes of n through PROSPECT model justified this approximation. We will discuss successively effects induced by variations of leaf area index, leaf mesophyll structure index, soil reflectance and leaf inclination.

- *Leaf area index.* Figure 10a shows that canopy reflectance obviously decreases when leaf absorptance increases. For low leaf area indices, canopy reflectance sensitivity to leaf optical properties is very small. Conversely, for dense canopies, canopy reflectance is much more sensitive. Further, the sensitivity is enhanced for small values of leaf absorptance. Variations of canopy reflectance with leaf absorptance are characterized by an exponential pattern. However, below leaf absorptance values of 0.6, it is almost linear.
- *Leaf mesophyll structure index.* An increase of leaf mesophyll structure index increases canopy reflectance sensitivity to leaf absorptance (figure 10b). The effect is more pronounced for high leaf area indices.
- *Soil reflectance.* For low leaf area indices, an increase in soil reflectance enhances canopy reflectance sensitivity to leaf absorptance. Conversely, no significant changes of canopy reflectance sensitivity are observed for high leaf area indices. We noticed that all the curves corresponding to various leaf area indices crossed each other almost at the same location characterized by a canopy reflectance close to the soil reflectance value.

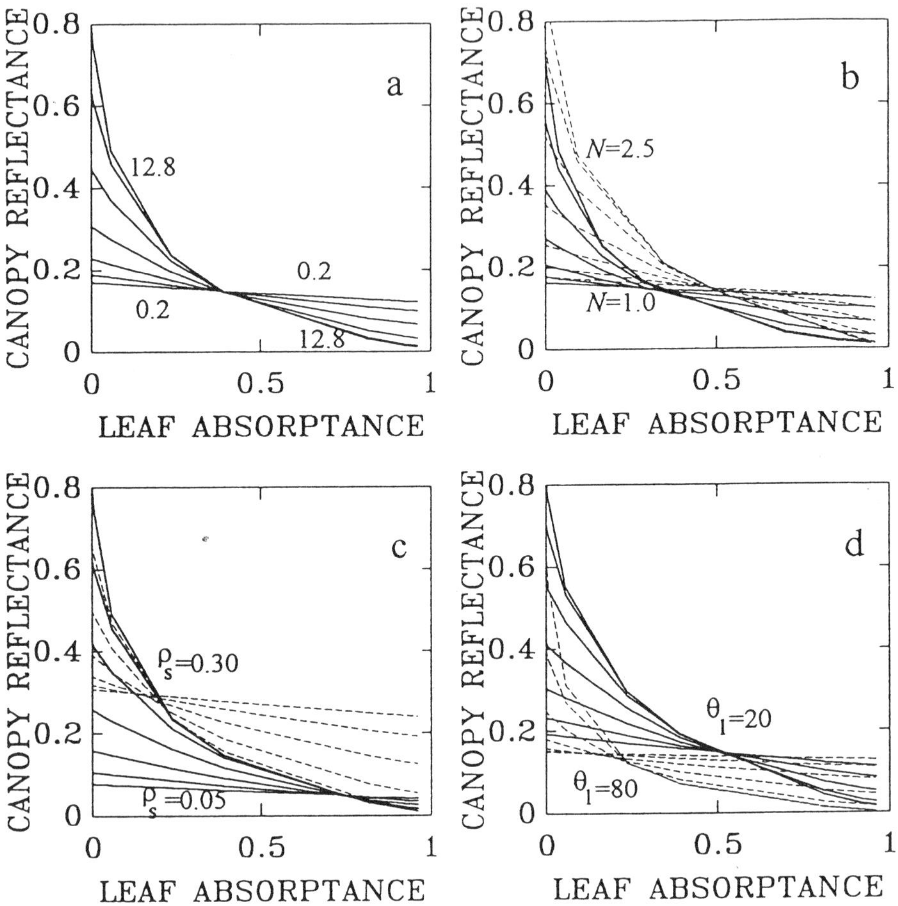

FIGURE 10. Canopy reflectance sensitivity to leaf absorptance. Results from SPECAN model simulation for nadir viewing, 45° sun zenith angle and no incoming diffuse radiation. Soil reflectance is assumed spectrally flat. First graph (a) corresponds to the standard case characterized by a leaf mesophyll structure index N = 1.5, soil reflectance ρ_s = 0.15, a range of leaf area indices (l = 0.2 0.4 0.8 1.6 3.2 6.4 12.8), and a quasi spherical leaf inclination distribution (θ_l = 60°). Other graphs present variations around this standard case: The second graph (b) corresponds to variations of the mesophyll structure index N = 1.0 (solid lines) and N = 2.5 (dashed lines). The third graph (c) presents variations due to soil reflectance ρ_s = 0.05 (solid lines) and ρ_s = 0.30 (dashed lines). The last one (d) presents variations with leaf inclination θ_l = 20° (solid lines), θ_l = 80° (dashed lines).

- *Leaf angle inclination.* Planophile canopies ($\theta_l = 20°$) are more sensitive to leaf absorptance changes than erectophile canopies ($\theta_l = 80°$) (figure 10d).

This brief sensitivity analysis shows that for low vegetation amount, 1% increase of leaf absorptance results in about 0.1% decrease of canopy reflectance. This result indicates that it should be difficult to estimate leaf biochemical composition from canopy reflectance in this case. For canopies with full background cover or dense canopies, and for low values of leaf absorptance, 1% leaf absorptance variation induces about 0.5-2% canopy reflectance variation. This sensitivity is in the range of what sensors could resolve. To get low values of leaf absorptance and increase canopy reflectance sensitivity, observations should focus on spectral domains characterized by low water specific absorption coefficient (figure 2) and moderate to high biochemical specific absorption coefficient values. However, the main problem still to be solved is the lack of information about biochemical specific absorption coefficients that could be used to model leaf reflectance sensitivity to biochemical composition.

4. Conclusions

Along this study, models caricaturing leaf, soil and canopy structures and optical properties appear to be very useful to help understanding the influence of factors governing canopy spectral response. Modeling could be used as a way to check the robustness of empirical methods to interpret high spectral resolution data. This was exemplified for the spectral shift of the red edge. The SPECAN model pointed out the sensitivity of this spectral index to various factors. Models could also be used to improve these indices or to create new indices with better performances. Therefore, the original empirical approach will revert into more general semi-empirical approaches. Usually, model will provide the key variables with which indices have to be related, as well as the general pattern of the relationships. However, these spectral indices should be tested on actual data expressing a large range of variation. In many cases, due to drastic simplifications made throughout modeling, experiments will provide data for the necessary adaptations and adjustments to insure that model predictions could be applied on actual canopies.

Spectral indices focus only on a restricted portion of the optical domain rather than taking advantage of the full coverage of the spectrum. To prevent such potential misuse of spectral information, we developed an alternative approach. It is based on model inversion that could retrieve canopy characteristics from canopy reflectance spectral variation. We showed that model inversion of measured canopy spectra allowed retrieval of leaf concentration of main absorbers such as chlorophyll and water. However, estimation of canopy structure was beyond the possibility of this method because of the instability of the inversion process for these variables. This was mainly due to the similarity of the effects of both leaf area index and leaf inclination on canopy reflectance. This instability problem could be solved by introducing some external information in terms of constraints on parameter values, or by reparameterizing canopy structure. It could be also solved by using concurrent measurements in few well chosen view directions. More research efforts should be directed towards the synergistic use of spectral and directional variations of canopy reflectance. Nevertheless, high spectral resolution data provides a unique source of information on canopies for two main reasons: (i) it allows to take advantage of the full range of sensitivity of the radiometric response to canopy variables, and (ii) it allows to extract specific absorption features that characterize leaf or soil biochemical composition.

Confrontation between modeling and experimental approaches points out gaps of knowledge. Along this study, we highlighted two main weaknesses that limit our understanding and the potential use of canopy reflectance spectral variation. First a more explicit description of soil spectral variations, with special attention to soil type and surface moisture effects. Second, the development of leaf optical properties models that take explicitly into account minor absorbers such as lignin, cellulose, nitrogen, starch, sugar, etc. The brief sensitivity analysis of canopy reflectance to leaf optical properties suggests that these minor absorbers could be detected on full cover canopies in domains where water absorption is not too strong such as in the near infrared plateau or in the 1550-1850 nm domain.

High spectral resolution sensors will soon routinely provide data from space. Transmission and processing of the large amount of data associated to this type of remote sensing information will certainly be the main technological limiting factor. These problems must direct avenues of research towards optimal specification of sensors in terms of the number, position and width of the wavelength bands, and inboard data processing algorithms. Further, one key point poorly addressed in this study, is the possibility to correct radiance values recorded by space borne spectro-imaging systems from atmospheric effects. This is also one specificity offered by the use of high spectral resolution data.

5. References

Adams J.B., M.O. Smith and A.R. Gillespie (1991) 'Imaging spectroscopy: data analysis and interpretation based on spectral mixture analysis', *in* P. A. Englert (eds.), *Remote Geochemical Analysis: Elemental and Mineralogical Composition.*

Allen W.A. and A.J. Richardson (1968) 'Interaction of light with a plant canopy', *J. Opt. Soc. Amer.*, 58(8), 1023-1028.

Allen W.A., H.W. Gausman, A.J. Richardson, and J.R. Thomas (1969) 'Interaction of isotropic light with a compact plant leaf ', *J. Opt. Soc. Amer.*, 59(10), 1376-1379.

Allen W.A., H.W. Gausman, and A.J. Richardson (1970) 'Mean effective optical constants of cotton leaves', *J. Opt. Soc. Amer.*, 60(4), 542-547.

Allen W.A., H.W. Gausman, and A.J. Richardson (1973) 'Willstater-Stoll theory of leaf reflectance evaluated by ray tracing'. *Appl. Opt.*, 12(10), 2448-2453.

Baret, F.(1988) 'Un modèle simplifié de réflectance et d'absorptance d'un couvert végétal', *in* ESA (ed.), *4ième Colloque International des Signatures Spectrales d'Objets en Télédétection.*, SP-187 113-120. Aussois, France: ESA.

Baret, F., and G. Guyot (1991) 'Potentials and limits of vegetation indices for LAI and APAR assessment', *Remote Sensing of Environment*,, 35, 161-173.

Baret F., S. Jacquemoud, and J.F. Hanocq (1992) 'The soil line concept in remote sensing', *Remote Sensing Reviews, (in press)*

Baret F., S. Jacquemoud, G. Guyot, and C. Leprieur (1992) 'Modeled analysis of the biophysical nature of spectral shifts and comparison with information content of broad bands', *Remote Sensing of Environment*, 41, 133-142.

Campbell, G.S.(1986) 'Extinction coefficients for radiation in plant canopies calculated using an ellipsoidal inclination angle distribution', *Agric. For. Meteorol.*, 36, 317-321.

Chandrasekhar, S.(1960) 'Radiative Transfer', New York, Dover Publications.

Conel J.E., R.O. Green, R.E. Alley, C.J. Bruegge, V. Carrere, J.S. Margolis, G. Vane, T.G. Chrien, P.N. Slater, S.F. Biggar, P.M.Teillet, R.D. Jackson, and S. Moran (1988) 'In-flight radiometric calibration of the airborne visible/infrared imaging spectrometer (AVIRIS)', *in* SPIE *(ed.) Recent advances in sensors, Radiometry and data processing for remote sensing*, 924,179-195. Orlando, Florida, USA.

Curran, P.J. (1989) 'Remote sensing of foliar chemistry', *Remote Sensing of Environment*, 30, 271-278.

Demetriades-Shah, T.H., M.D. Steven, and J.A. Clark (1990) 'High resolution derivative spectra in remote sensing', *Remote Sensing of Environment*, 33, 55-64.

Hall, F.G., K.F. Huemmrich, and S.N. Goward (1990) 'Use of narrow-band spectra to estimate the fraction of absorbed photosynthetically active radiation', *Remote Sensing of Environment*, 33, 47-54.

Hapke, B.(1981) 'Bidirectional reflectance spectroscopy, 1. Theory', *J. Geophys. Res.*, 86, 3039-3054.

Horler D.N.H., M. Dockray, and J. Barber (1983) 'The red edge of plant leaf reflectance', *Int. J. Remote Sens.*, 4(2), 273-288.

Huete, A.R.(1986) 'Separation of soil-plant spectral mixtures by factor analysis', *Remote Sensing of Environment*, 19, 237-251.

Jacquemoud, S., and F. Baret (1990) 'PROSPECT: A model of leaf optical properties spectra', *Remote Sens. Environ.*, 34: 75-91.

Jacquemoud, S., F. Baret, and J.F. Hanocq (1992) 'Modeling spectral and directional soil reflectance', *Remote Sensing of Environment*, 41, 123-132.

Jacquemoud, S.(1992) 'Utilisation de la haute résolution spectrale pour l'étude des couverts végétaux: Développement d'un modèle de réflectance spectrale', Université Paris VII (France)/INRA/CNES,1-92.

Johnson L.F., and D.L. Peterson (1991) 'AVIRIS observation of forest ecosystems along Oregon transect', *in* G. Vane (ed.), *Second JPL Airborne Geoscience Workshop*, Pasadena, CA, USA, JPL, 190-199.

Johnson L.F., F. Baret, and D.L. Peterson (1992) 'Oregon Transect: Comparison of leaf-level reflectance with canopy-level and modelled reflectance', *in* R.O. Green (ed.), *Third JPL Airborne Geoscience Workshop*, Pasadena, CA, USA, JPL, 113-115.

Kneizys, F., E. Shettle, G. Anderson, L. Abrew, J. Chetwynd, J. Shelby, and W. Gallery (1989) Atmospheric transmittance/radiance, Computer code Lowtran 7', Hanscom AFB, MA (USA).

Kubelka P., and Munk F.(1931) 'Ein Beitrag zur Optik der Farbanstriche', *Ann. Tech. Phys.*, 11, 593-601.

Kumar, R., and L. Silva (1973) 'Light ray tracing through a leaf cross section', *Appl. Opt.*, 12(12), 2950-2954.

Malthus, T.J.(1989) 'Anglo-French collaborative reflectance experiment. Experiment I, Broom's Barn experimental Station, July 1989', INRA Bioclimatologie, BP 91, 84143 Montfavet, France.

Marten G.C., J.S. Shenk, F.E. Barton II (eds.) (1989) 'Near infrared reflectance spectroscopy (NIRS): analysis of forage quality', *United States Department of Agriculture Research Series Handbook* Number 643.

Pinty, B., M.M. Verstraete, and R.E. Dickinson (1989) 'A physical model for predicting bidirectional reflectances over bare soils', *Remote Sensing of Environment*, 27, 273-288.

Price, J.C.(1990) 'On the information content of soil reflectance spectra', *Remote Sensing of Environment*, 33, 113-121.

Rock ,B.N., T. Hoshizaki, and J.R. Miller (1988) 'Comparison of in situ and airborne spectral measurements of the blue shift associated with forest decline', *Remote Sensing of Environment*, 24, 109-127.

Rondeaux, G., and V.C. Vanderbilt (1992) 'Estimation of photosynthetic capacity using polarization', Proc. of *IGARSS '92,* Houston, Texas, USA, IEEE, 1471-1473.

Smith, M.O., J.B. Adams, and D.E. Sabol (1994) 'Mapping sparse vegetation canopies', In: Hill, J. and J. Mégier (eds.) *'Imaging Spectrometry - a tool for environmental observations'*, Kluwer Academic Publishers, Dordrecht (this volume).

Tanré, D., C. Deroo, P. Duhaut, M. Herman, J.J. Morcrette, J. Perbos, and P.Y. Deschamps (1986) 'Simulation of the satellite signal in the solar spectrum: The 5S code', *Int. J. Remote Sens.* 11(4), 659-668.

Tucker, C.J., and M.W. Garratt (1977) 'Leaf optical system modeled as a stochastic process', *Appl. Opt.*, 16(3), 635-642.

Ustin S.L., M.O. Smith, and J.B. Adams (1991) 'Remote sensing of ecological processes: A strategy for developing and testing ecological models using spectral mixture analysis', *in* J. E. Field and C. Field (eds.), *Scaling Ecological Processes from Leaf to Landscape,* Academic Press.

Vanderbilt, V.C., S.L. Ustin, and J. Clark (1988) 'Canopy geometry changes due to wind cause red edge spectral shift', *in* Proc. of *IGARSS '88*, Edinburgh (Scotland), ESA SP-284, 835-836.

Verhoef, W. (1984) 'Light scattering by leaf layers with application to canopy reflectance modeling: the SAIL model', *Remote Sensing of Environment*, 16: 125-141.

Verhoef ,W. (1985) 'Earth observation modeling based on layer scattering matrices', *Remote Sensing of Environment*, 17: 165-178.

OPTICAL PROPERTIES OF LEAVES: MODELLING AND EXPERIMENTAL STUDIES

JEAN VERDEBOUT, STÉPHANE JACQUEMOUD, and GUIDO SCHMUCK
Institute for Remote Sensing Applications
Commission of the European Communities
Joint Research Centre
I-21020 Ispra (Va), Italy

ABSTRACT. This paper deals with the interpretation of leaves spectra following an approach based on modelling and laboratory studies. First, the leaves structure and principal constituents are described together with the way they interact with light. The effects of growth, senescence and environmental factors on the leaf optical properties are summarised. A laboratory study conducted on drought stress of maize (*Zea Mays*) plants is reported as an example. A succinct review of the existing models is then made: ray tracing, Kubelka-Munk and developments, plate models, and the stochastic model. The use of these models to determine leaf constituents and structure by inversion on reflectance spectra is then discussed with an emphasis on the research of good specific absorption coefficients for the constituents. The validation of the PROSPECT model (generalised plate model) on the basis of leaves spectra is presented. The problems linked with the application of these procedures to remote sensing data is evoked, and an example of inversion on experimental spectra of sugar beet (*Beta vulgaris L*) fields is briefly reported.

1. Introduction

The scientific community developed an increasing awareness that Earth functions as a single biogeochemical system in which terrestrial ecosystems play a key role. Air pollution along with climatic changes and other human impacts may alter Earth's regulating capability. International programmes are necessary to improve our understanding of these processes. New approaches in the use of remotely sensed measurements, especially as inputs to ecosystem models, hold the key to this problem (Ustin et al., 1991).

Up to now, broad band satellite sensors such as Landsat -MSS and -TM, SPOT-HRV, NOAA-AVHRR, proved their capacity to assess the extent, density or composition of vegetation. However, they have not permitted to describe in detail vegetation status and its functioning. The new generation of airborne or satellite imaging spectrometers (AIS, AVIRIS, DAIS 7919, MERIS, MODIS) already provides a considerable improvement. These sensors record a complete reflectance spectrum for each pixel in the image, revealing spectral features characteristic of a particular vegetation type or its environmental conditions. However, development of algorithms to interpret high spectral resolution data has not yet followed the technological advances of sensors.

Since the main connection between changes in the terrestrial ecosystem and the radiative transfer from a landscape is through changes in the spectral properties of the single leaves, one of the primary directions must be the development of leaf models, based on experimental techniques and

J. Hill and J. Mégier (eds.), Imaging Spectrometry - a Tool for Environmental Observations, 169–191.

reference methods. Fundamental understanding gained from laboratory studies will be required to scale to the canopy and landscape level. Special attention has to be paid to the influence of development (growth, senescence) and of changed environmental conditions on the reflectance properties of leaves.

2. Interaction of electromagnetic radiation with leaf tissue

Among terrestrial vegetation, one can distinguish between the Gymnosperms, a systematic old group composed of arborescent or bushy perennials, and the Angiosperms, a group which widely dominates the terrestrial flora, and which gathers herbaceous and ligneous species (Camefort et Boué, 1969). These two groups represent major differences in leaf anatomical organisation. Gymnosperms are mainly characterised by needle-shaped leaves such as pine, spruce, larch, etc. According to Camefort (1972), the cross section of a pine needle can be described by:

- *an epidermis* built of cells with very thick membranes, inwardly doubled by a layer of cells with thick and lignified cell walls.

- *a mesophyll*, homogeneous chlorophyllian parenchyma.

- *a central vascular cylinder* which ensures the circulation of crude (xylem) and elaborated (phloem) saps inside the leaf.

Angiosperms are more developed plants. One can distinguish the monocotyledons (graminae) and the dicotyledons (leguminous plants) whose leaves show the following fundamental tissues (Camefort, 1972):

- *an epidermis* covering the whole surface of the limb. It is made up of a layer of epidermal cells with no chloroplasts, topped by a cuticle of variable thickness according to the species, and sometimes doubled by a waxy layer. Monocotyledons show off stomata on both sides, dicotyledons only on the abaxial (lower) face.

- *a mesophyll structure*: monocotyledonous leaves have an homogeneous chlorophyllian parenchyma, with only few intercellular air-spaces. The mesophyll of dicotyledons is not homogeneous but it is differentiated between a palisade and a spongy mesophyll. Palisade cells are high, packed, and arranged in one to two layers; they contain the largest amount of the chloroplasts. The spongy mesophyll is made up of irregular cells, separated by large intercellular air-spaces; they contain less chloroplasts than the palisade cells.

- *a vascular system* corresponding to the main and secondary limb nerves.

Although plant leaves present numerous anatomical structures, the basis elements are the same, and the variability of the leaf optical properties (see Section 3) only results from their arrangement inside the leaf. The two main interactions of the electromagnetic radiation with these elements are diffusion and absorption.

2.1. SURFACE REFLECTION

The first boundary encountered by solar radiation is the interface air-epidermis. The leaf surface reflectance is a combination of diffuse and specular reflectance, and is not Lambertian: Breece and Holmes (1971), Brakke et al. (1989), Walter-Shea et al. (1989) have demonstrated the importance of specular reflectance for oblique incidence. The leaf can be compared to a Lambertian scatterer only for normal incidence. Moreover, Grant (1987), Grant et al. (1987, 1992) showed that all leaf surfaces polarise incident light and that the polarised component of the reflectance depends on the characteristics of the leaf surface. The physiological role of the leaf surface is very important because plants can modify it according to the external conditions. Thus, the leaves of many desert species are pubescent to decrease the amount of light and heat in the tissues: this is expressed by an increase of the diffuse reflectance. Another example is given by the waxy layer that may cover the cuticle: it protects the underlying cellular structures, but also leads to an increase of leaf reflectance (Grant, 1987).

The shape of epidermal cells may influence the path of the incident beam: convex cells of certain plants act as lenses that focus light within the upper region of the palisade which contains many chloroplasts adapted to high light. This phenomenon has been first presented as an adaptation to the low light environment on the tropical forest floor (Bone et al., 1985), but Martin et al. (1989) showed that, among cultivated plants, it could increase absorption of light from low angles: in that case, the diameter of the focal spot displaces laterally within the palisade while increasing.

2.2. DIFFUSION OF ELECTROMAGNETIC RADIATION

The cells of the main leaf tissues (protective tissues, parenchyma and conductive tissues) are surrounded by a cell wall, containing a cytoplasm with several organelles (nucleus, mitochondria, chloroplasts, amyloplasts, etc.) and a vacuole of large size. These cells are separated by intercellular air-spaces which permit the circulation of gases inside the tissues. The general term of scattering gathers complex phenomena of reflection, refraction and diffusion (Vogelmann and Björn, 1986).

- *at microscopic scale* (the size of particles is greater than the wavelength), refractive index differences between two different media induce optical boundaries; at each time the light changes medium, it is partly specularly reflected and partly refracted according to the Snell-Descartes law ($n_1 \sin\theta_1 = n_2 \sin\theta_2$). The cell walls-air interfaces are the main optical boundaries encountered by light. Most of the time, absorption at the interface is assumed to be negligible so that the refractive index of foliar constituents is real (Kumar and Silva, 1973). The measurement or evaluation of this index has been subjected to many studies, and it is approximately equal to 1.41 at 800 nm (Allen et al., 1969; Gausman et al., 1974; Woolley, 1975). In comparison, the refractive index of the air is unity, and that of water equals 1.328 at 1000 nm (Palmer and Williams, 1974).

- *at macroscopic scale* (the particle size is less than or equal to the wavelength), one observes phenomena of Rayleigh ($d \ll \lambda$) and Mie ($d \approx \lambda$) scattering. The dimensions of palisade cells (15 μm x 15 μm x 60 μm), or epidermal and spongy cells (18 μm x 15 μm x 20 μm) in view of the wavelength is too important to induce such phenomena (Gates et al., 1965; Sinclair et al., 1973). The cytoplasmic organelles (chloroplasts and grana) may produce them: their size is comparable to the wavelength and the Mie scattering dominates. In fact, these phenomena are very complex and not well known.

2.3. Absorption of Electromagnetic Radiation

2.3.1. Electronic transitions. In this case, the absorption of light corresponds to a disturbance of the valence electrons in the electron cloud of a molecule, resulting in the formation of an electronically excited state (Schanda, 1986). These electronic transitions consume a lot of energy so that only the short wavelengths (400-1000 nm) are concerned. The molecules which absorb light in this spectral domain are mainly the foliar pigments (chlorophylls a and b, carotenoids, xanthophylls, brown pigments).

- *chlorophylls*: the chlorophylls of higher plants consist of chlorophyll a as the major pigment, and of chlorophyll b as an accessory pigment (Lichtenthaler, 1987). Chlorophyll is located in the chloroplasts of green leaves; it is one of the most important biological compounds: it acts as a photoreceptor and catalysator for photosynthesis (conversion of sunlight into chemical energy for the reduction of CO_2 into carbohydrates). Chlorophylls *a* and *b* present absorption peaks around 430/450 nm and 660/640 nm, due to the presence of a metal atom (Mg) at the centre of the molecule.

  Recent work of Vogelmann et al. (1989) show that absorption mainly occurs in the first 60 μ m of the palisade parenchyma (the cytoplasm of each palisade cell may contain up to 50 chloroplasts in suspension). There is a gradient in the chloroplast morphology that corresponds to a gradient of their photosynthetic properties. Thus, 90% of the visible light is absorbed by the palisade parenchyma of *Medicago sativa* leaves (Vogelmann et al., 1989).

- *carotenoids*: this group of molecules can be divided into the oxygen-free carotenes (α - and β - form), and into the xanthophylls (zeaxanthin, lutein, violaxanthin, neaxanthin) which contain fixed oxygen in different forms. The position of the absorption peaks of these organic molecules depends on the number of functional groups in the chain (for example, C=C and C-OH bonds). β -carotene and xanthophylls present absorption peaks around 450 nm. The properties of β - carotene are twofold: energy transfer towards chlorophyll, and protection of chlorophyll *a* from photo-oxydation (Lichtenthaler, 1987).

- *polyphenols and brown pigments*: during senescence, plant leaves are subjected to a sequence of metabolic and structural changes. Polyphenols of cellular walls (cathecols, anthocyanins, flavonoids...) may induce enzymatic (action of polyphenoloxydase) or non enzymatic (interaction of heavy metals like iron) reactions. The resulting products are high molecular weigh compounds which absorb visible light.

- *other pigments*: phytochrome is a photoreceptor that controls morphogenesis; it absorbs in the red (660 nm) and far-red (730 nm). Flavoproteins, nucleic acids absorb mainly in the blue region of the spectrum (400-450 nm).

2.3.2. Vibrations of polyatomic molecules. The atoms of a molecule can hardly displace around their balance position, inducing vibrational processes. These processes are translations, rotations (kinetic energy), and vibrations (kinetic and potential energy). This system can be compared to a linear harmonic oscillator whose energy is characterised by the vibrational quantum number *vi*. Let ξ_{v1v2v3} be the total energy:

$$\xi_{v1v2v3} = \xi_{translation} + \xi_{rotation} + \xi_{vibration} = \sum_{1}^{3} h\, \nu_i \left(v_i + 1/2 \right) \quad (1)$$

where h is Planck's constant and ν_i the oscillator frequency. The absorption results from transitions between neighbouring states under the effect of a light beam whose frequency is close to those of the oscillator (Schanda, 1986; Goetz, 1992). This kind of absorption concerns the near and middle infrared (1000-2500 nm). Curran (1989) and Elvidge (1990) have made the inventory of the main chemical compounds involved:

- *water* represents from 40% to 90% of the leaf fresh weight. The measurement of the leaf water status is of some interest for the description of cell expansion in growing tissues and the physiological state of the plant. Water absorbs at 970 nm (weakly), 1200 nm, 1450 nm, 1940 nm, and 2500 nm.

- *cellulose* is a polysaccharide build of β -*D*-glucose. It is mainly located in cell walls of all plants where it acts to strengthen and protect plant structure. It absorbs at 1220 nm, 1480 nm, 1930 nm, 2100 nm, 2280 nm, 2340 nm, and 2480 nm.

- *lignin* is another structural component which gives the wood its hardness and rigidity. This polymer is built of complex units with encrusted cellulosic materials. Lignin has an intense absorption in the ultraviolet at 280 nm, and in the middle infrared at 1450 nm, 1680 nm, 1930 nm, and 2100 nm.

- *starch*, principal food storage molecule of plants, is a polysaccharide formed of α -*D*-glucose units. It absorbs at 990 nm, 1220 nm, 1450 nm, 1560 nm, 1700 nm, 1770 nm, 1930 nm, 2100 nm, 2320 nm, and 2400 nm.

- *other compounds*: one can also quote pectins, waxes, tannins and nitrogen compounds whose amounts in the leaf, and consequently whose absorption in the middle infrared, are weaker.

3. Variability of leaf optical properties

Classically, three spectral domains are distinguished (Jacquemoud, 1992):

- *the visible region* (VIS: 400-700 nm), characterised by a strong absorption of the foliar pigments leading to low reflectance and transmittance values.

- *the near infrared region* (NIR: 700-1300 nm) also called the infrared plateau because of high and almost constant values of reflectance and transmittance.

• *the middle infrared region* (MIR: 1300-2500 nm), the main absorption domain of water and other foliar biochemical components.

3.1. THE VISIBLE DOMAIN (VIS)

The visible light acts as a source of energy for photosynthesis when absorbed by chlorophyll and carotenoids; it controls photomorphogenesis when absorbed by other photoreceptors such as the phytochrome. The contribution of these last pigment systems to the leaf optical properties is negligible; therefore, only photosynthetic pigments will be considered in this paper. The relationships between the leaf optical properties and the pigment concentrations have been intensively studied in view of finding an indicator for the physiological state of vegetation. The chlorophyll concentration may indeed be affected by soil salinity, nitrogen or mineral deficiencies, heavy metal stress and atmospheric pollution (Belanger, 1990).

The reflectance and transmittance spectrum of a green leaf classically shows minima around 420 nm, 490 nm and 670 nm, and one maximum around 550 nm. The red edge (transition to high reflectance from 670 to 780 nm) corresponds to the wavelength interval at which the chlorophylls cease to absorb. The amplitude of the reflectance spectral features varies as a function of the concentration of the associated pigments in the leaf. On average, chlorophylls *a* and *b* are ten times more concentrated than carotenoids so that the effect of carotenoids in a green leaf is masked by that of chlorophylls. During leaf senescence, chlorophylls degrade faster than carotenoids (Sanger, 1971) which become the main pigments: the leaf turns yellow. Finally, at the leaf death, brown pigments appear: the leaf reflectance and transmittance regularly decrease between 750 nm and 400 nm (Boyer et al., 1988).

Classical methods for chlorophyll or carotenoids analysis are destructive; spectroradiometry offers a possibility for non-destructive evaluation of the concentration of these pigments. So far, empirical relationships have been proposed: Thomas and Gausman (1977) have described a good correlation between the concentration of these pigments and the reflectance levels at 550 nm (minimum absorption). Chappelle et al. (1992) found very good relationships between the reflectance ratio R675/R700 and chlorophyll *a* concentration, R760/R500 and carotenoids concentration, R675/(R650*R700) and chlorophyll *b* concentration. Instead of studying the leaf reflectance (or transmittance) values at a given wavelength, one can reason in terms of shape of reflectance spectra which typically present sharp variations between high and low absorptance levels. The position of the inflection point of the red edge is also related to chlorophyll or other absorbing pigments concentration, as well as to the mesophyll internal structure (Gates et al., 1965; Horler et al., 1983; Guyot and Baret, 1989; Belanger, 1990; Baret et al., 1992, etc.). It has been used as a general indicator of stress (figure 1): heavy metal stress, lack of nutrients, and exposure of plants to air pollutants, could be monitored *via* a shift of the red edge towards shorter wavelengths (Horler et al., 1983; Rock et al., 1986; Belanger, 1990). Therefore, this "blue shift" is not specific for particular types of environmental stresses. Well controlled laboratory experiments conducted on homogeneous sets of plants and isolating a single stress factor, will be useful to better document the effects on the spectral signatures. The reflectance measurements have to be accompanied by classical plant physiological work (pigment analysis, gas exchange measurements, stomata conductance determination) in order to determine biological parameters necessary for the correct interpretation of the optical data.

An example of such a study is found in Maracci et al. (1991): these authors showed that some disturbances in the photosynthetic functioning of the plant may occur without a change in the pigment concentration. Maize plants (*Zea mays*) were cultivated in a phytochamber, and submitted to drought stress for varying periods of time. With increasing water deficiency, the net photosynthetic activity decreased by a factor of about 100, while the chlorophyll concentration of the samples, as well as the VIS reflectance, remained almost unchanged (table 1). The decrease of photosynthetic activity, measured by the CO_2 uptake, is here due to the closure of the stomata.

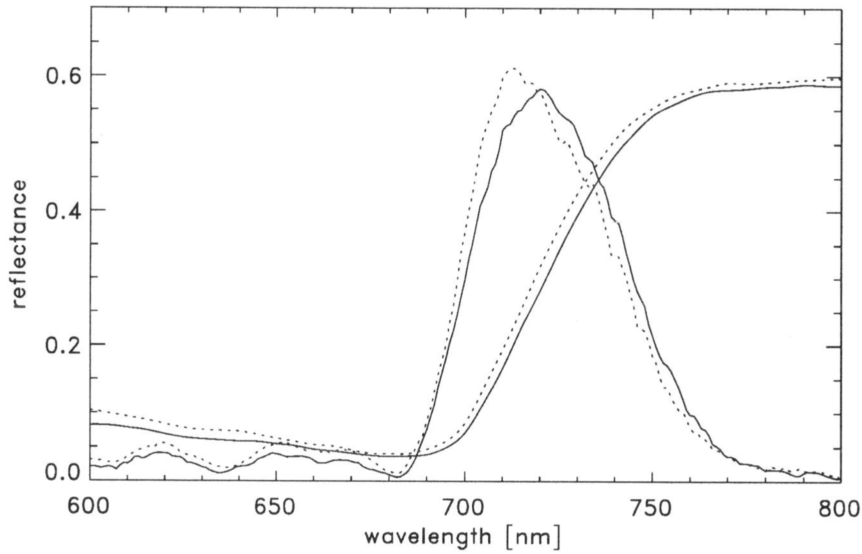

FIGURE 1. Comparison of leaf reflectance curves and their first derivative for plants from high- (dotted lines) and low-damaged (solid lines) sites in Vermont (from Rock et al., 1988).

3.2. THE NEAR INFRARED DOMAIN (NIR)

In this region, the absorption is low and light penetrates deeper in the leaf tissue. This wavelength range therefore *a priori* contains more information on the leaf structure.

As light penetrates inside the leaf, it encounters optical boundaries which control its distribution within the different tissues. For example, the boundary between the palisade layer and the air-rich spongy mesophyll layer in dicotyledonous leaves acts as an internal reflector that bounces light back into the chloroplast-rich palisade (Vogelmann and Björn, 1986). Scattering inside the leaf is mainly due to intercellular air-spaces: Gausman et al. (1970) have related the near infrared reflectance levels to the number of air-spaces between cells. Then, the leaf optical properties in this spectral region are determined by the internal leaf structure (Sinclair et al., 1973; Grant et al., 1987; Grant, 1987). The scattering is more correctly described as being due to multiple refractions and reflections induced by the steps of refractive index ($n \approx 1.4$ for hydrated cellular walls, $n = 1.326$ for water at 1000 nm, and $n = 1.0$ for air).

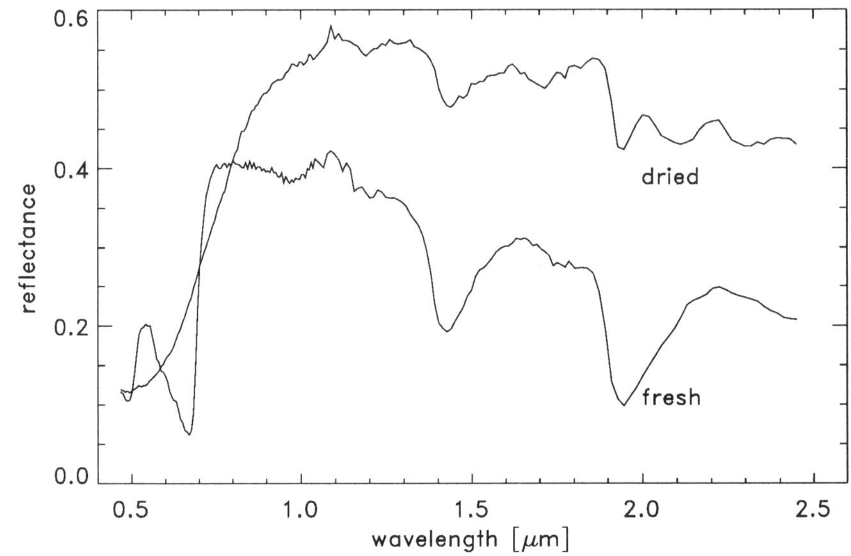

FIGURE 2. Comparison between reflectance spectra of green and dried platanus leaves.

For the same thickness, monocotyledons whose mesophyll is compact have a lower near infrared reflectance than dicotyledons which present a palisade and a spongy mesophyll; on the other hand, their transmittance is higher. This can be simply explained: the more lacunous structure of dicotyledonous leaf allows for greater multiple reflection within the leaf, and, as a result, leaf reflectance is greater (Grant, 1987). Recent studies have highlighted the role of there cavities: they trap light so that there can be from three to four times more light inside plants than is present on the outside (Vogelmann and Björn, 1986). This property may improve photosynthesis efficiency.

During the plant growth, the near infrared reflectance of a given species is almost constant (genetic determinism); the most important changes appear during maturation and senescence (Sinclair et al., 1971; Allen et al., 1971). Young, immature leaves have a compact mesophyll: the reflectance is lower than the transmittance. As leaves mature, air-spaces in the spongy mesophyll of dicotyledonous leaves increase, as well as the reflectance. As leaves senesce, they drop in moisture and cell geometry changes: the reflectance greatly increases. A similar effect is induced by severe drought stress: by comparing the reflectance spectra of fresh and dried leaves (figure 2), Maracci et al. (1991) showed that a strong decrease in water content results in an increase of the NIR reflectance. This is due to changes in the leaf internal structure, leading to stronger scattering (Ripple, 1984; Boyer et al., 1988).

3.3. THE MIDDLE INFRARED DOMAIN (MIR)

3.3.1. Water. The main absorption bands are located around 1200 nm, 1450 nm, 1940 nm, and 2500 nm, but the overall shape of the MIR spectrum is largely influenced by water. Dehydration increases the reflectance and the transmittance in the middle infrared by lowering the absorption and increasing the scattering because of structural modifications. The detection of a water stress by remote sensing has been the subject of many studies: a major factor limiting plant growth and productivity by reducing photosynthesis is the lack of water (Tucker, 1980; Hunt et al., 1987; Bowman, 1989; Maracci, 1991). In fact, stomatal conductance and water potential are the physiological variables which characterise the leaf water status but, the link between these variables and radiometric measurements is not direct. Nevertheless, water potential can be related to intermediate variables such as the leaf Relative Water Content (RWC) which is the ratio RWC= (WF-WD)/(WTF-WD) where WF, WD, and WTF are the masses of the leaf during the measurement, after drying, and in the fully turgid state. Tucker (1980), Hunt et al. (1987), Hunt and Rock (1989) found a relationship between RWC and a spectral index, the Leaf Water Content Index (LWCI). Other indices such as the Normalised Difference Infrared Index (NDVI) or the Moisture Stress Index (MSI) have been proposed in the literature.

	PN [$\mu g \cdot m^{-2} \cdot s^{-1}$]	**g H_2O** [$mmol \cdot m^{-2} \cdot s^{-1}$]	**Chl. a+b** [$\mu g \cdot cm^{-2}$]	**WP** [-MPa]
control	15.00	165.10	30.01	0.13
3 days stress	4.03	29.5	30.84	0.48
4 days stress	0.88	9.35	32.04	0.93
5 days stress	0.17	2.81	32.07	1.55

TABLE 1. Net photosynthesis (PN), stomata conductance (g H_2O) and chlorophyll concentrations of controlled and water-stressed leaves of wheat and maize (from Maracci et al., 1991).

The application to remote sensing of water stress is however not immediate in the short wavelengths. Some investigators (Hunt and Rock, 1989; Bowman, 1989) even concluded that these indices, and more generally the use of middle infrared reflectance, were insufficient to estimate the leaf water status because reflectance changes within a biologically meaningful range are too insignificant. The work of Maracci et al. (1991) partly confirmed this assertion: during the applied drought stress period, the reflectance of the maize leaves did not show a significant increase over the MIR spectral region (figure 3); differences between the control and stressed plants could not be observed until the fifth day of the experiment, while in the same time, the water content decreases from 91% to 75%, and the water potential (Ψ) decreases from -0.13 MPa in the well watered control plants to -1.55 MPa in the stressed plants (table 1). However, the curve fitting techniques (Goetz et al., 1990), and the model based approach (Jacquemoud and Baret, 1990), which take into account the combined effects of the changes in scattering and absorption offer renewed hope for canopy moisture detection.

3.3.2. Other foliar constituents. When the leaf dehydrates, cellulose, lignin, starch, etc. whose absorption was masked by water appear (Peterson et al., 1988; Wessman et al., 1988; Elvidge, 1990; Peterson, 1992). Absorption spectra of dried leaves reveal some absorption peaks due to vibrational processes of foliar constituents. Correlations have been established between these peaks and concentrations of organic compounds. Such procedures are commonly used by the United States Department of Agriculture to analyse forage quality. The accuracy of these measurements are comparable to those obtained by laboratory chemical methods (Curran, 1989). Unfortunately, a direct application of these methods is not possible with fresh leaves, mainly because of water which

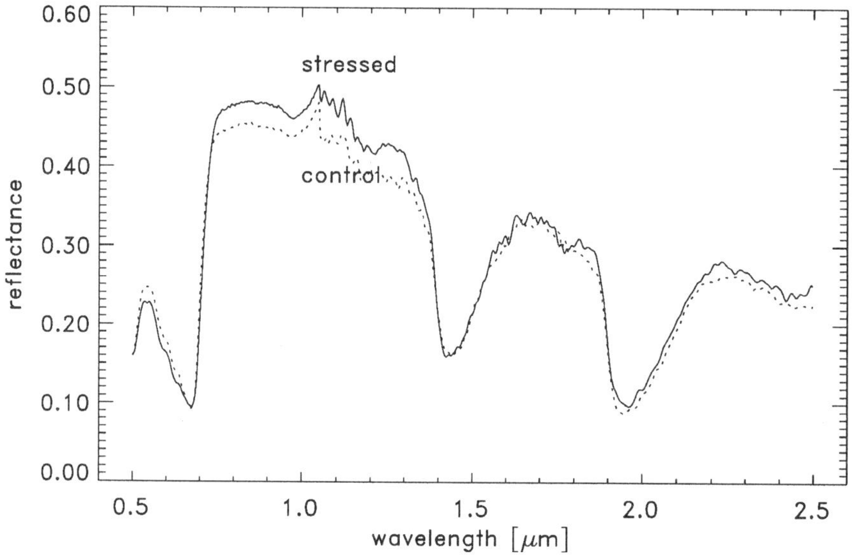

FIGURE 3. Comparison between leaf reflectance of drought-stressed and control maize plants (from Maracci et al., 1991).

represents from 60% to 90% of the fresh weight. Moreover, organic compounds absorb in similar wavebands so that a given wavelength is never associated with a unique compound. For instance, the strong O-H bond is a component of the absorption spectra of water, cellulose, sugar, starch and lignin (Curran, 1989). In spite of these difficulties, Curran et al. (1992) and Goetz et al. (1990) showed that an accurate estimation of lignin, nitrogen, cellulose, and starch was conceivable from fresh leaves.

4. Modelling of leaf reflectance

4.1. Ray Tracing

Given the complex structure of the leaf tissue, and the fact that internal diffusion is mainly due to reflections and refraction at the cell walls, the only rigorous approach to compute the reflectance and transmittance of plant leaves is ray tracing. The principle is to start with a detailed description of the leaf internal structure, for example deduced from a microscope photography of the leaf section or mimicked using geometrical shapes. In this way, individual cells and their particular arrangement are taken into account. The optical constants of the various materials (cell tissue, air cavities, etc.) are defined. Using the laws of reflection, refraction and absorption, it is then possible to simulate the propagation of individual light rays incident on the leaf surface. Once a sufficient number of rays have been simulated, statistically valid estimates of the macroscopic leaf optical properties are obtained (*i.e.* reflectance, transmittance). Such calculations are also able to reproduce the angular distributions of the light field and its polarisation state (Brakke et al., 1989).

The technique can be applied with a number of variants: Allen et al. (1973) used a leaf structure described by 100 circular arcs and considered two media (air and a medium characterised by a complex index of refraction to take into account the leaf tissue optical properties). They used their model to test the specular or diffuse nature of the reflection at the cell walls. In any case, their calculation led to an underestimation of the reflectance and an overestimation of the transmittance in the near infrared plateau. Shortly afterwards, Kumar and Silva (1973) showed that the actual reflectance and transmittance can be reproduced by including four media in the model: cell walls, chloroplasts, cell sap and air, thereby increasing the internal diffusion.

Ray tracing calculations are a tool of choice to test the assumptions made on the way the light interacts with the various leaf elements. However, if the goal is to use a model in order to retrieve information on the leaf constituents from, let us say, the reflectance spectrum, this technique is not practical. Indeed, it requires long computational times and the structure description is too complex. Other models, sufficiently simple to be inverted, were therefore developed.

4.2. SIMPLE MODELS

4.2.1. The K-M model and its developments. This model consists in considering the leaf as being a slab of diffusing and absorbing material (the medium is described by means of an absorption coefficient, *k*, and a scattering coefficient, *s*). The theory of Kubelka and Munk (two stream solution of a simple case of the radiative transfer equation) is then used, which yields simple analytical formulas for the diffuse reflectance and transmittance. This model was proposed by Allen and Richardson (1968); Yamada and Fujimura (1988, 1992) later developed a more sophisticated version in which the leaf is described as composed of four layers (upper cuticle, palisade parenchyma, spongy mesophyll and lower cuticle). The Kubelka-Munk theory is used to model the radiative transfer with different parameters in each layer, and the solutions are coupled with suitable boundary conditions to yield the leaf reflectance and transmittance. They used this model to perform non destructive measurements of chlorophyll concentration.

Ma et al. (1990) published a model describing the leaf as a slab of water with an irregular surface containing randomly distributed spherical particles. The angular distribution and the degree of polarisation of the reflected and transmitted light is computed using vector radiative transfer and Kirchoff rough surface scattering theories. The model is shown to reproduce well data measured on laurel, potato and corn leaves.

4.2.2. The plate models. The first plate model introduced in 1969 by Allen et al. consisted in representing the leaf as a non diffusing but absorbing plate with rough surfaces giving rise to Lambertian diffusion. The parameters are here an index of refraction and an absorption coefficient (more exactly the plate transmission coefficient). The model was successful in reproducing the reflectance spectrum of a compact leaf (corn) in which there are few air-cell wall interfaces. It was later extended (Allen et al., 1970) to the case of non compact leaves which are then regarded as a pile of N plates separated by N-1 air spaces, where N does not need to be an integer (generalised plate model). The additional parameter N in fact describes the internal structure and plays a role similar to that of the scattering coefficient in the K-M model. The latest development of the plate model is probably the PROSPECT model, of which a result is illustrated in figure 4 (Jacquemoud and Baret, 1990).

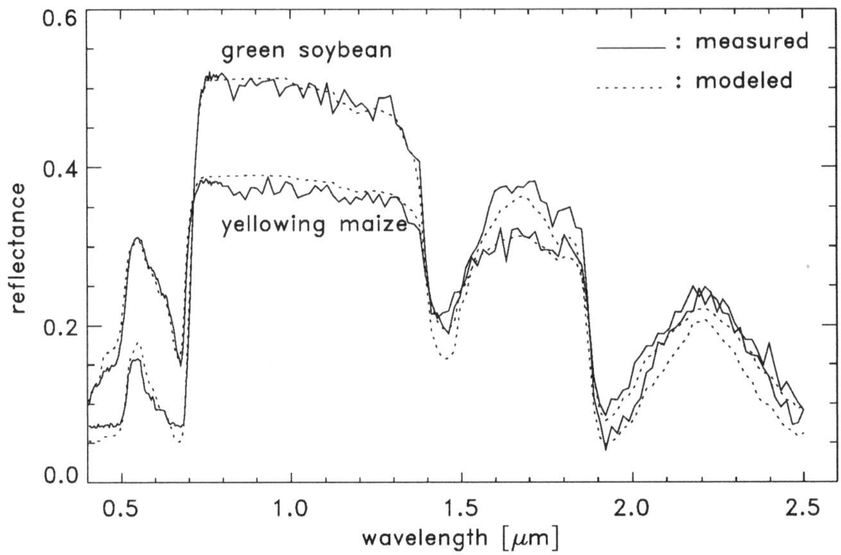

FIGURE 4. Comparison of the spectral reflectance modelled (dotted lines) and measured (solid lines) for green soybean and yellowing maize (from Jacquemoud and Baret, 1990).

4.2.3. The stochastic model. Tucker and Garatt (1977) have proposed a model in which the palisade parenchyma and the spongy mesophyll are explicitly taken into account as two layers of independently variable thicknesses. The reflectance calculation is performed using a Markov chain to model the radiation transfer. The chain makes use of ten compartments representing the radiation states (solar, reflected, absorbed and transmitted), the two cellular layers and the occurrence of scattering in each of these. Transition probabilities from compartment to compartment are estimated on the basis of the optical properties of the leaf material. Starting with an initial state vector representing the incident radiation, the steady state is computed by iteratively applying the one-step transition matrix, and yields both the reflectance and transmittance. This process was able to simulate the spectrum of maple leaves.

4.3. THE USE OF SIMPLE MODELS IN LEAF SPECTROMETRY - ROLE OF LABORATORY STUDIES

All the models presented above contain, in one way or another, the absorption coefficient of the leaf materials. If the assumption is made that leaf materials can be considered as an homogeneous mixture of its various components, one can write for the absorption coefficient:

$$k(\lambda) = \sum_i a_i \, k_i(\lambda) \qquad (2)$$

where $k_i(\lambda)$ and a_i are respectively the specific absorption coefficients and the concentrations of the constituents (water, chlorophyll, carotenoids, biochemicals, etc.). By adjusting a_i in order to reproduce the optimal reflectance spectrum, one can hope to determine the constituent concentrations. In order to perform this operation, one must know the spectral variation of the specific absorption coefficients k_i, and the quality of the result will depend on the correctness of these base functions. It will also depend on how well the functional dependence of the reflectance with respect to the absorption coefficient is reproduced by the model. In a similar way, the adjustment of the model other parameters will provide information on the leaf internal structure. This may prove to be important to monitor growth or stress.

4.3.1. Determination of the specific absorption coefficients. A first obvious idea is to deduce them from optical measurements performed on the pure substances. The infrared absorption coefficient of distilled water has been carefully measured by Curcio and Petty (1951); the measurement has been repeated and confirmed later by several authors. The water index of refraction can be found in Palmer and Williams (1974). The data for chlorophylls and, to a lesser extent, for non-photosynthetic pigments (carotenoids, xantophylls) are also available from the literature. In this case, though, a problem arises: the absorption spectrum obtained from extracts of chlorophyll in a solvent does not correspond exactly to the *in vivo* measurement; spectral shifts of the order of 10 nm are observed (Lichtenthaler, 1987). These differences are attributed both to the influence of the solvent and to the fact that the chlorophylls inside the leaf tissue are complexed with other pigments and proteins. Similar effects have been reported for the other pigments, for water, and for the biochemical components, for which spectral signature changes due to the interactions with the leaf tissue cannot be excluded.

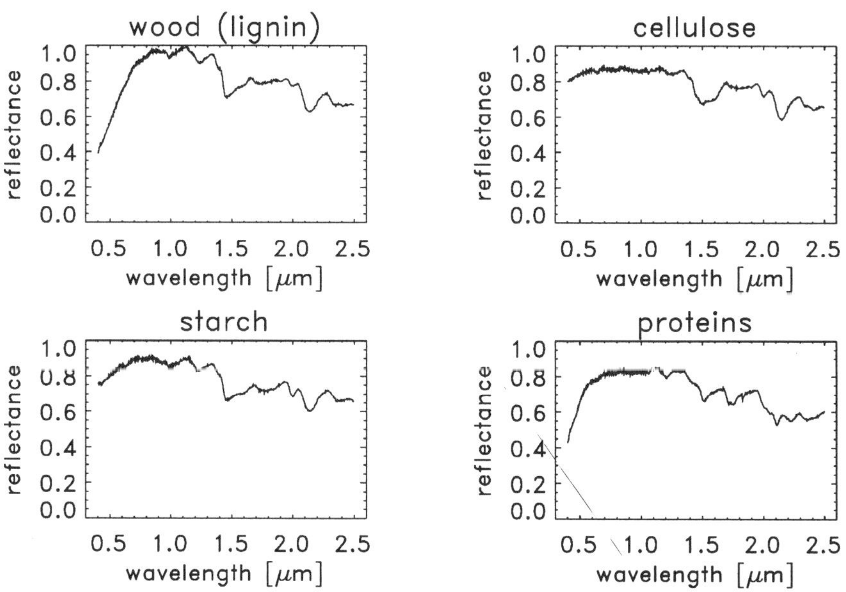

FIGURE 5. Reflectance spectra of four biochemical components found in leaves, obtained on "pure" substances.

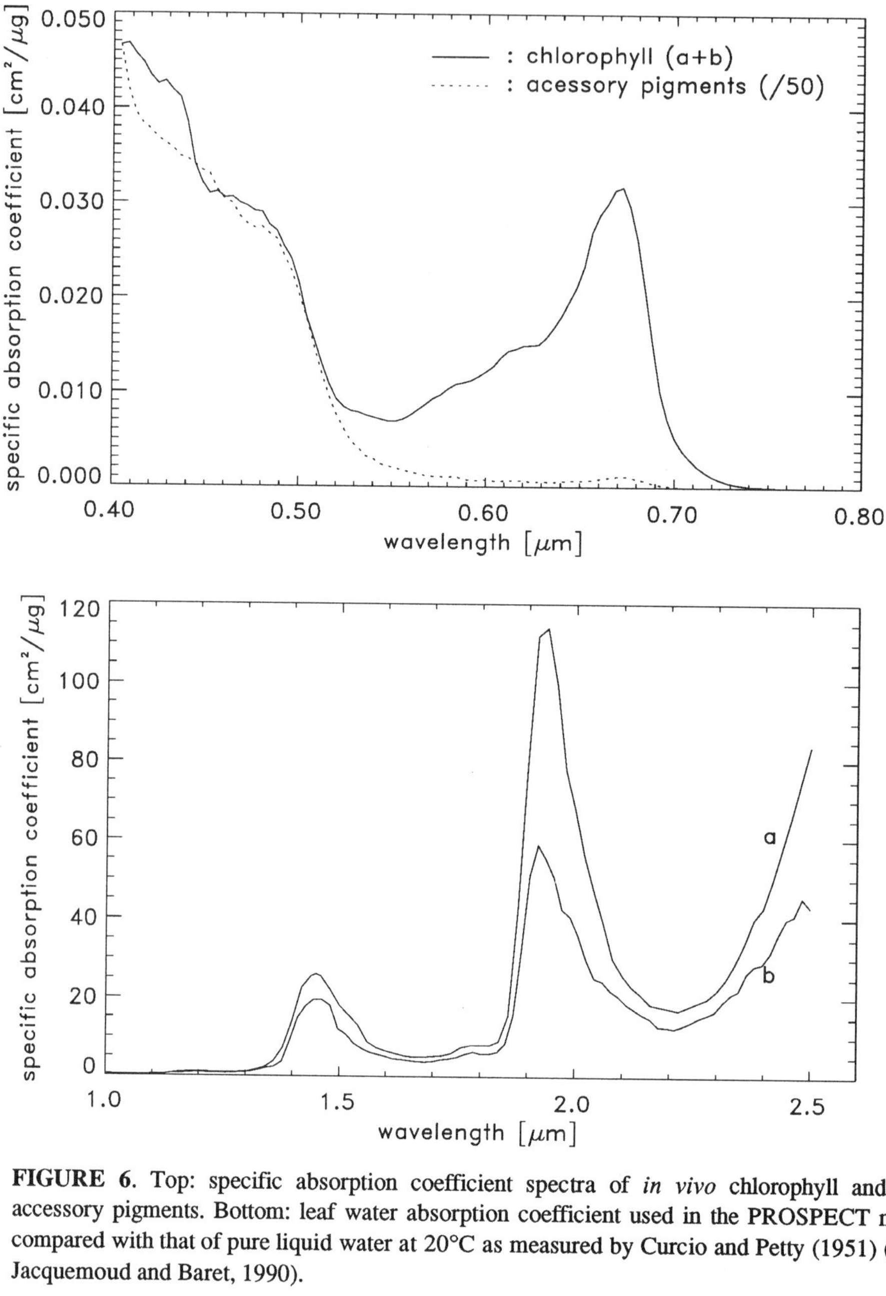

FIGURE 6. Top: specific absorption coefficient spectra of *in vivo* chlorophyll and *in vivo* accessory pigments. Bottom: leaf water absorption coefficient used in the PROSPECT model (a) compared with that of pure liquid water at 20°C as measured by Curcio and Petty (1951) (b) (from Jacquemoud and Baret, 1990).

Today, one question is to investigate the possibility of evaluating the vegetation content in such products as lignin, cellulose, nitrogen (mainly contained in proteins), starch, sugar, etc. These biochemicals have, in common, absorption features in the middle infrared part of the spectrum (1200-2500 nm). As already stated, these features are fundamentally associated with certain chemical bonds (C-H, N-H, C-O, O-H), the absorption bands and overtones of which combine to form the product signature. These signatures are rather complex because the large molecules contain many different bonds in various proportions.

Here again, the spectra of pure substances are used in practice, but their correctness is disputable:

- starch and cellulose are well defined substances and can be obtained pure.
- proteins and sugars can be isolated but both are families of substances, and this induces a variability of the signature due to the exact composition in the leaf.
- the situation is worse for lignin which cannot be really isolated and is not even precisely defined. In practice, the spectra measured on wood are used to substitute the pure lignin specific curve.

Figure 5 shows the reflectance of optically thick samples of these "pure" substances. The specific absorption coefficient can be deduced from these measurements by using the Kubelka-Munk theory.

In order to bypass these difficulties, another strategy can be used to determine the specific absorption coefficients: deduce them from measurements on the leaves themselves. For instance, the PROSPECT model (Jacquemoud and Baret, 1990) makes an extensive use of this approach: it uses specific absorption coefficients for water, non photosynthetic pigments, and chlorophylls (figure 6), all deduced from the adjustment of the model on measurements performed on specially prepared leaves (albino leaf, dried leaf, etiolated leaf) corresponding to limiting cases of the model. The use of such semi-empirical curves is likely to improve the way the calculation reproduces the observed spectra, but it may be argued that the model can loose generality.

The research for the best specific absorption curves is certainly not concluded, especially concerning the biochemicals.

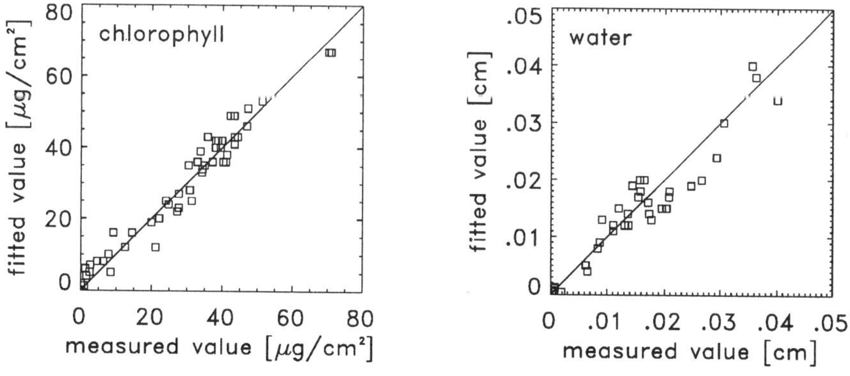

FIGURE 7. Estimation of chlorophyll a+b concentration, and leaf water equivalent thickness by inversion of the PROSPECT model (from Jacquemoud and Baret, 1990)

4.3.2. Validation. Before a model can be used with any confidence, it has to be validated. This means that its performance in determining some leaf characteristic (for instance the constituents concentration) has to be evaluated by comparing the model-derived value with an independent measurement of the same parameter. As an example, we shall briefly present the validation of the PROSPECT model (Jacquemoud and Baret, 1990) for determination of chlorophyll and water concentrations.

The validation was carried out with four data sets, consisting in reflectance and transmittance measurements performed on four species of leaves (maize,soybean,wheat and sugar beet). The optical data were accompanied by measurements of the water and chlorophyll content (using weighing and wet chemical analysis). In figure 7, the values provided by the model inversion on optical data are plotted against the measured values: the high correlation shows that the procedure is indeed successful in retrieving the water and chlorophyll content from optical data. This experiment also allowed to validate the interpretation of the structural parameter N (number of stacked elementary plates): the value of N was found to be close to 2 for a fresh soybean leaf, and close to 1 for a yellowing maize leaf; the internal structure of these two leaves can indeed be roughly described as being made of 2 and 1 layers respectively.

A similar work was conducted previously by Allen et al. (1970). They compared the performance of three models (the generalised plate model, the K-M model and the Melamed theory in reproducing the reflectance spectra of a set of 200 mature cotton leaves. All three models were found to perform well, and it was shown that they could yield a water equivalent thickness close to the leaf physical thickness. Note: the Melamed theory is based on the reflectance of a pile of identical spheres characterised by optical constants n and k; the interaction of light with the spheres is based on Fresnel and Lambert laws (Melamed, 1963).

4.4. APPLICATION OF THE MODELS TO REMOTE SENSING

The spectrum measured by a remote sensor over a canopy is not directly interpretable in terms of the leaf reflectance. When studying spatial data, a first necessary step is to deduce the ground reflectance from the radiance signal by performing atmospheric corrections. This ground reflectance has then to be unfolded to take into account the respective contributions of vegetation and soil. This problem can be approached by the radiative transfer modelling (this subject is developed in another lecture in this course).

So far, the canopy models essentially aim at explaining the bi-directional reflectance function with the goal of retrieving canopy structural parameters; the effect of the canopy on the spectral signature has been studied to a much lesser extent. The possibility of inverting a coupled leaf/canopy reflectance model has been recently tested both on synthetic and real spectra (Jacquemoud, 1992; 1993; Jacquemoud and Baret, 1992): in this study, the leaf reflectance and transmittance were simulated by the PROSPECT model, and were directly introduced in the SAIL model describing the canopy (Verhoef, 1984); the soil spectrum was assumed to be known. The leaf is described by a chlorophyll concentration (C_{ab}), an equivalent water thickness (C_w), and the structural parameter N; the canopy by the leaf area index (LAI) and the mean leaf inclination angle (θ_l). The sensitivity studies showed:

- that the canopy reflectance was almost insensitive to the leaf structural parameter N.
- that the effects of LAI and θ_l were essentially unseparable by using the spectral information.

The inversion on real spectra was therefore conducted with a restricted model in which the parameter *N* was fixed to a reasonable value, and the leaf inclination angle to the mean measured value. A data set obtained on sugar beet plots, and containing reflectance spectra measured with a GER IRIS field spectroradiometer together with measurements of the *LAI*, chlorophyll concentration, and equivalent water thickness was used (Malthus, 1989). The variability between the plots was ensured by different types of soil, and by inducing chlorosis in some plots. The results are illustrated in figure 8. It can be seen that the *LAI* and chlorophyll concentration are correctly predicted, while the equivalent water thickness shows a larger dispersion, probably due to the variability of the soil moisture (This subject is developed in another lecture in this course).

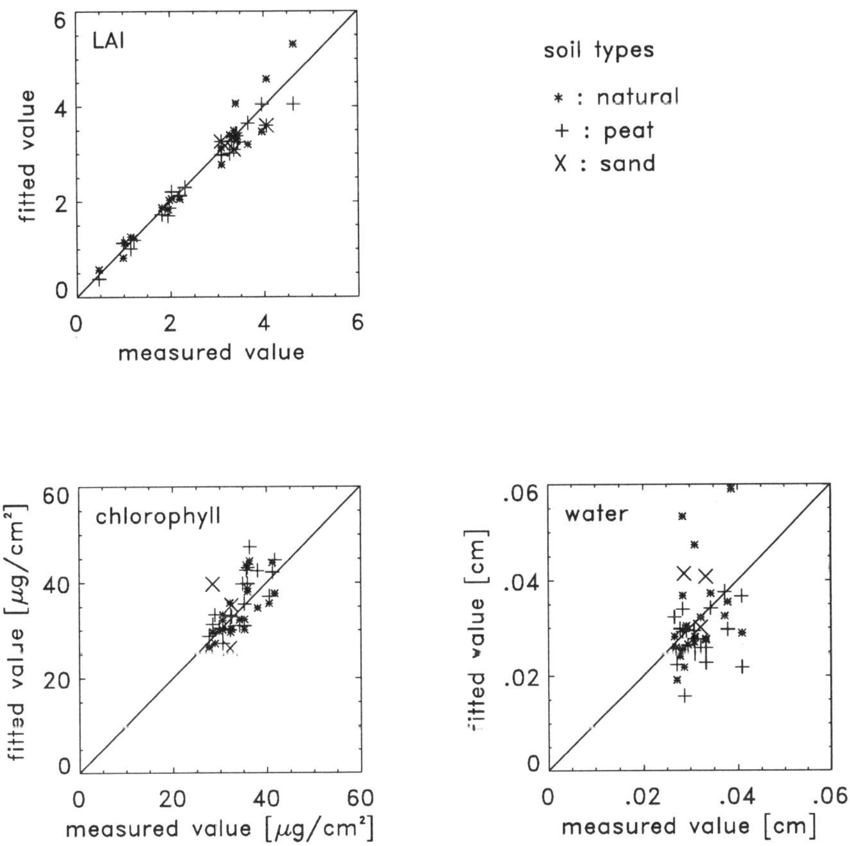

FIGURE 8. Comparison between the measured and fitted chlorophyll a+b concentration, water equivalent thickness and Leaf Area Index for beet plots with different soil types (from Jacquemoud and Baret, 1992).

This experiment demonstrated that the results obtained by inversion of models on leaf spectra are at least partially transferable to the canopy level.

5. Conclusion

This paper has reviewed the main factors controlling the leaf optical properties, and most of the models used to mathematically describe the variability of the reflectance and transmittance spectra. The results obtained at leaf level are now sufficiently robust to undertake the application of this approach at canopy level.

For this purpose a further validation of canopy models, regarding in particular the spectral effects, is necessary. Further work has to be done on the inversion procedures of the models in order to adapt them to the constraints of remote sensing data such as a lower spectral resolution, the large number of spectra to be processed, and the fact that some spectral ranges are highly disturbed by atmospheric effects.

Further work is also needed at leaf and canopy levels in order to include the biochemical constituents in the description, and to better assess the practical use of the information retrieved through the modelling approach.

6. References

Allen, W.A. and Richardson, A.J. (1968) 'Interaction of light with a plant canopy', *J. Opt. Soc. Am.*, 58 (8), 1023-1028.

Allen, W.A., Gausman, H.W., Richardson, A.J. and Thomas, J.R. (1969) 'Interaction of isotropic light with a compact leaf', *J. Opt. Soc. Am.*, 59 (10), 1376-1379.

Allen, W.A., Gayle, T.V., Richardson, A.J. (1970) 'Plant-canopy irradiance specified by the Duntley equations', *J. Opt. Soc. Am.*, 60 (3), 372-376.

Allen, W.A., Gausman, H.W., Richardson, A.J., Cardenas R. (1971) 'Water and air changes in grapefruit, corn and cotton leaves with maturation', *Agron. J.*, 63, 392-394.

Allen, W.A., Gausman, H.W. and Richardson, A.J. (1973) 'Willstätter-Stoll theory of leaf reflectance evaluation by ray tracing', *Appl. Opt.*, 12 (10), 2448-2453.

Baret, F., Jacquemoud, S., Guyot, G. and Leprieur, C. (1992) 'Modelled analysis of the biophysical nature of spectral shifts and comparison with information content of broad bands', *Remote Sens. Environ.*, 41, 133-142.

Belanger, B. J. (1990) 'A seasonal perspective of several leaf developmental characteristics as related to the red edge of plant leaf reflectance', PhD thesis, York University, Ontario (Canada).

Bone, R.A., Lee, D.W. and Norman, J.M. (1985) 'Epidermal cells functioning as lenses in leaves of tropical rain forest shade plants', *Appl. Opt.*, 24 (10), 1408-1414.

Bowman, W.D. (1989), 'The relationships between leaf water status, gas exchange, and spectral reflectance in cotton leaves', *Remote Sens. Environ.*, 30,249-255.

Boyer, M., Miller, J., Belanger, M., Hare, E. and Wu, J. (1988) 'Senescence and spectral reflectance in leaves in Northern Pin Oak (Quercus palustris Muenchh.)', *Remote Sens. Environ.*, 25, 71-87.

Brakke, T.W., Smith, J.A. and Harnden, J.M. (1989) 'Bi-directional scattering of light from tree leaves', *Remote Sens. Environ.*, 29, 175-183.

Breece, H.T. and Holmes, R.A. (1971) 'Bi-directional scattering characteristics of healthy green soybeans and corn leaves in vivo', *Appl. Opt.*, 10 (1), 119-127.

Camefort, H. (1972) 'Morphologie des végétaux vasculaires (Cytologie, anatomie, adaptation)', Doin, Paris.

Camefort, H. and Boué, H. (1969) 'Reproduction et biologie des principaux groupes végétaux (Les Cormophytes ou Archégoniates)', Doin, Paris.

Chappelle, E., Kim, E.S. and McMurtrey III, J.E. (1992) 'Ratio analysis of reflectance spectra (RARS): an algorithm for the remote estimation of the concentrations of chlorophyll A, chlorophyll B, and carotenoids in soybean leaves', *Remote Sens. Environ.*, 39, 239-247.

Curcio, J.A. and Petty, C.C. (1951) 'The near infrared absorption spectrum of liquid water', *J. Opt. Soc. Am.*, 41 (5), 302-304.

Curran, P.J. (1989) 'Remote sensing of foliar chemistry', *Remote Sens. Environ.*, 30, 271-278.

Curran, P.J., Dungan, J.L., Macler, B.A., Plummer, S.E. and Peterson, D.L. (1992) 'Reflectance spectroscopy of fresh whole leaves for the estimation of chemical concentration', *Remote Sens. Environ.*, 39, 153-166.

Elvidge, C.D. (1990) 'Visible and near infrared reflectance characteristics of dry plant materials', *Int. J. Remote Sens.*,11 (10), 1775-1795.

Gates, D.M., Keegan, H.J., Schleter, H.R. and Weidner, V.R. (1965) 'Spectral properties of plants', *Appl. Opt.*, 4 (1),11-20.

Gausman, H.W., Allen, W.A., Cardenas, R. and Richardson, A.J. (1970) 'Relation of light reflectance to histological and physical evaluation of cotton leaf', *Appl. Opt.*, 9 (3), 545-552.

Gausman, H.W., Allen, W.A. and Escobar, D.E. (1974) 'Refractive index of plant cell walls', *Appl. Opt.*, 13 (1), 109-111.

Goetz, A.F.H., Gao, B.C., Wessman, C.A. and Bowman, W.D. (1990) 'Estimation of biochemical constituents from fresh, green leaves by spectrum matching techniques', in Proc. Int. Geosci. and Remote Sens. Symp. (IGARSS'90), Washington DC, 20-24 May 1990, 971-974.

Goetz, A.F.H. (1992) 'Imaging spectrometry for earth remote sensing', *in* F. Toselli and J. Bodechtel (eds.) *Imaging spectroscopy: fundamentals and prospective applications*, ECS, EEC, EAEC, Brussels and Luxembourg, 1-19.

Grant, L. (1987) 'Diffuse and specular characteristics of leaf reflectance', *Remote Sens. Environ.*, 22, 309-322.

Grant, L., Daughtry, C.S.T. and Vanderbilt, V.C. (1987) 'Variations in the polarised leaf reflectance of sorghum bicolor', *Remote Sens. Environ.*, 21, 333-339.

Grant, L., Daughtry, C.S.T. and Vanderbilt, V.C. (1992) 'Polarised and specular reflectance variation with leaf surface features', *Physiol. Plant.* (submitted).

Guyot, G. and Baret, F. (1989) 'La haute résolution spectrale. Déterminisme des déformations spectrales entre le rouge et le proche infrarouge', *in* Bernier, Bonn and Gagnon (eds.)*Télédétection et Gestion des Ressources*, Vol VI, 197-209.

Horler, D.N.H., Dockray, M. and Barber, J. (1983) 'The red edge of plant leaf reflectance', *Int. J. Remote Sens.*, 4 (2), 273-288.

Hunt, E.R., Rock, B.N. and Nobel, P.S. (1987) 'Measurement of leaf relative water content by infrared reflectance', *Remote Sens. Environ.*, 22, 429-435.

Hunt, E.R. and Rock, B.N. (1989) 'Detection of changes in leaf water content using near and middle-infrared reflectances', *Remote Sens. Environ.*, 30, 43-54.

Jacquemoud, S. and Baret, F. (1990) 'PROSPECT: a model of leaf optical properties spectra', *Remote sens. Environ.*, 34, 75-91.

Jacquemoud, S. (1992) 'Utilisation de la haute résolution spectrale pour l'étude des couverts végétaux: développement d'un modèle de réflectance spectrale', Thèse de doctorat de l'université Paris 7 (INRA/CNES).

Jacquemoud, S. and Baret, F. (1992) 'Estimating vegetation biophysical parameters by inversion of a reflectance model on high spectral resolution data', in Proc. Coll. "Structure du couvert végétal et climat lumineux: méthodes de caracterisation et applications", Saumane (France), 23-27 September 1991 (in press).

Jacquemoud, S. (1993) 'Inversion of the PROSPECT+SAIL canopy reflectance model from AVIRIS equivalent spectra. 1. Theoretical study', *Remote Sens. Environ.* (under press).

Knipling, E.B. (1970) 'Physical and physiological basis for the reflectance of visible and near infrared radiation from vegetation', *Remote Sens. Environ.*, 1, 155-159.

Kumar, R. and Silva, L. (1973) 'Light ray tracing through a leaf cross section', *Appl. Opt.*, 12 (12), 2950-2954.

Lichtenthaler, H.K. (1987) 'Chlorophylls and carotenoids: pigments of photosynthetic biomembranes', *Methods Enzymol.*, 148, 350-382.

Ma, Q., Ishimaru, A., Phu, P. and Kuga, Y. (1990) 'Transmission, reflection, and depolarisation of an optical wave for a single leaf', *IEEE Trans. Geosci. Remote Sens.*, 28 (5), 865-872.

Malthus, T.J., (1989) 'Anglo-French collaborative reflectance experiment, July 1989', Brooms Barn Experimental Station internal report

Maracci, G., Schmuck, G., Hosgood, B. and Andreoli, G. (1991) 'Interpretation of reflectance spectra by plant physiological parameters', in Int. Geosci. and Remote Sens. Symp. (IGARSS'91), Espoo (Finland), 3-6 June 1991, 2303-2306.

Martin, G., Josserand, S.A., Bornman, J.F. and Volgemann, J. (1989) 'Epidermal focusing and the light microenvironment within leaves of Medicago sativa', *Physiol. Plant.*, 76, 485-492.

Melamed, N.T. (1963) 'Optical properties of powders. Part I. Optical absorption coefficients and the absolute value of the diffuse reflectance. Part II. properties of luminescent powders', *J. Appl. Phys.*, 34, 560-570.

Palmer, K.F. and Williams, D. (1974) 'Optical properties of water in the near infrared', *J. Opt. Soc. Am.*, 64 (8), 1107-1110.

Peterson, D.L., Aber, J.D., Matson, P.A., Card, D.H., Swanberg, N., Wessman, C. and Spanner, M. (1988) 'Remote Sensing of Forest canopy and leaf biochemical contents', *Remote Sens. Environ.*, 24, 85-108.

Peterson, D.L. (1992) 'Report on the workshop Remote sensing of plant biochemical content: theoretical and empirical studies', Marshall (CA), 18-20 March 1991.

Ripple, W.J. (1986), 'Spectral reflectance relationships to leaf water stress', Photogramm. Eng. Remote Sens., 52 (10), 1669-1675.

Rock, B.N., Hoshizaki, T. and Miller, J.R. (1983) 'Comparison of in situ and airborne spectral measurements of the blue shift associated with forest decline', *Remote Sens. Environ.*, 24, 109-127.

Sanger, J.E. (1971) 'Quantitative investigation of leaf pigments from their inception in buds through autumn coloration to decomposition in falling leaves', *Ecology*, 52 (6), 1075-1089.

Schanda, E. (1986) 'Physical fundamentals of remote sensing', Springer-Verlag, Berlin.

Sestak, Z. (1985) 'Chlorophylls and carotenoids during leaf ontogeny' (Z. Sestak, ed), Academia Praha.

Sinclair, T.R., Hoffer, R.M. and Schreiber, M.M. (1971) 'Reflectance and internal structure of leaves from several crops during a growing season', *Agron. J.*, 63, 864-867.

Sinclair, T.R., Schreiber, M.M. and Hoffer, R.M. (1973) 'Diffuse reflectance hypothesis for the pathway of solar radiation through leaves', *Agron. J.*, 65, 276-283.

Thomas, J.R., Namken, L.N., Oerther,G.F. and Brown, R.G. (1971) 'Estimating leaf water content by reflectance measurements', *Agron. J.*, 63, 845-847.

Thomas, J.R. and Gausman, H.W. (1977) 'Leaf reflectance versus leaf chlorophyll and carotenoids concentration for eight crops', *Agron. J.*, 63, 845-847.

Tucker, C.J. and Garratt, M.W. (1977) 'Leaf optical properties as a stochastic process', *Appl. Opt.*, 16 (3), 635-642.

Tucker, C.J. (1980) 'Remote sensing of leaf water content in the near infrared', *Remote Sens. Environ.*, 10, 23-32.

Ustin, S.L., Wessman, C.A., Curtiss, B., Kasischke, F., Way, J. and Vanderbilt, V.C. (1991) 'Opportunities for using the EOS imaging spectrometers and synthetic aperture radar in ecological models', *Ecology*, 72 (6), 1934-46.

Verhoef, W. (1984) 'Light scattering by leaf layers with application to canopy reflectance modelling: the SAIL model', *Remote Sens. Environ.*, 16, 125-141.

Vogelmann, T.C. and Björn, L.O. (1986) 'Plants as light traps', *Physiol. Plant.*, 68, 704-708.

Vogelmann, T.C., Bornman, F.J. and Josserand, S. (1989) 'Photosynthesis gradients and spectral regime within leaves of Medicago sativa', *Phil. Trans. R. Soc. Lond. B*, 323, 411-421.

Walter-Shea, E.A., Norman, J.M. and Blad, B.L. (1989) 'Leaf bi-directional reflectance and transmittance in corn and soybean', *Remote Sens. Environ.*, 29, 161-174.

Wessman, C.A., Aber, J.D., Peterson D.L. and Melillo, J.M. (1988) 'Foliar analysis using near infrared reflectance spectroscopy', *Can. J. For. Res.*, 18, 6-11.

Wessman, C.A., Aber J.D. and Peterson, D.L. (1989) 'An evaluation of imaging spectroscopy for estimating forest canopy chemistry', Int. J. Remote Sens., 10 (8), 1293-1316.

Willstätter, R. and Stoll, A. (1918) 'Untersuchungen uber die Assimilation der Kohlensäure', Springer, Berlin.

Woolley, J.T. (1975) 'Refractive index of soybean leaf cell walls', *Plant Physiol.*, 55, 172-174.

Yamada, N. and Fujimura, S. (1988) 'A mathematical model of reflectance and transmittance of plant leaves as a function of chlorophyll pigment content', in Int. Geosci. and Remote Sens. Symp. (IGARSS'88), Edinburgh (Scotland), 13-16 Sept 1988, 833-834.

Yamada, N. and Fujimura, S. (1992) 'Non destructive measurement of chlorophyll pigment content in plant leaves from three-color reflectance and transmittance', *Appl. Opt.* (submitted).

IMAGING SPECTROMETRY IN AGRICULTURE - PLANT VITALITY AND YIELD INDICATORS

JAN G.P.W. CLEVERS
Wageningen Agricultural University
Dept. Landsurveying and Remote Sensing
P.O. Box 339
6700 AH Wageningen
The Netherlands

ABSTRACT. For monitoring agricultural crop production, growth of crops has to be studied, e.g. by using crop growth models. Estimates of crop growth often are inaccurate for non-optimal growing conditions. Remote sensing can provide information on the actual status (e.g. its vitality) of agricultural crops. This information can be used to initialize, calibrate or update crop growth models, and it can yield parameter estimates to be used as direct input into growth models: (1) leaf area index (LAI), (2) leaf angle distribution (LAD) and (3) leaf colour (optical properties in the PAR region). LAI and LAD determine the amount of light interception. Leaf (or crop) colour influences the fraction of absorbed photosynthetically active radiation (APAR) and the maximum (potential) rate of photosynthesis of the leaves. A framework is described for integrating optical remote sensing data from various sources in order to estimate the mentioned parameters. Emphasis is on the importance of the red edge index as a measure for plant vitality. Imaging spectrometry data are needed for an accurate estimation of this red edge index.

The above concepts for crop growth estimation were elucidated and illustrated with a case study for sugar beet using groundbased and airborne data obtained during the MAC Europe 1991 campaign. A simple reflectance model was used for estimating LAI. Quantitative information concerning LAD was obtained by measurements at two viewing angles. The red edge index was used for estimating the leaf optical properties. Finally, a crop growth model (SUCROS) was calibrated on time-series of optical reflectance measurements to improve the estimation of beet yield.

1. Introduction

Remote sensing (RS) techniques can provide information about agricultural crops over a large area, quantitatively and non-destructively. A lot of research has been devoted to land cover classification and acreage estimation with considerable success. Within the present chapter crop growth monitoring and crop yield prediction are focused at. In order to obtain a reliable yield prediction, growth and production of agricultural crops have to be modelled by means of, e.g., crop growth models taking account of actual growing conditions and plant vitality. Emphasis is put on the possible role of optical RS, particularly imaging spectroscopy.

J. Hill and J. Mégier (eds.), Imaging Spectrometry - a Tool for Environmental Observations, 193–219.

1.1 CROP GROWTH MODELS

Crop growth models describe the relationship between physiological processes in plants and environmental factors such as solar irradiation, temperature and water and nutrient availability (de Wit, 1965; Penning de Vries & van Laar, 1982; Spitters et al., 1989). These models compute the daily growth and development rate of a crop, simulating the dry matter production from emergence till maturity. Finally, a simulation of yield at harvest time is obtained. The basis for the calculations of dry matter production is the rate of gross CO_2 assimilation of the canopy.

The main driving force for crop growth in these models is absorbed solar radiation, and a lot of emphasis is given to the modelling of the solar radiation budget in the canopy. Incoming photosynthetically active radiation (PAR $\approx$ 400-700 nm) is first partly reflected by the top layer of the canopy. The direct reflectance of the canopy is a function of solar elevation, leaf area index (LAI), leaf angle distribution and optical properties of the leaves. The complementary fraction is potentially available for absorption by the canopy. Subsequently, the absorptance by the canopy is a function of LAI, scattering coefficient (which may be derived from the direct reflectance) and extinction coefficient. The extinction coefficient is a function of solar elevation, leaf angle distribution and scattering coefficient. The product of the amount of incoming photosynthetically active radiation (PAR) and the absorptance yields the amount of absorbed photosynthetically active radiation (APAR). The rate of CO_2 assimilation (photosynthesis) is calculated from the APAR and the photosynthesis-light response of individual leaves. The maximum rate of photosynthesis at light saturation is highly correlated to the leaf nitrogen content. The assimilated CO_2 is then reduced to carbohydrates which can be used by the plant for growth.

Because of this detailed modelling of the solar radiation budget, crop growth models are especially suitable for the linkage with optical RS through the use of optical reflectance models.

1.2 OPTICAL REMOTE SENSING

Crop growth models as described above were developed to formalize and synthesize knowledge on the processes that govern crop growth. When applied to operational uses such as yield estimation, these models often appear to fail when growing conditions are non-optimal (caused by stresses, e.g., fertilizer deficiency, pest and disease incidence, severe drought, frost damage). Therefore, for yield estimation, it is necessary to 'check' modelling results with some sort of information on the actual status of the crop throughout the growing season (Bouman, 1991). Optical RS can provide such information. From section 1.1 it may be concluded that there are three 'key-factors' useful in crop growth models which may be derived from optical RS data: (a) LAI, (b) leaf angle distribution and (c) leaf optical properties in the PAR region (determining the scattering coefficient). This is illustrated in figure 1.

1.2.1 Leaf area index (LAI). The LAI during the growing season is an important state variable in crop growth modelling. Moreover, the LAI is a major factor determining crop reflectance and is often used in crop reflectance modelling (e.g. Suits, 1972; Bunnik, 1978; Verhoef, 1984). The estimation of LAI from RS measurements has received much attention. Much research has been aimed at determining combinations of reflectances, so-called Vegetation Indices (VIs), to correct for the effect of disturbing factors on the relationship between crop reflectance and crop characteristics such as LAI (Richardson and Wiegand, 1977; Tucker, 1979; Clevers, 1988, 1989; Bouman, 1992a). A sensitivity analysis revealed that the main parameter influencing the relationship between many VIs and green LAI is the leaf angle distribution (e.g., Clevers and Verhoef, 1990; Clevers, 1992).

1.2.2 Leaf angle distribution (LAD). LAD (leaf angle distribution) affects the process of crop growth because it has an effect on the interception of APAR by the canopy (e.g., Clevers et al., 1992). Moreover, as seen before, it is the main parameter influencing the relationship between VI and LAI. With optical RS techniques it has been more difficult to obtain quantitative information on LAD than on LAI. A solution may be found by performing measurements at different viewing angles. Goel & Deering (1985) have shown that measurements at two viewing angles for fixed solar zenith and view azimuth angles are enough to allow estimation of LAI and the LAD by the near-infrared reflectance.

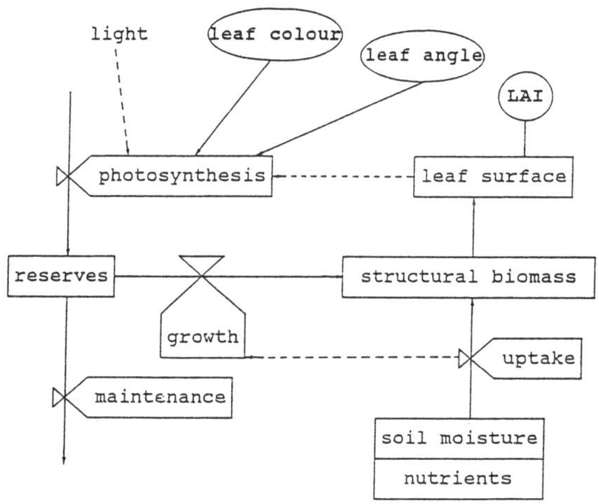

FIGURE 1. Possible links between optical remote sensing information and a crop growth model.

1.2.3 Leaf optical properties in the PAR region. Leaf optical properties (leaf colour) are important in the process of crop growth because: (1) they influence the fraction of absorbed PAR, and (2) they can be indicative for the nitrogen status (or chlorophyll content) of leaves which affects the maximum rate of photosynthesis. Leaf optical properties in the PAR region may be ascertained by spectral measurements in the visible region (VIS) of the electromagnetic (EM) spectrum. However, at low soil cover the measured signal will be confounded by soil influence. At complete coverage spectral measurements in the VIS offer information only on leaf colour. However, since the signal in VIS at complete coverage is relatively low, it may be heavily confounded by atmospheric effects for which must be corrected. Another measure of chlorophyll content may be offered by the so-called red edge index (blue shift): a decrease in leaf chlorophyll content results into a shift of the red edge towards the blue (cf. section 1.3). A decreased leaf chlorophyll content may be the result of a decreased vitality caused by abnormal or non-optimal environmental conditions. The possible role of imaging spectroscopy in detecting plant vitality will be analysed first in the next section.

1.3 IMAGING SPECTROSCOPY OF VEGETATION

The interaction of electromagnetic radiation with plant leaves is determined by their chemical and physical properties.

In the visible region (VIS) from 400 to 700 nm, various pigments such as chlorophyll, xanthophyll and carotene influence this interaction. They absorb energy in this region to a large extent and use it for the displacement of electrons by which the synthesis of carbohydrates proceeds with atmospheric CO_2 and absorbed groundwater.

In the near-infrared (NIR) region from 700 to 1300 nm, the interaction is mainly determined by the absence of absorption by pigments. Approximately 50% of the NIR energy is reflected by the leaf. Reflection takes place in the leaf at the transition of air and cellulose cell walls.

In the shortwave infrared (SWIR) region from about 1300 to 2500 nm, a lot of energy is absorbed by water in the cells. There exist strong absorption peaks of water at about 1400 and 1900 nm. Weak absorption bands of water occur at about 960 and 1100 nm.

The above-mentioned features are apparent when measuring and analysing vegetation reflectance with rather broad spectral bands (20 - 50 nm). Laboratory spectral measurements using spectrometers showed that specific absorption features of individual dried, ground leaves may be found when the spectral resolution is high. In this way, in addition to the above-mentioned main absorption features, a large number of minor absorption features were found (Curran, 1989). These minor features are correlated to concentrations of leaf organic compounds, such as cellulose, lignin, protein, sugar and starch. Absorption is most pronounced below 400 nm and above 2400 nm. The absorption features of these organic compounds are quite weak in the range 400-2400 nm.

These specific absorption features may also be found when moving such a spectrometer into an aeroplane (or even satellite) and using it as an RS technique (Goetz, 1991). For the five major absorption features (caused by chlorophyll and leaf water) this is quite well possible. However, up to now the remote sensing of foliar chemical concentrations (other than chlorophyll and water) has not been very successful. The presence of water in living leaf tissue almost completely masks these biochemical absorption features (Vane & Goetz, 1988). A number of airborne spectrometers have been developed, operating in the 400-2400 nm spectral range. These instruments do not operate in the region where the absorption features of leaf chemicals are most pronounced: below 400 nm atmospheric influence disturbs the remote recording of such features; above 2400 nm not enough solar radiation reaches the earth's surface to allow recording from a remote platform in narrow spectral bands.

Vegetation response to stress varies with both the type and the degree of stress. On the one hand, stress may cause biochemical changes at the cellular and leaf level, which have an influence, e.g., on pigment systems and the canopy moisture content. On the other hand, stress may cause changes in canopy structure, coverage or biomass. Essentially, alterations of leaf chemistry also may be used to detect subtle changes in the vitality of vegetation.

Up to now, however, most promising results for detecting the occurrence of plant stress (decrease in vitality) are obtained by studying the sharp rise in reflectance of green vegetation between 670 and 780 nm (Horler et al., 1983). This region is called the red edge. Both the position and the slope of the red edge change under stress conditions, resulting into a blue shift of the red edge. The position of the red edge is defined as the position of the main inflexion point of the red infrared slope. Reliable detection of the blue shift requires sampling at about 10 nm intervals or less, requiring high resolution spectral measurements.

1.4 MAIN OBJECTIVES

The overall framework, within which the present study was carried out, is the integration of optical RS data from various sources for estimating LAI, LAD and leaf colour, and the subsequent linking of this RS information with crop growth models for growth monitoring and yield prediction. Emphasis in the present chapter is on the possible role of high resolution imaging spectroscopy in addition to traditional broad-band spectral information. The presented concepts were evaluated in a case study using data from the MAC Europe 1991 campaign over the Dutch test site Flevoland.

2. Theoretical Study on the Red Edge Index

2.1 INTRODUCTION

From literature (see Clevers and Büker, 1991) it may be concluded that the red edge shift is related to the leaf chlorophyll content and to the LAI. This means that this red edge index and a vegetation index for estimating LAI (cf. section 1.2) contain complementary information. In order to use the red edge index for estimating the leaf chlorophyll content (or the leaf colour), first (or simultaneously) the LAI has to be estimated. The so-called WDVI-concept has proven to be a very useful concept for estimating the LAI of various agricultural crops under practical conditions (Uenk et al., 1992; Bouman et al., 1992a, 1992b).

2.1.1 *The WDVI Concept.* A simplified, semi-empirical reflectance model for estimating LAI of a green canopy (vegetative stage) was introduced by Clevers (1988, 1989). For estimating LAI a "corrected" (adjusted) NIR reflectance was calculated by subtracting the contribution of the soil in line of sight from the measured reflectance of the composite canopy-soil scene. This corrected NIR reflectance was ascertained as a weighted difference between the measured NIR and red reflectances (called WDVI = weighted difference vegetation index), assuming that the ratio of NIR and red reflectances of bare soil is constant, independent of soil moisture content (which assumption is valid for many soil types). Subsequently, this WDVI was used for estimating LAI according to the inversion of an exponential function.

The simplified reflectance model derived by Clevers (1988, 1989) consists of two steps. Firstly, the WDVI is calculated as:

$$WDVI = r_{ir} - C\,r_r \tag{1}$$

with r_{ir} total measured NIR reflectance
r_r total measured red reflectance

and

$$C = r_{s,ir} / r_{s,r} \tag{2}$$

with $r_{s,ir}$ NIR reflectance of the soil
$r_{s,r}$ red reflectance of the soil.

Secondly, the relation between WDVI and LAI is modelled as:

$$LAI = -1/\alpha \cdot \ln\left(1 - WDVI/WDVI_{\infty}\right) \tag{3}$$

with α as a combination of extinction and scattering coefficients describing the rate with which the function of equation (3) runs to its asymptotic value, and $WDVI_{\infty}$ as the asymptotic limiting value for the WDVI. Parameters α and $WDVI_{\infty}$ have to be estimated empirically from a training set, but they have a physical interpretation (Clevers, 1988). Bouman et al. (1992) arrived at the same formulation of the relationship between LAI and WDVI through a similar line of reasoning. They empirically found consistent parameters for various years, locations, cultivars and growing conditions for some main agricultural crops (Uenk et al., 1992).

2.1.2 *Red Edge Sensitivity Analysis*. One of the interesting features of this red edge index is that it seems to be independent of soil reflectance. Moreover, the atmosphere seems to have only a minor influence on the position of the red edge. Both the soil background and the atmospheric influence hamper the use of solely a spectral band in the VIS for estimating leaf colour (leaf chlorophyll content). In this section a sensitivity analysis is described using theoretical leaf and canopy reflectance models in order to study the influence of leaf and canopy parameters and of external parameters on the relationship between red edge index and leaf chlorophyll content. The atmospheric influence on the red edge index will be given attention in an empirical way when analysing the available imaging spectroscopy data.

2.2 COMBINATION OF LEAF REFLECTANCE MODEL (PROSPECT) WITH CROP REFLECTANCE MODEL (SAIL)

In order to perform a theoretical study towards the possibilities of imaging spectroscopy for agricultural applications, a leaf reflectance model (PROSPECT) and a canopy reflectance model (SAIL) were used. Both models were implemented on personal computer, linked to one another and adapted for relevant simulation exercises. Since it was concluded from literature that the information provided by the red edge index (only obtainable from narrow spectral bands in the red-NIR region) is most promising for agricultural applications, in addition to traditional broad-band spectral information, emphasis was put on this red edge index. Moreover, up to now no leaf reflectance model is available that takes explicitly into account main absorption features of biochemical constituents (except chlorophyll).

2.2.1 *SAIL Model*. The one-layer SAIL radiative transfer model (Verhoef, 1984) simulates canopy reflectance as a function of canopy parameters (leaf reflectance and transmittance, LAI and LAD), soil reflectance, ratio diffuse/direct irradiation and solar/view geometry (solar zenith angle, zenith view angle and sun-view azimuth angle). Recently, the SAIL model has been extended with the hot spot effect (Looyen et al., 1991). Leaf inclination distribution functions used with the SAIL model are given by Bunnik (1978) and Verhoef & Bunnik (1981). The SAIL model has been used in many studies and validated with various data sets (e.g. Goel, 1989).

2.2.2 *PROSPECT Model.* Recently, Jacquemoud & Baret (1990) developed a leaf model that simulates leaf reflectance and leaf transmittance as a function of leaf properties: the PROSPECT model. The PROSPECT model is a radiative transfer model for individual leaves. It is based on the generalized "plate model" of Allen et al. (1969, 1970), which considers a compact theoretical plant leaf (without air cavities) as a transparent plate with rough plane parallel surfaces. An actual leaf is assumed to be composed of a pile of N homogeneous compact layers separated by $N-1$ air spaces. The compact leaf ($N = 1$) has no intercellular air spaces or the intercellular air spaces of the mesophyll have been infiltrated with water. The discrete approach can be extended to a continuous one where N need not be an integer. PROSPECT allows to compute the 400-2500 nm reflectance and transmittance spectra of very different leaves using only three input variables: leaf mesophyll structure parameter N, pigment content and water content. All three are independent of the selected wavelength. The output of the PROSPECT model can be used directly as input into the SAIL model. As a result, these models can be combined into one combined model.

2.3 RED EDGE INDEX

Since the position of the red edge mostly is defined as the inflexion point of the red infrared slope, an accurate determination requires a large number of spectral measurements in very small bands in this region. For practical reasons, the inflexion point often is approximated by fitting a curve to fewer measurements. First, a polynomial function may be fitted to the data (Clevers and Büker, 1991). Secondly, a so-called inverted Gaussian fit to the red infrared slope may be applied (Bonham-Carter, 1988). Finally, Guyot and Baret (1988) applied a simple linear model to the red infrared slope. A comparison of the three methods yielded comparable results (Clevers and Büker, 1991; Büker and Clevers, 1992). Since the method of Guyot and Baret (1988) uses only four wavelength bands, calculations are performed very fast. Therefore it was chosen to use this latter method for further simulations.

Guyot and Baret (1988) used only four wavelength bands for ascertaining the position of the red edge. They used reflectance measurements at 670, 700, 740 and 780 nm. First of all, they estimated the reflectance value at the inflexion point halfway minimum (at 670 nm) and maximum (at 780 nm) reflectance (figure 2). Secondly, they applied a linear interpolation procedure between the measurements at 700 and 740 nm for estimating the wavelength corresponding to the estimated reflectance value at the inflexion point. This method, which will be called the "method of Guyot", can be described in the following way:

(1) Calculation of the reflectance at the inflexion point:

$$R_{red\ edge} = \left(R_{670} + R_{780} \right)/2 \qquad (4)$$

(2) Calculation of the red edge wavelength:

$$\lambda_{red\ edge} = 700 + 40 \cdot \left(\left(R_{red\ edge} - R_{700} \right) / \left(R_{740} - R_{700} \right) \right) \qquad (5)$$

$R_{red\ edge}$ is the estimated reflectance value at the main inflexion point. R_{670}, R_{700}, R_{740}, and R_{780} are the reflectance values at 670, 700, 740 and 780 nm, respectively.

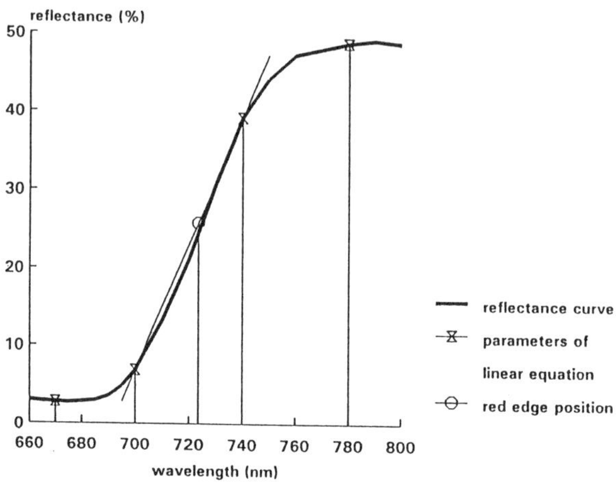

FIGURE 2. Illustration of the linear method of Guyot (Guyot and Baret, 1988).

2.4 Sensitivity Analysis

The subject of this section deals with the influence of leaf and plant properties and external factors on the position of the red edge, as calculated by means of the method of Guyot, and on the relationship between the red edge index and the leaf chlorophyll content (as a measure for leaf colour). An extensive sensitivity analysis is given by Büker and Clevers (1992).

In this study, simulations were performed for a standard crop under standard irradiation and viewing conditions unless otherwise indicated. The input parameters for this standard crop are:

chlorophyll content of 34.24 $\mu g \cdot cm^{-2}$
N parameter of 1.8320
water content of 0.0137 cm

example of a dicotyledonous plant as given by Jacquemoud and Baret (1990)

leaf area index (LAI) of 4
spherical leaf angle distribution
hot-spot size parameter of 0
soil reflectance of 20%
only direct solar irradiation
solar zenith angle of 45°
nadir viewing.

2.4.1 *Influence of LAI.* The effect of varying LAI on simulated red edge index for various leaf chlorophyll contents is presented in figure 3. The LAI has a significant influence on the red edge position. There is a distinct shift of the red edge to longer wavelength positions with increasing LAI. This shift is most pronounced at the lower LAI values. Of course, the influence of leaf chlorophyll content is also quite obvious from this figure.

2.4.2 *Influence of mesophyll structure.* The effect of changing the N parameter at a given LAI (illustrations are for an LAI of 4.0) on simulated red edge index is presented for the standard crop in figure 4. At low chlorophyll contents an increasing N value (up to N=3) caused a small shift of the red edge position to longer wavelengths whereas higher N values do not change the red edge position. At high chlorophyll contents the red edge position changed up to N values of 4. So, the influence of variations in the N parameter will be largest at low values of N. However, for a given crop type the N parameter does not exhibit a large variation (Jacquemoud, 1992).

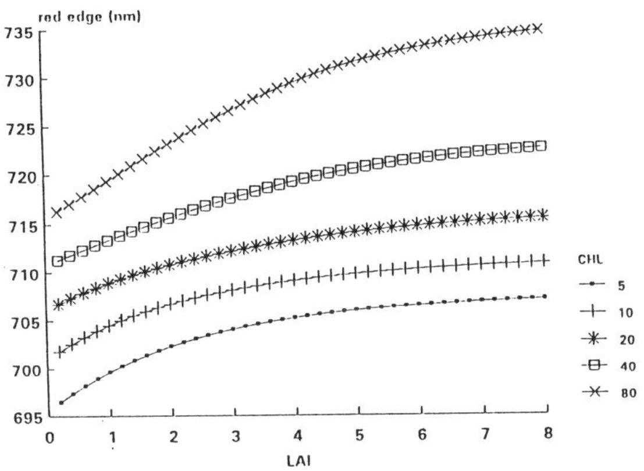

FIGURE 3. Influence of LAI on simulated red edge values for several chlorophyll contents (CHL in μg ·cm-2) of the standard crop.

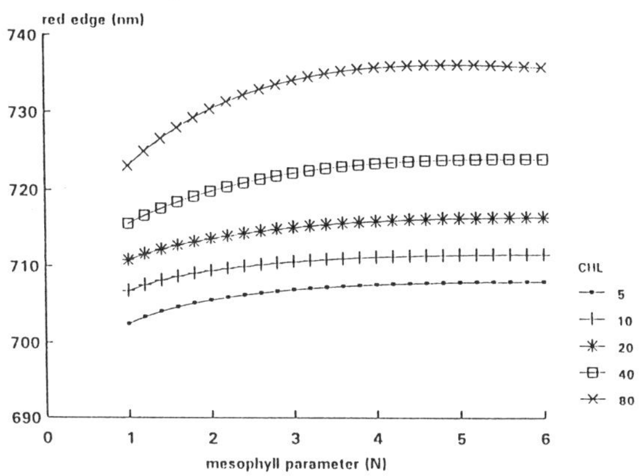

FIGURE 4. Influence of leaf mesophyll structure on simulated red edge values for several chlorophyll contents (CHL in μg ·cm-2) of the standard crop.

2.4.3 *Influence of leaf inclination angle*. The effect of varying leaf inclination angle on simulated red edge index is presented in figure 5. To perform these calculations a discrete distribution of leaf inclination angles was assumed. Since the leaf inclination angle is defined as the angle with the horizontal plane, an erectophile LAD coincides with large leaf angles. The leaf inclination angle has only a small influence on the red edge position. A small shift of the red edge index to longer wavelength positions was found with increasing leaf inclination angle from planophile to more erectophile leaves. Whereas hardly any effect at low chlorophyll contents is measured, the influence is more pronounced at high chlorophyll contents.

2.4.4 *Influence of soil reflectance*. The effect of varying soil reflectance on simulated red edge index is presented in figure 6 (for an LAI of 1.0). The soil reflectance has hardly any influence on the red edge position. With increasing soil reflectance a small shift of the red edge to longer wavelength positions was found.

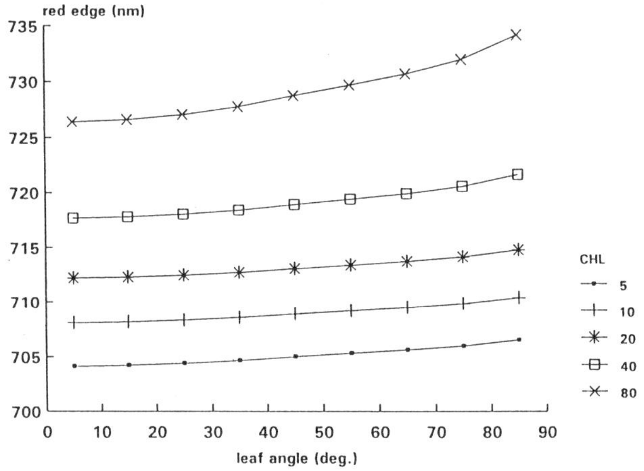

FIGURE 5. Influence of leaf inclination angle on simulated red edge values for several chlorophyll contents (CHL in $\mu g \cdot cm^{-2}$) of the standard crop.

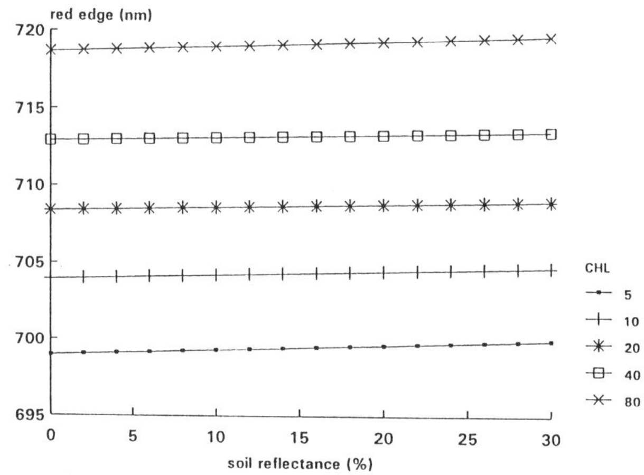

FIGURE 6. Influence of soil reflectance on simulated red edge values for several chlorophyll contents (CHL in $\mu g \cdot cm^{-2}$) of the standard crop with an LAI of 1.0.

2.4.5 *Influence of solar zenith angle*. The effect of varying solar zenith angle on simulated red edge index is presented in figure 7. The solar zenith angle has only a small influence on the red edge position. A small shift of the red edge index to shorter wavelength positions is found at solar zenith angles above 60° which increases with increasing solar zenith angle. The influence is most pronounced at high chlorophyll contents.

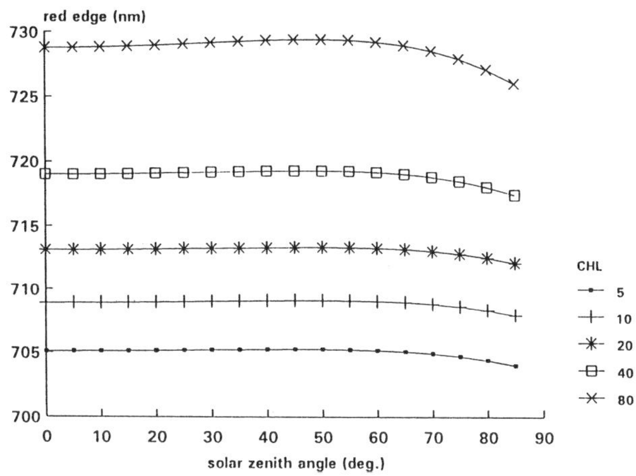

FIGURE 7. Influence of solar zenith angle on simulated red edge values for several chlorophyll contents (CHL in $\mu g \cdot cm^{-2}$) of the standard crop.

2.5 Conclusions

The LAD is the main parameter influencing the estimation of LAI by means of the WDVI. So, information on the LAD is required to obtain LAI estimates from optical RS data. As stated in section 1.1, the dual look concept will be applied for deriving information on both LAI and LAD.

The LAI and the leaf chlorophyll content, and the LAD to a lesser extent, are the main parameters determining the position of the red edge. So, after estimating LAD and LAI, the red edge index can be applied for estimating the leaf chlorophyll content. This approach will be elucidated further in section 5 and tested with data from the MAC Europe 1991 campaign (as described in section 3).

3. Experimental Data Flevoland 1991

The potential of using imaging spectroscopy for agricultural applications was tested in a case study using data of the MAC Europe 1991 campaign from the Flevoland test site in The Netherlands.

In the MAC Europe campaign, initiated by the National Aeronautics and Space Administration (NASA) and the Jet Propulsion Laboratory (JPL), both radar and optical airborne measurements were made over selected test sites during the growing season of 1991. One of the test sites was Flevoland in the Netherlands. In the optical remote sensing domain, NASA executed one overflight with the AVIRIS scanner (for system description, see Vane et al., 1984). In addition, the Dutch experimenters flew three flights with the Dutch CAESAR scanner (for system description, see Looyen et al., 1991). The Joint Research Centre financed an additional flight with the GER imaging spectrometer. The radar observations made during MAC Europe do not form part of this case study and will not be considered here (the synergism between RS data from different domains of the EM spectrum is the topic of another study).

An extensive description of the collected ground truth and of the airborne optical data during the 1991 season over Flevoland is provided by Büker et al. (1992a) and Büker et al. (1992b), respectively.

3.1 TEST SITE

The test site was located in Southern Flevoland in the Netherlands, an agricultural area with very homogeneous soils reclaimed from the lake "IJsselmeer" in 1966. The test site comprised ten different agricultural farms, 45 to 60 ha in extension. Main crops were sugar beet, potato and winter wheat. Due to hailstorms and night-frost damage of the sugar beet in April '91 some of the sugar beet fields were sown for a second time in late April resulting into quite some growth differences.

3.2 CROP PARAMETERS

Crop parameters concerning acreage, variety, planting date, emergence date, fertilization, harvest date, yield and occurring anomalities were collected for the main crops. During the growing season, additional parameters were measured in the field. The selected parameters were the estimated soil cover by the canopy, the mean crop height, row distance, plants per m^2, the soil moisture condition and comments about plant development stage.

3.3 METEOROLOGICAL DATA

Daily meteorological data are needed as input for crop growth simulation models. For the 1991 growing season these were obtained from the Royal Dutch Meteorological Service (KNMI) for the station Lelystad. Data consisted of daily minimum and maximum temperature, daily global irradiation and daily precipitation.

3.4 SPECTRA OF SINGLE LEAVES

Leaf optical properties were investigated with a LI-COR laboratory spectroradiometer at the Centre for Agrobiological Research (CABO) in Wageningen. The reflectance or transmittance signature of the upper and lower surface of several leaves was recorded continuously from 400 to 1100 nm wavelength in 5 nm steps. The instrument was calibrated with a white barium sulphate plate.

3.5 GROUND-BASED REFLECTANCE MEASUREMENTS

Field reflectance measurements were obtained during the 1991 growing season with a portable CROPSCAN radiometer. Eight narrow-band interference filters with photodiodes were oriented upwards to detect hemispherical incident radiation and a matched set of interference filters with photodiodes were oriented downwards to detect reflected radiation. Spectral bands were located at 490, 550, 670, 700, 740, 780, 870 and 1090 nm with a bandwidth of 10 nm. The sensor head of the radiometer was mounted on top of a long metal pole and positioned three metres above the ground surface. The distance to the crop was 2.5 to 1.5 m depending on the crop height. As the diameter of the field of view (FOV 28°) was half the distance between sensor and measured surface, the field of view varied from 1.23 m^2 to 0.44 m^2.

3.6 CAESAR

The CAESAR (CCD Airborne Experimental Scanner for Applications in Remote Sensing) applies linear CCD arrays as detectors. It has a modular set-up and it combines the possiblities of a high spectral resolution with a high spatial resolution. For land applications three spectral bands are available in the green, red and NIR part of the EM spectrum. One of the special options of CAESAR is the capability of acquiring data according to the so-called dual look concept. This dual look concept consists of measurements performed when looking nadir and under the oblique angle of 52°. Combining these measurements provides information on the directional reflectance properties of objects (Looyen et al., 1991). Successful overflights over the test site were carried out on July 4th, July 23rd and August 29th, 1991.

3.7 AVIRIS

The ER-2 aircraft of NASA, carrying the airborne visible-infrared imaging spectrometer (AVIRIS), performed a successful overflight over the Flevoland test site on July 5th, 1991. AVIRIS acquires 224 contiguous spectral bands from 0.41 to 2.45 μm. However, because during the recording of the Flevoland test site the last spectrometer in the SWIR range yielded only noise data, spectral information was available only in the 0.4 µm to 1.86 μm wavelength range. The ground resolution is 20 m as it is flown at 20 km altitude.

3.8 GERIS

The GERIS scanner (Geophysical Environmental Research Corporation Imaging Spectrometer) yielded spectral images with a pixel size of about 10x10 m from 3 km flight altitude. The sensor consists of three independent spectrometers measuring the spectral range from 0.43 to 2.45 μm wavelength in 63 bands. Due to battery problems the cooling device was not operational during the flight over the Flevoland test site and therefore no values are available for bands 32 to 35 (between 1.44 and 1.80 μm) which represent the second spectrometer. The first spectrometer measures the visible and NIR wavelength range (0.43 to 1.15 μm; band 1 to 31) with a spectral resolution of about 25 nm whereas the third spectrometer (1.96 to 2.45 μm; band 36 to 63) has a resolution of roughly 18 nm. It must be noted that the spectral resolution is rather coarse for an accurate determination of the red edge index.

3.9 CALIBRATION

CAESAR data were radiometrically calibrated using calibrated reference targets placed at a little airfield in the test area (Büker et al., 1992b). Since it was not possible to detect the reference targets in the AVIRIS and GERIS images due to their spatial resolution, an empirical procedure for atmospheric correction was applied. Groundbased reflectance measurements of large homogeneous agricultural fields were used as reference objects. This procedure yielded satisfactory results.

Since it was concluded from an investigation of Clevers & Büker (1991) that the red edge position is rather insensitive for atmospheric influence, the calculation of red edge positions from radiance values should yield almost the same results as those from calibrated reflectance values. Figure 8 illustrates the comparison of red edge index values based on radiances and on reflectances for AVIRIS. These results confirm the statement that the atmospheric influence plays a minor role in the calculation of the red edge position.

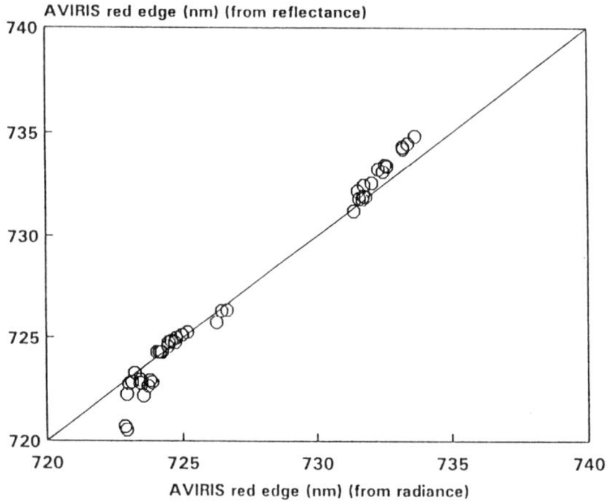

FIGURE 8. Comparison of the red edge index based on radiances and the one based on reflectances for AVIRIS data, July 5th 1991, Flevoland test site.

4. Optimal Band Selection from Imaging Spectroscopy Data

4.1 PRINCIPAL COMPONENTS ANALYSIS

In order to select the optimal set of spectral bands from a large number of bands as in imaging spectroscopy, a principal components analysis (a simplified form of factor analysis) was performed first. Factor analysis is a statistical technique used to identify a relatively small number of factors that can be used to represent relationships among sets of many interrelated variables. In the case of imaging spectroscopy of vegetation the (observed) variables are the responses in the individual spectral bands, whereas (unobservable) factors could be common sources of variation like leaf chlorophyll content, leaf structure, water content, LAI or LAD at canopy level.

In a principal components analysis (used for the factor extraction), linear combinations of the observed variables are formed. The first principal component is the combination that accounts for the largest amount of variance in the sample. The second principal component is uncorrelated with the first one and accounts for the next largest amount of variance. Successive components explain progressively smaller portions of the total sample variance, and all are uncorrelated with each other. To help us decide how many principle components (factors) we need to represent the data, it is helpful to examine the percentage of total variance explained by each.

Although the factor matrix obtained in the principal components analysis indicates the relationship between the factors and the individual variables, it is usually difficult to identify meaningful factors based on this matrix. Often the variables and factors do not appear correlated in any interpretable pattern. Most factors are correlated with many variables. Since one of the goals of factor analysis is to identify factors that are substantially meaningful, a factor rotation attempts to transform the initial factor matrix (from the principal components analysis) into one that is easier to interpret. If each factor would have high loadings for only some of the variables, this would help the interpretation. Moreover, if many variables would have a high loading on only one factor, the factors could be differentiated from each other. Most rotation procedures (e.g. the varimax procedure) try to realize such a simple structure. It should be noted that the explained variance is redistributed over the individual factors, while the total variance explained by the chosen number of factors does not change.

Finally, to identify the factors, an interpretation has to be given to groups of variables that have large loadings for the same factor.

4.2 RESULTS FROM THE ANALYSIS OF AVIRIS - DATA

A principal components analysis and factor rotation was applied to the AVIRIS data of July 5th 1991. As stated in section 3, the fourth spectrometer (from 1830 nm onwards) was not functioning. Moreover, measurements near the water absorption bands yielded only noisy data. As a results, spectral bands from 410 nm till 1350 nm and from 1480 nm till 1800 nm were used in the analysis (a total of 135 spectral bands). A selection of the spectral signatures of 101 pixels within the test site was made. All crops and bare soil were included in the data set, whereas for each object type pixels were randomly selected.

Applying the usual criteria, the principal components analysis resulted in three factors explaining 96.8% of the total variance in the selected data set. Subsequently, a factor rotation was performed. Figure 9 illustrates the relationship between the initial spectral bands and the three rotated factors. Depicted are the factor loadings which equal the correlation coefficients between the spectral bands and the respective factors. Figure 9 shows that factor 1 is highly correlated to the NIR region (from 730 nm up to about 1350 nm) and little to all other bands. This figure also shows that this factor 1 may be described as one broad band in this NIR region. Factor 2 appears to be highly correlated with the visible region (from about 500 nm up to 700 nm) and with the SWIR region (from about 1500 nm onwards). As a result, factor 2 may be described as a combination of two broad bands, one in the VIS region and one in the SWIR region (up to 1800 nm). Finally, factor 3 does not exhibit high correlations with any spectral band at all. However, it should be noticed that factor 3 shows the highest correlation with a few spectral bands around 717 nm (see figure 9). This is exactly the region of the red edge, not covered by factors 1 or 2.

It may be concluded that the principal components analysis on AVIRIS data confirms that the investigated data set can best be described by one broad spectral band in the NIR region, one broad band in the VIS region and one broad band in the SWIR region between the two main water absorption features. However, the bands in the VIS and in the SWIR appear to be highly correlated. The results also indicate that some extra information may be provided by spectral measurements around 717 nm (the red edge region), not covered by the information provided by a combination of an NIR and a VIS broad spectral band. Concerning high spectral resolution data it seems to be most promising to pay particular attention to this red edge region.

In judging the factors resulting from this analysis, factor 1 may be related to the leaf mesophyll structure and the LAI (NIR reflectance). Factor 2 may be related to the leaf chlorophyll content and the LAI (VIS reflectance). Moreover, factor 2 may be related to the leaf water content (SWIR reflectance). It is noticeable that the VIS and SWIR reflectances are highly correlated for the analysed AVIRIS data. The negative correlation between factor 1 and factor 2 may be explained by the effects of LAI on VIS and NIR reflectances. Generally, NIR reflectance increases with increasing LAI, whereas VIS reflectance decreases.

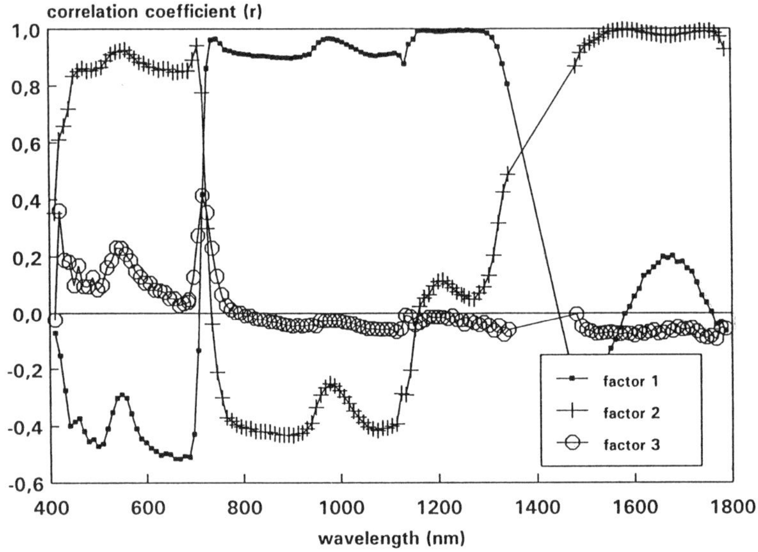

FIGURE 9. Factor loadings (correlation coefficients) for the main factors resulting from a principal components analysis and factor rotation for an agricultural data set based on spectral bands of AVIRIS spectrometers 1, 2 and 3. Flevoland test site, July 5th 1991.

5. Concept for Crop Parameter Estimation

5.1 Measurements of Leaf Optical Properties

During July and August 1991 individual leaves of sugar beet were measured in the laboratory with a LI-COR LI-1800 portable spectroradiometer. During this period leaf properties were rather constant. The measurements yielded for an NIR band (at 870 nm) an average reflectance of 46.0% and an average transmittance of 48.4%. These values were respectively 7.3% and 0.6% for a red (at 670 nm) and 15.8% and 13.8% for a green band (at 550 nm). The average scattering coefficient was 0.144 for the whole PAR region.

5.2 Estimating Leaf Angle Distribution (LAD)

Since the LAD is one of the main parameters influencing the relationship between WDVI and LAI, information on this parameter is very important. Figure 10 shows a nomogram of the simulated WDVI (SAIL model) at an oblique viewing angle (52°) plotted against the simulated WDVI at nadir viewing for several LADs and LAI values. By plotting measured WDVI values into this nomogram, an estimate of both LAI and LAD is obtained.

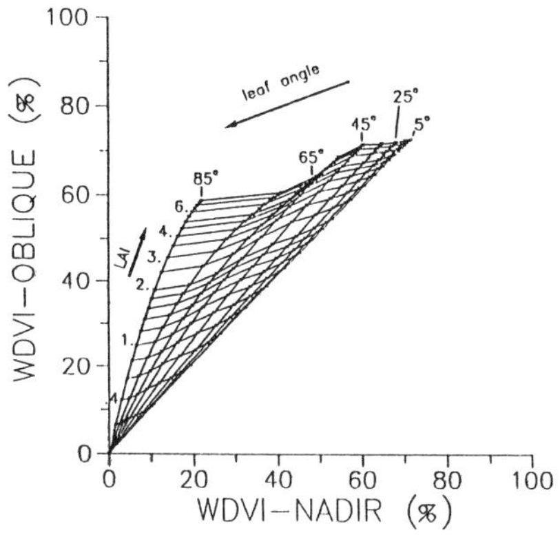

FIGURE 10. Nomogram illustrating the influence of LAI and LAD (in this graph each LAD consists of just one leaf angle) on the WDVI measured from nadir and the WDVI measured at an oblique viewing angle of 52°. Solar zenith angle of 36° and azimuth angle between plane of observation and sun of 7° as simulated with the SAIL model (simulations are for MAC Europe 1991, CAESAR overflight July 4th [Julian day 185], 13.30 GMT).

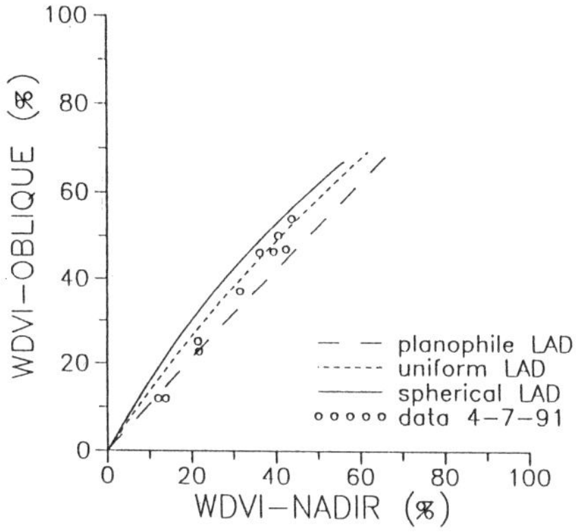

FIGURE 11. Relationship between the simulated WDVI from nadir and the simulated WDVI at an oblique viewing angle of 52° for a spherical, uniform and planophile LAD (SAIL model with a hot spot size-parameter of 0.5 for sugar beet) and measurements obtained with CAESAR, July 4th [Julian day 185], 1991 (13.30 GMT). Solar zenith angle of 36° and azimuth angle between plane of observation and sun of 7°.

Figure 11 gives the results of July 4th for the CAESAR scanner (note: CROPSCAN measurements over bare soil yielded an estimate for C in equation (1) of 1.15) together with simulated curves for a spherical, uniform and planophile LAD (LADs as defined by Verhoef & Bunnik, 1981). In this figure more realistic LADs are shown as opposed to figure 10 with LADs consisting of just one angle.

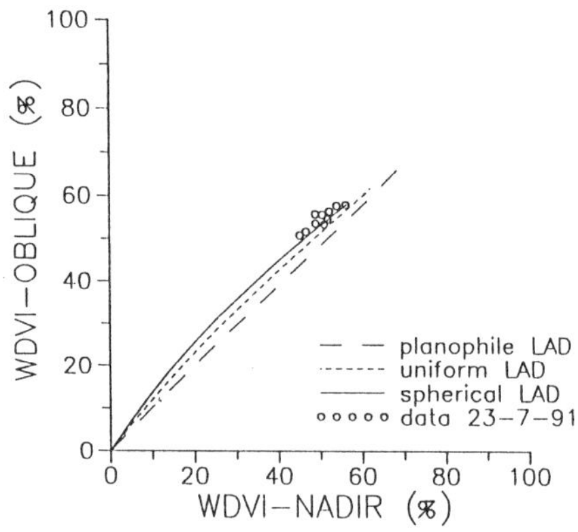

FIGURE 12. Relationship between simulated WDVI nadir and WDVI oblique (cf. figure 11) and measurements obtained with CAESAR, July 23rd [Julian day 204], 1991 (10.45 GMT). Solar zenith angle of 34° and azimuth angle between plane of observation and sun of 78°.

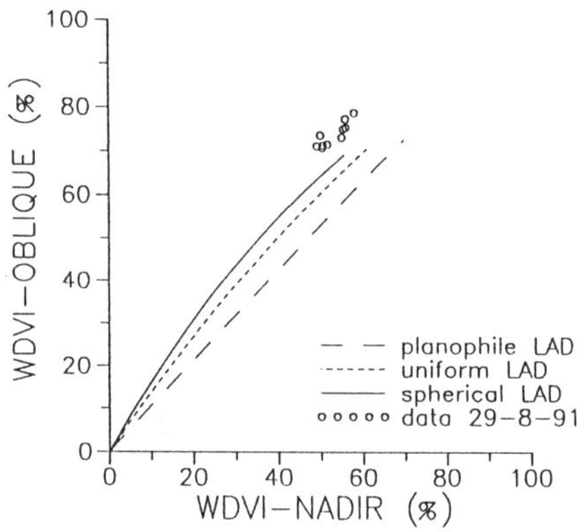

FIGURE 13. Relationship between simulated WDVI nadir and WDVI oblique (cf. figure 11) and measurements obtained with CAESAR, August 29th [Julian day 241], 1991 (12.50 GMT). Solar zenith angle of 45° and azimuth angle between plane of observation and sun of 28°.

Figures 12 and 13 present results obtained with the CAESAR scanner on July 23rd and August 29th, respectively. Concerning the data of August 29th, it must be noted that the WDVI-oblique (forward looking) may be overestimated because the NIR forward looking reflectances were extrapolations from the values of the reference panels (cf. Büker et al., 1992b). As a result, the absolute level of the WDVI-oblique values may be biased. Results for all three dates showed that sugar beet mostly matched the curve for a spherical LAD rather well, except for the beginning of the growing season (LAI<1.5) when the LAD was more planophile. This information is important for determining the relationship between WDVI and LAI. It can also be used to derive extinction coefficients that are input for crop growth models.

5.3 ESTIMATING LEAF AREA INDEX (LAI)

Using the optical leaf properties found in section 5.1 and the spherical LAD in section 5.2, the relationship between WDVI and LAI was simulated using the SAIL model. The regression of LAI on WDVI (eq. 3) yielded for a an estimate of 0.418 and for $WDVI_\infty$ an estimate of 57.5 (spherical LAD). For sugar beet Bouman et al. (1992) estimated empirically an a of 0.485 and a $WDVI_\infty$ of 48.4, whereby the WDVI was based on green reflectance instead of red reflectance. CROPSCAN measurements and SAIL simulations yielded a ratio between WDVIs based on green and red reflectances, respectively, for sugar beet of 1.16. As a result, a value of 48.4 for $WDVI_\infty$ (based on green reflectances) corresponds to a value of 56.3 for a $WDVI_\infty$ based on red reflectances. Figure 14 illustrates that the simulated relationship and the empirical relationship match rather well.

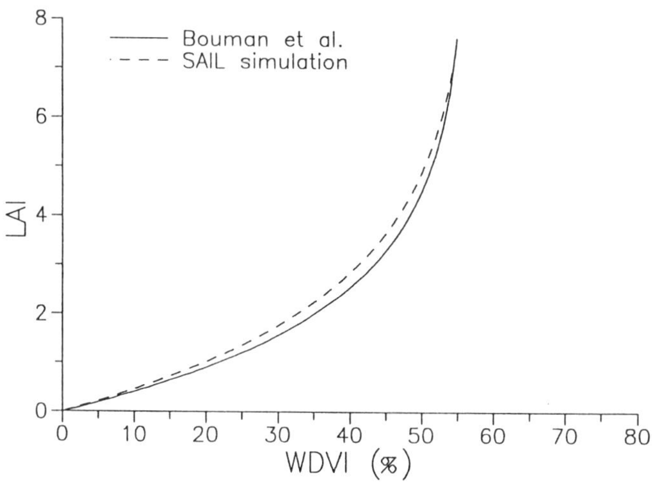

FIGURE 14. Theoretical (SAIL model) and empirical (Bouman et al., 1992) relationship between WDVI (based on NIR and red reflectances) and LAI for sugar beet.

5.4 ESTIMATING LEAF OPTICAL PROPERTIES

As stated before, in practice it will be very difficult to ascertain leaf colour unless leaves are analysed in the laboratory. A more practical measure may be offered by the red edge index. However, in section 2 it was concluded that this index is determined by both LAI and leaf colour (related to leaf chlorophyll content). Two (independent) measurements are therefore needed, one more related to LAI (like WDVI) and one more related to chlorophyll content (like red edge index). Figure 15 illustrates the simulated influence of LAI and leaf chlorophyll content on the position of the red edge and the WDVI (using a combined PROSPECT-SAIL model). As stated before, the method of Guyot & Baret (1988) for determining the position of the red edge was applied, using only four wavelength bands. Measurements of the WDVI and the position of the red edge may be combined for ascertaining the leaf chlorophyll content.

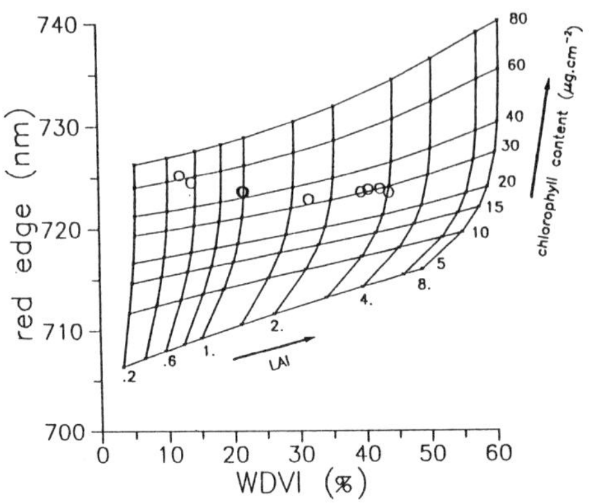

FIGURE 15. Nomogram illustrating the influence of LAI and leaf chlorophyll content on the WDVI (from nadir) and the position of the red edge as simulated with a combined PROSPECT-SAIL model (measurements are of MAC Europe 1991, CAESAR overflight July 4th [Julian day 185] (WDVI), and AVIRIS overflight July 5th (red edge)).

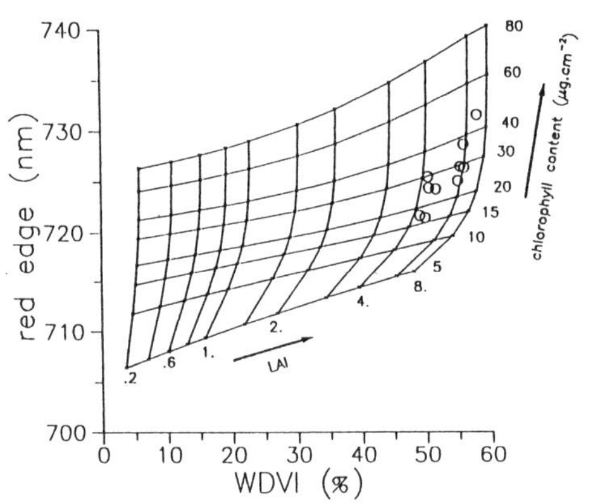

FIGURE 16. Relationship between the WDVI (from nadir) and the position of the red edge as simulated with a combined PROSPECT-SAIL model and measurements obtained with CAESAR (WDVI) and GERIS (red edge index), August 29th [Julian day 241] 1991.

By plotting both the measured WDVI (acquired with CAESAR) and the red edge values (acquired with AVIRIS) into such a nomogram for actual recording conditions, an estimate of both LAI and leaf chlorophyll content is obtained. Since AVIRIS data were calibrated accurately up to radiances and only an empirical calibration towards reflectances was applied, WDVI values of CAESAR were used. For calculating red edge values, radiances may be used instead of reflectances (cf. section 3.9, figure 8). Results for sugar beet yielded an estimated chlorophyll content of about 30 $\mu g \cdot cm^{-2}$, except for the beginning of the growing season (LAI<1.5) when the chlorophyll content was somewhat higher. By using the PROSPECT model this leaf chlorophyll content yielded an average leaf scattering coefficient in the PAR region of 0.158 (N value of 1.832). This value is comparable to the one found in section 5.1.

Figure 16 presents the results of the WDVI (acquired with CAESAR) and the red edge values (acquired with GERIS) for August 29th 1991. Results for sugar beet now yielded an estimated chlorophyll content varying between 15 and 50 $\mu g \cdot cm^{-2}$. Using PROSPECT this yielded an average leaf scattering coefficient in the PAR region varying between 0.281 and 0.099 (N value of 1.832). As already stated in section 3.8, red edge index estimates with GERIS must be used with caution since the spectral resolution is rather course in the red-NIR region.

5.5 CONCLUDING REMARKS

By using the bidirectional reflectance characteristics of a red and NIR spectral band (yielding, e.g., bidirectional WDVI values), both LAI and LAD of agricultural crops may be estimated. In practice, the observation geometry (relative to the position of the sun) is important for discrimination between different LADs. A combination of nadir viewing and viewing towards the hot spot yielded good results.

In order to acquire additional information on leaf colour (in terms of leaf chlorophyll content), the red edge index may be used. Since this red edge index is influenced by LAI and leaf chlorophyll content (and LAD to a lesser extent), application of the red edge index should be combined with use of the WDVI at two viewing angles. With regard to this red edge index determination high resolution imaging spectrometers are most important.

6. Linking Optical Remote Sensing Data with Crop Growth Models

6.1 FRAMEWORK

Two methods can be distinguished to link optical remote sensing data with crop growth models. In the first method, called 'model initialization', crop parameters are estimated from optical remote sensing and 'fed' into a growth model as input or forcing function. Mostly, crop parameters that have been used succesfully so far are measures for the fractional light interception by the canopy, namely LAI and soil cover (Steven et al., 1983; Kanemasu et al., 1984; Maas, 1988; Bouman & Goudriaan, 1989). However, parameters like LAD and leaf colour can also be used as input in more elaborate growth models.

In the second method, called 'model calibration', crop growth models are calibrated on time-series of remote sensing measurements. Maas (1988) presented a method in which crop growth model parameters were adjusted in such a way that simulated values of LAI by the growth model matched LAI values that were estimated from reflectance measurements. Bouman (1992b) developed a procedure in which remote sensing models (a.o. optical reflectance) were linked to crop growth models so that canopy reflectance was simulated together with crop growth. The growth model was then calibrated to match simulated values of canopy reflectance to measured values of reflectance. In a case study for sugar beet, the simulation of (above-ground) biomass was more accurate after calibration than before calibration on reflectance measurements. The calibration in this procedure is governed by the parameters which link the crop growth model and the remote sensing model (LAI, LAD and leaf colour).

6.2 SUCROS

In this study the used crop growth model was SUCROS (Simple and Universal CROp growth Simulator; Spitters et al., 1989). It is a mechanistic growth model describing the potential growth of a crop as a function of irradiation, air temperature and crop characteristics. The light profile within a crop canopy is computed on the basis of the LAI and the extinction coefficient. At selected times during the day and at selected depths within the canopy, photosynthesis is calculated from the photosynthesis-light response of individual leaves. Integration over the canopy layers and over time within the day gives the daily assimilation rate of the crop. Assimilated matter is used for maintenance respiration and for growth (cf. figure 1). The newly formed dry matter is partitioned to the various plant organs. An important variable that is simulated is the LAI, since the increase in leaf area contributes to next day's light interception and thus rate of assimilation.

6.3 RESULTS CALIBRATION SUCROS

The crop growth model SUCROS, extended to simulate WDVI from the growing crop (Bouman, 1992b), was run to estimate the final beet yield for ten selected farmers in the test area. Input for the model were the location parameters, weather data for the 1991 growing season, sowing date, harvest date and crop-specific model parameters. The difference between simulated yield and attained yields by the farmers is given in figure 17.

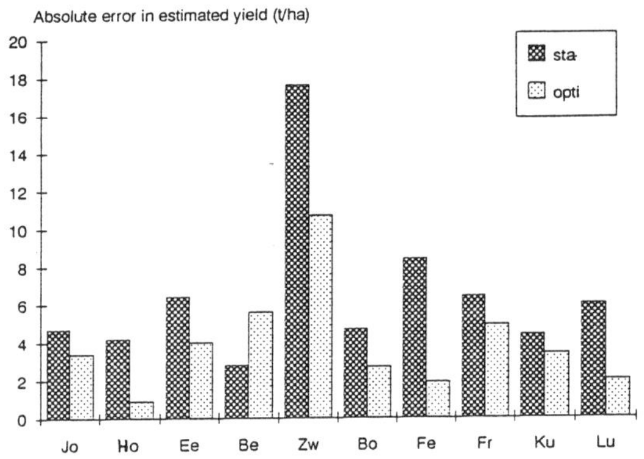

FIGURE 17. Absolute error (tons/hectare) in estimated beet yield using standard SUCROS (sta) and using SUCROS calibrated to measured time-series of WDVI (opti). The different pairs of bars relate to different farmers.

Next, SUCROS was calibrated so that simulated WDVI during the growing season matched measured WDVI (as measured with the CROPSCAN) as close as possible for all ten fields individually. Figure 18 gives the simulated beet yields after model calibration versus actually obtained yields, and the differences between simulated and actual yields are again given in figure 17. For all except one fields, the simulated yield after calibration was closer to actually obtained yields than before calibration. On the average, the simulation error of (fresh) beet yield decreased from 6.6 t/ha (8.6%) using 'standard' SUCROS, to 4.0 t/ha (5.2%) with SUCROS calibrated to time-series of WDVI.

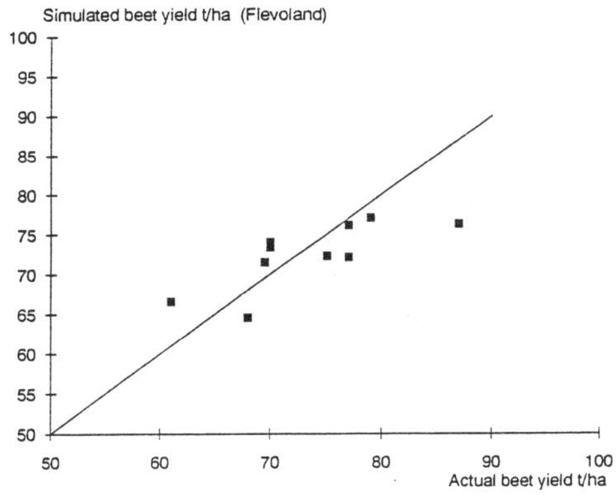

FIGURE 18. Simulated beet yield using SUCROS calibrated to measured WDVI versus actually obtained beet yields for ten farmers in 1991 in the Flevopolder test site.

7. Discussion

A framework was presented to integrate crop canopy information derived from optical remote sensing with crop growth models for the purpose of growth monitoring and yield estimation. First, crop parameters that play an important role in both the processes of crop growth and canopy reflectance were estimated from optical remote sensing data. These crop parameters were LAI, LAD and leaf colour. For each parameter, an estimation methodology was developed or taken from literature. Secondly, a crop growth model called SUCROS was extended with a canopy reflectance model to calculate remote sensing signals from the growing crop. The estimated parameter values can be used as direct input into crop growth models or they can be used for calibrating crop growth models. In the latter case use can be made of time-series of remote sensing data.

The framework was applied to data gathered during the MAC Europe 1991 campaign over the Dutch test site Flevoland. Results for sugar beet indicated the feasibility of estimating LAI, LAD and leaf colour from optical reflectance measurements. A critical point to consider is the precision and additional value of the parameter values derived from remote sensing compared to the standard values already used in the growth model. For instance, much relative benefit might be obtained from the estimation of leaf colour expressed in leaf nitrogen or chlorophyll content. Especially the modelling of leaf nitrogen status in canopies is extremely complicated (but equally important through its effect on maximum leaf photosynthesis rate) and actual information derived from optical reflectance would be valuable. However, it will take more research and dedicated experiments together with crop physiologists to investigate the potentials of optical remote sensing for the assessment of leaf (or canopy) nitrogen status.

The method of model calibration was tested on sugar beet. For nine out of ten fields, the simulated yield was better in agreement with actually obtained yields after model calibration than without model calibration. Since the calibration procedure mainly concerned the calibration of the simulated LAI, these results indicate the importance of LAI for accurate growth simulation.

8. Acknowledgements

W. Verhoef is acknowledged for providing the SAIL model and B. Bouman and J. Goudriaan are acknowledged for providing the SUCROS model. We are very grateful to S. Jacquemoud and F. Baret (INRA, Montfavet - France) for providing the PROSPECT model. NASA is acknowledged for providing the AVIRIS data in the framework of MAC Europe 1991. This chapter describes a study that was carried out in the framework of the NRSP-2 under responsibility of the Netherlands Remote Sensing Board (BCRS) and under contract no. 4530-91-11 ED ISP NL of the Joint Research Centre.

9. References

Allen, W.A., H.W. Gausman, A.J. Richardson & J.R. Thomas (1969) 'Interaction of isotropic light with a compact plant leaf', *J. Opt. Soc. Am.*, 59, 1376-1379.

Allen, W.A., H.W. Gausman and A.J. Richardson (1970) 'Mean effective optical constants of cotton leaves', *J. Opt. Soc. Am.*, 60, 542-547.

Bonham-Carter, G.F. (1988) 'Numerical procedures and computer program for fitting an inverted gaussian model to vegetation reflectance data', *Computers & Geosciences*, vol. 14 (3), 339-356.

Bouman, B.A.M. (1991) 'Linking X-band radar backscattering and optical reflectance with crop growth models', Thesis Agricultural University Wageningen, Wageningen, The Netherlands.

Bouman, B.A.M. (1992a) 'Accuracy of estimating the leaf area index from vegetation indices derived from reflectance characteristics, a simulation study', *International Journal of Remote Sensing*, 13, 3069-3084.

Bouman, B.A.M. (1992b) 'Linking physical remote sensing models with crop growth simulation models, applied for sugar beet', *International Journal of Remote Sensing*, 13, 2565-2581.

Bouman, B.A.M. and J. Goudriaan (1989) 'Estimation of crop growth from optical and microwave soil cover', *International Journal of Remote Sensing*, 10, 1843-1855.

Bouman, B.A.M., H.W.J. van Kasteren and D. Uenk (1992) 'Standard relations to estimate ground cover and LAI of agricultural crops from reflectance measurements', *Eur. J. Agronomy* (in press).

Bunnik, N.J.J. (1978) 'The multispectral reflectance of shortwave radiation by agricultural crops in relation with their morphological and optical properties', Ph.D. Thesis, Mededelingen Landbouwhogeschool Wageningen 78-1, 175 pp.

Büker, C. and J.G.P.W. Clevers (1992) 'Imaging spectroscopy for agricultural applications', Report LUW-LMK-199206, Dept. Landsurveying & Remote Sensing, Wageningen Agricultural University.

Büker, C., J.G.P.W. Clevers, H.J.C. van Leeuwen, B.A.M. Bouman, and D. Uenk (1992a) 'Optical component MAC Europe, Ground truth report, Flevoland 1991', Report LUW-LMK-199204, Dept. Landsurveying & Remote Sensing, Wageningen Agricultural University, 54 pp.

Büker, C., J.G.P.W. Clevers, and H.J.C. van Leeuwen (1992b) 'Optical component MAC Europe, Optical data report, Flevoland 1991', Report LUW-LMK-199205, Dept. Landsurveying & Remote Sensing, Wageningen Agricultural University.

Clevers, J.G.P.W. (1988) 'The derivation of a simplified reflectance model for the estimation of leaf area index', *Remote Sensing of Environment*, 25, 53-69.

Clevers, J.G.P.W. (1989) 'The application of a weighted infrared-red vegetation index for estimating leaf area index by correcting for soil moisture', *Remote Sensing of Environmen,* 29, 25-37.

Clevers, J.G.P.W. (1992) 'Modelling and synergistic use of optical and microwave remote sensing. Report 4: Influence of leaf properties on the relationship between WDVI and LAI: a sensitivity analysis with the SAIL and the PROSPECT model', BCRS report 92-14, 36 pp.

Clevers, J.G.P.W. and C. Büker (1991) 'Feasibility of the red edge index for the detection of nitrogen deficiency', *Proc. 5th Int. Coll. on Physical Measurements and Signatures in Remote Sensing*, Courchevel, France, ESA SP-319, 165-168.

Clevers, J.G.P.W. and W. Verhoef (1990) 'Modelling and synergistic use of optical and microwave remote sensing. Report 2: LAI estimation from canopy reflectance and WDVI: a sensitivity analysis with the SAIL model', BCRS report 90-39, 70 pp.

Clevers, J.G.P.W., W. Verhoef and H.J.C. van Leeuwen (1992) 'Estimating APAR by means of vegetation indices: a sensitivity analysis', *International Archives of Photogrammetry and Remote Sensing*, Vol. XXIX, Part B7, Comm. VII, XVIIth ISPRS Congress, Washington D.C., 1992, pp. 691-698.

Curran, P.J. (1989) 'Remote sensing of foliar chemistry', *Remote Sensing of Environment,* 29, 271-278.

Goel, N.S. and D.W. Deering (1985) 'Evaluation of a canopy reflectance model for LAI estimation through its inversion', *IEEE* GE-23, 674-684.

Goel, N.S. (1989) 'Inversion of canopy reflectance models for estimation of biophysical parameters from reflectance data', *in* G. Asrar (ed.) *'Theory and applications of optical remote sensing',* J. Wiley & Sons, Inc., New York. 205-251.

Goetz, A.F.H. (1991) 'Imaging spectrometry for studying Earth, air, fire and water', *EARSeL Advances in Remote Sensing*, 1, 3-15.

Guyot, G. and F. Baret (1988) 'Utilisation de la haute resolution spectrale pour suivre l'etat des couverts vegetaux', *Proc. 4th Int. Coll. on Spectral Signatures of Objects in Remote Sensing*, Aussois, France, 18-22 January 1988. ESA SP-287, 279-286.

Horler, D.N.H., M. Dockray and J. Barber (1983) 'The red edge of plant leaf reflectance', *International Journal of Remote Sensing*, 4, 273-288.

Jacquemoud, S. (1992) 'Utilisation de la haute resolution spectrale pour l'etude des couverts vegetaux: devellopement d'un modele de reflectance spectrale', Ph.D. Thesis, University of Paris VII, 92 pp.

Jacquemoud S. and F. Baret (1990) 'PROSPECT: a model of leaf optical properties spectra', *Remote Sensing of Environment*, 34, 75-91.

Kanemasu, E.T., G. Asrar and M. Fuchs (1984) 'Application of remotely sensed data in wheat growth modelling', *in* Day, W. and R.K. Atkin (ed.) *'Wheat growth modelling'*, Plenum Press, New York and London, Published in cooperation with NATO Scientific Affairs Division, 357-369.

Looyen, W.J., W. Verhoef, J.G.P.W. Clevers, J.T. Lamers and J. Boerma (1991) 'CAESAR: evaluation of the dual-look concept', BCRS report 91-10, 144 pp.

Maas, S.J. (1988) 'Use of remotely sensed information in agricultural crop growth models', *Ecological modelling,* 41, 247-268.

Penning de Vries, F.W.T. and H.H. van Laar (1982) 'Simulation of plant growth and crop production', Simulation Monographs, PUDOC, Wageningen, The Netherlands, 308 pp.

Richardson, A.J. and C.L. Wiegand (1977) 'Distinguishing vegetation from soil background information', *Photogr. Engineering and Remote Sensing*, 43, 1541-1552.

Spitters, C.J.T., H. van Keulen and D.W.G. van Kraalingen (1989) 'A simple and universal crop growth simulator: SUCROS87', *in* Rabbinge, R., S.A. Ward and H.H. van Laar (eds.) *'Simulation and systems management in crop protection'*, Simulation Monographs 32, PUDOC, Wageningen, The Netherlands, pp. 147-181.

Steven M.D., P.V. Biscoe and K.W. Jaggard (1983) 'Estimation of sugar beet productivity from reflection in the red and infrared spectral bands', *International Journal of Remote Sensing*, 2, 117-125.

Suits, G.H. (1972) 'The calculation of the directional reflectance of a vegetation canopy', *Remote Sensing of Environment*, 2, 117-125.

Tucker, C.J. (1979) 'Red and photographic infrared linear combinations for monitoring vegetation', *Remote Sensing of Environment*, 8, 127-150.

Uenk, D., B.A.M. Bouman and H.W.J. van Kasteren (1992) 'Reflectiemetingen aan landbouwgewassen, Handleiding voor het meten van gewasreflectie', Standaardlijnen voor de bepaling van bodembedekking en LAI (in Dutch). CABO-DLO report 156, 56 pp.

Vane, G., M. Chrisp, H. Enmark, S. Macenka and J. Solomon (1984) 'Airborne Visible/Infrared Imaging Spectrometer: An advanced tool for Earth remote sensing', *Proc. IGARSS '84*, SP215, 751.

Vane, G. and A.F.H. Goetz (1988) 'Terrestrial imaging spectroscopy', *Remote Sensing of Environment*, 24, 1-29.

Verhoef, W. (1984) 'Light scattering by leaf layers with application to canopy reflectance modelling: the SAIL model', *Remote Sensing of Environment*, 16, 125-141.

Verhoef, W. and N.J.J. Bunnik (1981) 'Influence of crop geometry on multispectral reflectance determined by the use of canopy reflectance models', *Proc. Int. Coll. on Signatures of Remotely Sensed Objects*, Avignon, France, 273-290.

Wit, C.T. de (1965) 'Photosynthesis of leaf canopies', Agricultural Research Report 663, PUDOC, Wageningen, The Netherlands.

MAPPING SPARSE VEGETATION CANOPIES

MILTON O. SMITH, JOHN B. ADAMS, and DON E. SABOL
Department of Geological Sciences AJ-20
University of Washington
Seattle, Washington, USA 98195

ABSTRACT. Mapping sparse vegetation communities is routinely applied using techniques such as band ratios, the normalized difference vegetation index (NDVI) and spectral mixture analysis (SMA). The uncertainty of these vegetative mapping techniques is examined using the soil spectral variability defined by the spectral reference endmembers from three Landsat Thematic Mapper images: Owens Valley, California, USA; Gran Desierto, Sonora, Mexico, and Fayyum, Egypt. We find that band ratios and NDVI are not optimized for detecting vegetation given soil spectral variability. For SMA, the detection of sparse vegetation is optimized when it is detected as a residual component. Depending on the uncertainty model utilized from two to four fold improvement in mapping sparse vegetation is possible compared to NDVI and band ratios.

1. Introduction

In a first paper (Spectral Mixture Analysis - New Strategies for the Analysis of Multispectral Data) procedures to minimize uncertainties of endmember abundances used in spectral mixture analysis (SMA) were introduced. These procedures were used to estimate effects of endmember variability and spectral contrast between endmembers on the uncertainty in abundances. In this paper, we investigate the uncertainties of some commonly used vegetation mapping techniques and explore strategies to optimize mapping sparse vegetation canopies using SMA and Landsat Thematic Mapper (TM) images. While the objectives of the first paper investigated techniques to maintain a consistent quantitative measure of vegetation, the focus here is directed toward maximizing vegetation detection.

Given todays controversy over changes in global carbon and the resulting effect on climate, it is important to understand the uncertainties of spectral estimates of vegetation independent of ground measures of cover. Although the sparse vegetation communities, typical of deserts and semiarid regions, make a small contribution to the total carbon pool, these communities perhaps are the most sensitive to changes in climate. Thus, if we are to monitor global climate change using remote sensing techniques perhaps a key element is to monitor changes of sparse vegetation communities as regional indicators of climate change. Our approach is to compare uncertainties of vegetation abundance obtained from image measurements to estimates derived from laboratory spectral measurements of endmembers.

The high spectral contrast of vegetation relative to soils has provided many successful studies measuring sparse vegetation communities using a variety of remote sensing techniques (Tucker

J. Hill and J. Mégier (eds.), Imaging Spectrometry - a Tool for Environmental Observations, 221–235.

1979; Smith et al.; 1990a). However, it has not been possible with indices, spectral matching or classification mapping techniques to estimate the uncertainty in these vegetation estimates or to validate these estimates in a regional or global context. Validation of remote sensing studies using ground truth measurements alone has been inadequate to guarantee extension of results from local studies to regional and or global scales. In remote sensing studies there is increasing concern of the effect of the background on quantifying vegetation in multispectral images (e.g. Tucker and Miller 1977; Satterwhite and Henley, 1987; Huete et al., 1985; Smith et al., 1990).

Early in the history of satellite remote sensing, vegetation has been mapped using the relative contrast in the visible wavelengths ratioed to the infrared wavelengths (e.g. ratio of Landsat TM bands 4/3). This characterization is consistent with green vegetation which contains active chlorophyll and water in the foliage. However, in the mapping of vegetation it is not always the case that the plants are alive, especially in areas with sparse cover. Many grasslands for example have a relative short active period of growth and exist in a dormant or senescent state most of the time. How do ratios and indices such as the normalized difference vegetation index (NDVI) function under conditions when most of the plant biomass is dead stems and litter? What is the reliability of these estimates in the context of spectral variability of soil?

Analytical techniques in remote sensing are for the most part deficient in accessing the validity of abundance estimates of vegetation in the context of spectral contrast to the background. For example, Smith et al. (1990) indicate separate relationships between vegetative ground cover estimates and spectral abundance estimates for the bajada and valley floor in Owens Valley. A review of mapping techniques by Ustin et al. (1988) indicates that maximum likelihood vegetative classes are associated with edaphic variation on the bajada in Owens Valley. In contrast, SMA indicated that the vegetative classes delineated by maximum likelihood classification utilized spectral differences of the bajada soils to separate the different vegetation types. Such correlations are common in multispectral data sets and often result in erroneous interpretations.

In SMA, using two bands similar to NDVI and band ratios, it is possible to uniquely separate at most three endmembers. For areas with sparse vegetation these endmembers are typically a soil, shade/shadow, and vegetation. Spectral variation in the soil and vegetation are often interpreted as changes in abundance estimates of these endmembers. Similarly, for NDVI and band ratios spectral variation in soils and vegetation is also interpreted as changes in the vegetation abundance. In selecting the three endmembers for two bands the instrumental error is folded into the abundance estimates. In addition, the soil endmember for NDVI and band ratios is fixed such that we define the soil as having no slope for Landsat TM bands 3 and 4. Vegetation abundance is maximum when the slope in bands 3 and 4 is maximum. The analogy between indices and SMA is necessary to apply uncertainties determined from the SMA framework to indices.

We divide the problem of mapping vegetation into two parts, namely, 1) detection and 2) quantification of abundance. Although in this paper we focus on detection, optimizing either objective requires knowledge of the background and its spectral contrast to vegetation. If, for example, vegetation is in the shadowed area of an image, then its detection is reduced in comparison to sunlit areas. Similarly, if our objective is to map dry grass against a soil background, the detection of dry grass will intuitively be lower than to map green grass over the same area because of the difference in spectral contrast. To the extent possible we desire to obtain images such that the contrast between spectral scene components is maximized. Thus, if we wish to map vegetation we would prefer that the vegetation was green foliage. However, we cannot completely control contrast between vegetation and natural backgrounds. Thus, the strategy of analysis must focus on an analytical framework that allows interpretation to proceed under conditions of variable confidence in abundance estimates resulting from changes in spectral contrast.

Two implicit assumptions commonly part of SMA are that 1) there is a single set of endmembers applicable to an entire image and 2) mimicry of endmember mixtures is negligible. If these two assumptions are valid, the uncertainties of vegetation abundances are as described by Sabol et al. (1992). Instrumental noise is the dominant factor affecting abundance uncertainties. Using the same notation as in the first paper, the uncertainty of the green vegetation fraction is < 0.01 against any combination of soil backgrounds short of the input dimensionality and for any of the three areas (e.g. Fayyum, Owens Valley, or Gran Desierto). This uncertainty is applicable to both AVIRIS and Landsat TM multispectral images. The problem of detecting vegetation reduces to quantifying the effect of instrument noise given the contrast between soil and a single green vegetation endmember. Because the spectral contrast between green vegetation and soil is high, the choice of the green vegetation endmember or the soil does not significantly affect the uncertainty of sparse vegetation abundances.

If, however, the soil endmembers are variable at spectral resolutions of Landsat TM then the instrumental noise is inappropriate to determine uncertainty. For each of the three areas we investigate, previous studies have found that the spectral dimensionality of the soils is greater than the input dimensionality of Landsat TM. However, to expand the number of soil endmembers to include the full input dimensionality of TM results in vegetation being mimicked by mixtures of substrate types. Uncertainties from both the instrument and endmember variability are expressed as mixtures of soils. The problem in measuring sparse vegetation is to determine what strategies are possible to provide both an optimal and realistic measure of vegetation when we only approximately know the conditions of spectral variability.

For example, we may have a good estimate of the soil types, but we may not know what specific mixtures of soils occur. We desire to make estimates of vegetation abundance with an incomplete knowledge of background reference spectra such as is performed with indices such as NDVI. Also we wish to determine how simple we can make the analyses and still achieve meaningful results. We know it is possible to select areas and measure the soil and vegetation spectral endmembers so that the abundance uncertainties converge to that arising from instrumental noise. However, for large scale regional mapping of vegetation it may not be feasible or affordable to make such an intensive effort. For these cases we desire to know an optimal set of endmembers which are applicable to large areas and the resulting uncertainties.

2. Methods

For each of the four TM images, SMA, a simple ratio (band 4 / band 3), and NDVI was used to estimate the vegetation abundance. For each image, the data were nominally calibrated to radiance and corrected for atmospheric effects as described by Smith et al. (1990a) prior to applying each mapping technique. A local calibration consisted of adjusting nominal gains and offsets to optimize the fit between image and reference endmembers. As a result of applying SMA we obtain a list of reference endmembers applicable for each of the three areas. The vegetation abundances from SMA are derived by using the reference endmembers applicable for each area independently.

Owens Valley, California, U.S.A., is scmi arid and the vegetation is part of the Great Basin Sage Community. Two Landsat TM images (December 1982 and May 1985) have overlapping coverage of the alluvial fans near Independence, California. It has been the objective of previous remote sensing studies to characterize the soils and vegetation in this area (Gillespie 1990; Smith et al., 1990a; Smith et al., 1990b). The semiarid Owens Valley is bounded by the Sierra Nevada on the west and by the Inyo and White Mountains on the east. The granitic alluvial fans emanating from the Sierra Nevada encompass a range in elevation from ~1200 to 1800 m. Most of the alluvial fan vegetation is part of the Great Basin sagebrush semi-desert ecosystem. The dominant spectral contrast in the soils on the alluvial fans is due to the degree of soil development as a function of age. One soil spectral endmember corresponds to young camborthids and torriorthents, poorly developed on bouldery granitic franglomerate. A second soil endmember is characterized by older haplargids which are tan colored by iron oxides.

Another area, also sparsely vegetated, is the Gran Desiertio in Sonora, Mexico. Blount et al. (1990) used spectral mixture analysis to determine the regional aeolian dynamics in this area from a single Landsat TM image acquired January 1985. We use the same TM image subset to determine the uncertainty of vegetation estimates given the spectral variability of the background soils and sands. This region is characterized by active sands juxtaposed on both volcanic and crystalline source rocks. To the east of the Gran Desierto is the volcanic complex of the Sierra Pinacate containing a variety of mesquite, palo verde, and acacia along arroyos. Surrounding the Sierra Del Rosario Mountains are ephemeral grass communities. In the active sand sea there is no vegetation cover.

A third image of Fayyum, Egypt, also has a spectral diversity of substrates and no vegetation except for some irrigated agricultural areas. Buck et al. (1986) in a study of this area used SMA to indicate the likelihood of archeological artifacts. This TM image (Scene ID V5121807552Xo, July 2, 1987) contains a significant amount of noise in band 3. The endmembers consist of limestone, dune sand, chert, shade and a vegetation spectrum. The endmember suite is similar in spectral contrast to that of the Gran Desierto.

In mapping vegetation we alter three components in the SMA analytical framework to determine how these components affect abundance uncertainty. These include selection of: 1) bandpasses, 2) image subsets, and 3) the number and type of endmembers. For the case of band ratios and NDVI it is not possible to alter these components. The reason to make a connection between the indices and SMA is to have a method to evaluate the uncertainties of indices.

Utilizing the same notation as in the first paper, the uncertainty of vegetation abundance is separated into three parts, namely: 1) σ_i, the effect of the noise from the instrument, 2) σ_m, the effect of soils not part of the endmember set that mimic sparse vegetation, and 3) σ_v the effect of soil variation which does not fit mixtures of soil endmembers. In this paper, the cases considered are such that σ_i is typically small compared to either σ_m or σ_v and is not included as part of the analyses results. The vegetation abundance uncertainty is the sum of all three components.

2.1 Conceptual Uncertainty Models

If the objective is to spectrally measure the amount of vegetation which has a high probability of not being present, then we propose to examine whether it is optimal to assume the absence or presence of vegetation in the context of the analytical framework. Both NDVI and band ratios assume vegetation is present by including it as an endmember, e.g., one end of the index range. We examine the uncertainty of vegetation estimates from these two perspectives using SMA. To include vegetation as part of the analysis we include it as an endmember. For this case we wish to determine the optimal number of bands for analysis and the best set of endmembers which minimize uncertainties in vegetation abundance. Table 1 lists reference endmembers that are applicable for each of the three study areas. We utilize the reference data set to examine the uncertainty of mapping sparse vegetation. We compute a single vegetation uncertainty for all three areas using a single set of endmembers.

In the second case vegetation is assumed absent by not including it as an endmember. However, in this case we know that the dimensionality of the soils span the input dimensionality of the Landsat TM images for each area. Thus, if we include all the soils there are no residuals left to detect vegetation. Results from Roberts et al. (1992) indicate the ability to separate multispectral images into regions that are mixtures of two endmembers. We find that the Landsat TM images for each area described above can also be separated into regions that are mixtures of two endmembers. The endmember pairs are not just the reference endmembers listed in table 1 but also mixtures of these endmembers. We, therefore assume that the substrate endmembers in table 1 are representative of the magnitude and direction of spectral variation due to substrate variability. For each two endmember pair we pose the question of how much vegetation is needed to detect it as a residual given that the residual may also be due to any other substrate reference

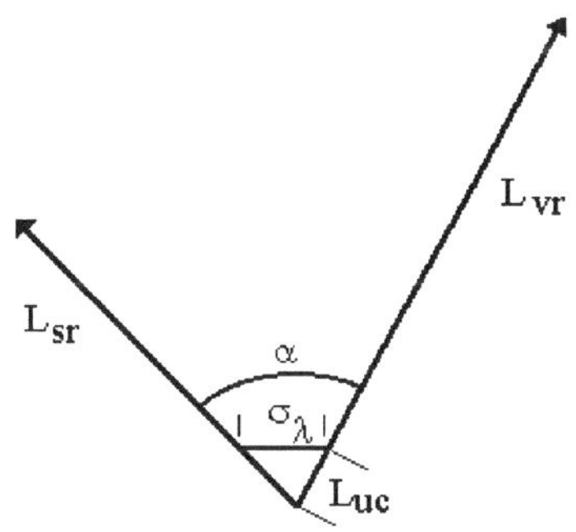

FIGURE 1. To identify a vegetation residual L_{vr} from a soil residual vector L_{sr} , when the spectral variation due to the dominant soil endmembers has been removed, requires determining the length of the vegetation residual vector L_{uc}. The amount of vegetation required to uniquely identify the residual as vegetation is dependent on the instrumental noise σ_λ.

spectrum. To answer this question we take all combinations of two soil endmembers and measure the abundance of vegetation required to uniquely identify the residual as vegetation and not any other substrate. The length of the vegetation residual vector L_{uc} required to exceed the noise threshold σ_λ for a given soil residual vector L_{sr} not part of the endmember pair is determined by:

$$L_{uc} = 0.5\,\sigma_\lambda \,/\, \sin(\alpha/2) \tag{1}$$

where α is the angle between the vegetation and substrate residual vectors that result from a soil endmember pair and σ_λ the instrumental noise over all bands. The minimum fraction of vegetation F_{vd} required to be detected from a given soil is determined by

$$F_{vd} = L_{uc} / L_{vr} \tag{2}$$

where L_{vr} is the length of the vegetation residual vector. To keep the problem simple we use a single noise value σ_λ for all TM bands. To detect vegetation from a specified substrate in the residuals that is affecting a two endmember substrate pair requires knowledge of the similarity in direction and magnitude of the soil and vegetation residual vectors. For cases when α is zero then vegetation and soil vectors have the same direction and the fraction of vegetation needed to be absolutely certain that the residual is due to vegetation is:

$$F_{vd} = \left(\sigma_\lambda + L_{rs} \right) / L_{vr} \quad when \quad \alpha = 0 \tag{3}$$

where L_{rs} is the length of the residual soil vector.

inactive-sand-gd	0.11	0.11	0.13	0.17	0.32	0.28
active-sand-gd/fy	0.27	0.38	0.51	0.66	0.75	0.74
red-basalt-gd	0.09	0.08	0.12	0.20	0.54	0.56
duri-crust-gd	0.15	0.13	0.20	0.23	0.41	0.35
basalt-gd	0.13	0.13	0.16	0.17	0.15	0.13
granite-sand-gd	0.23	0.27	0.35	0.40	0.44	0.37
carbonate-gd	0.28	0.32	0.40	0.44	0.51	0.40
granite-gd	0.22	0.24	0.29	0.38	0.41	0.40
chert-fy	0.12	0.15	0.17	0.21	0.39	0.34
gravel-fy	0.23	0.30	0.34	0.37	0.54	0.50
limestone	0.28	0.34	0.42	0.51	0.54	0.49
tan-soil-ov	0.11	0.15	0.20	0.29	0.57	0.55
gray-soil-ov	0.22	0.27	0.29	0.30	0.28	0.26
salt-ov	0.64	0.71	0.74	0.77	0.59	0.43
Populus	0.07	0.12	0.09	0.62	0.28	0.12
Artemisia	0.24	0.28	0.26	0.74	0.38	0.24
Coleogyne	0.09	0.11	0.13	0.33	0.23	0.14
Drygrass	0.22	0.30	0.35	0.41	0.54	0.41
Plant Litter	0.24	0.31	0.37	0.49	0.57	0.45

TABLE 1. Reflectances of reference endmembers for three separate areas. A two letter abbreviation is appended to each spectrum to indicate the area where that spectrum was applicable. The abbreviations are as follows: gd - Gran Desierto, fy - Fayyum, and ov - Owens Valley. The last five spectra were acquired from plant samples.

Application of the above equations must be performed given all combinations of two endmember soil vectors. For each two endmember soil pair we compute the fraction of vegetation required to be detected to each soil not part of the two endmember pair. Each soil not part of the endmember pair is potentially equally likely to cause measured residuals in addition to vegetation. For the 14 soils in table 1 there are 1095 separate estimates of uncertainty. The interpretation of these uncertainties is made in the context of their distribution rather than any single uncertainty. Although we cannot separate 1095 two endmember pairs in the image the distribution of these pairs provide specific information that can be used to focus efforts to minimize uncertainty in estimating vegetation.

3. Results

The results of computing vegetation abundances from the described image data sets are listed in table 2 for 18 separate areas. The abundances have been ordered by the abundance estimates derived from SMA. The first seven areas contain no vegetation except for the east side Owens Valley which has a halophytic community consisting entirely of nonphotosynthetic vegetation (NPV) with less than 15 percent cover. The NDVI of soils in table 1 has a mean of 0.095, a σ of 0.065, and range from 0.017 to 0.254. The band 4/3 ratio for the soils in table 1 has a mean of 1.22 with a σ of 0.17 and a range from 1.03 to 1.66. These values are nominally equivalent to that obtained from the image. The variability of these indices in the image for the first seven areas (table 2) is consistent with that measured directly from the soils.

Figure 2 indicates similar responses to estimating vegetation over the full range of measurements for NDVI and a band 4/3 ratio for the 18 areas listed in table 2. However, a plot of SMA versus NDVI estimates (figure 2) show less similarity especially with areas of no vegetative cover. Abundance estimates up to 0.20 (SMA) indicate low correlations between NDVI and SMA. Independent of accessing the correct estimate, we find remote sensing techniques do not provide similar estimates of sparse vegetation. At low vegetation cover both NDVI and ratios are close to being a linear metric and thus for sparse cover we would expect the best comparisons between scales. However, it is at low covers where differences are most pronounced.

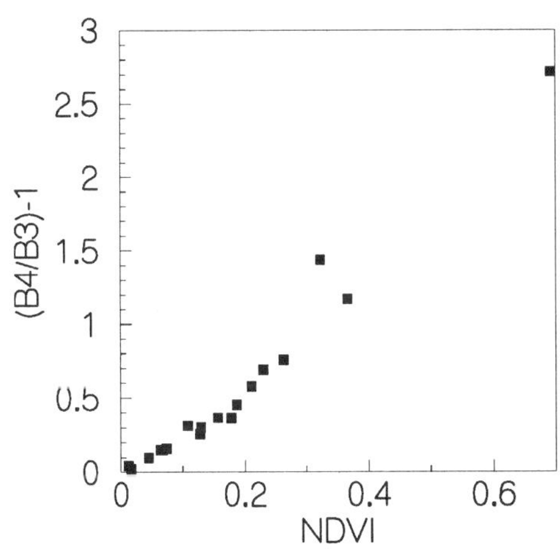

FIGURE 2. A comparison of the variation in vegetation estimates between NDVI and a simple ratio for 18 areas over 4 separately calibrated TM images. Both NDVI and ratio provide a similar estimate of vegetation over the range of soil types characteristic of these arid and semi arid areas.

Figure 3 is used to provide an estimate of the uncertainty by scaling the range independently for each technique. Qualitatively, the variation in the first seven areas (a - g) is an indication of the variability given the range in the y-axis. For both NDVI and B3/B4 ratio the variability covers 20% of the range in the y-axis. Thus, qualitatively this data set has a resolution of roughly 1 part in 5 that may be considered common between cover estimates. SMA, in contrast, indicates a range of 0.06 abundance for green vegetation cover in areas with no vegetation. There is nearly a factor of four difference in the sensitivity of SMA to estimates based on two band indices in areas with no vegetation.

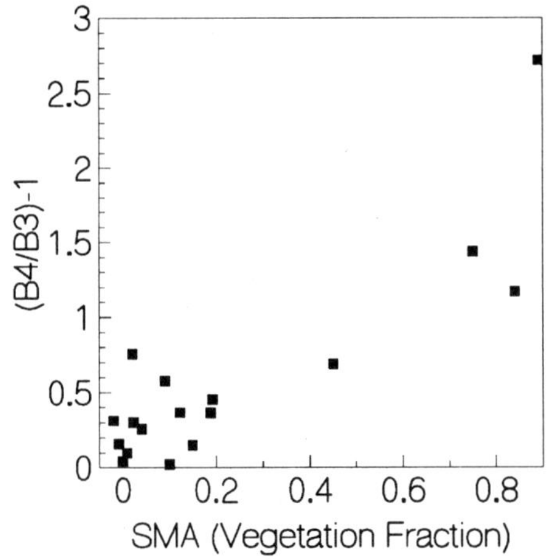

FIGURE 3. The variation in vegetation estimates between NDVI and SMA for 18 areas over 4 separately calibrated TM images. Unlike NDVI and a band 4/band 3 ratio, the SMA estimate is nearly uncorrelated to estimates of NDVI at the low range of vegetative cover.

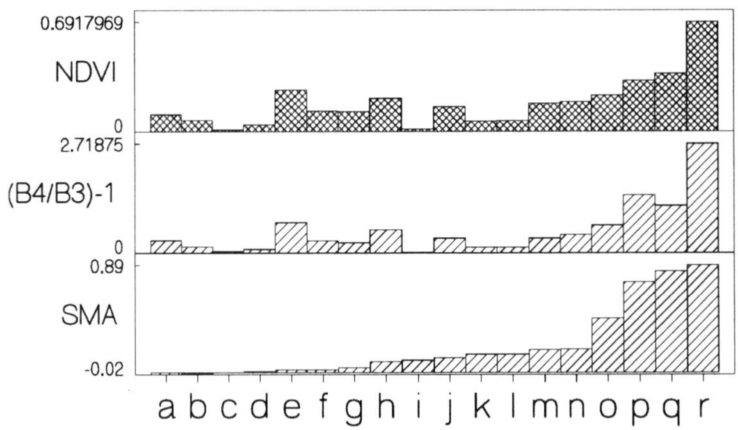

FIGURE 4. A comparison of the estimates of vegetation over 4 TM images of NDVI, a simple band ratio, and SMA. The scaling of the y-axis has been performed relative to each index using 3 areas with near 100 percent cover. The areas have been ordered by the relative magnitudes in

SMA. The area descriptions are provided as part of table 1. Areas labeled (a - g) have no vegetation cover, the spectral variability is due solely to the spectral variability of the soils.

For the Owens Valley images acquired in different seasons, we find a difference in the change of vegetation abundance at the top of the alluvial fans compared to the bottom of the alluvial fans. However, the magnitude of the difference of vegetation estimates is significantly different for the indices and SMA. In the case of the two band indices a decline in the vegetation abundances is indicated from December to May (e.g. table 2, OVD-high - OVD-low compared to OVM-high - OVM-low). In contrast, SMA indicates that the minimal vegetation is in December rather than May. The variability of the indices with seasonal changes in vegetation can potentially lead to erroneous conclusions regarding change if one does not consider the uncertainty of the estimate.

Area	Description	B4/B3	SMA	NDVI
a	GD-low	1.21	-0.02	0.108
b	FY-high	1.16	-0.01	0.074
c	OVM-east valley	1.04	0.00	0.012
d	FY-low	1.10	0.01	0.045
e	GD-basalt	1.76	0.02	0.263
f	OVD-east valley	1.30	0.02	0.130
g	GD-high	1.26	0.04	0.128
h	GD-granite	1.57	0.09	0.211
i	OVM-low	1.02	0.10	0.017
j	OVD-low	1.36	0.11	0.157
k	OVD-high	1.15	0.15	0.064
l	OVM-argsoil	1.15	0.15	0.068
m	OVD-argsoil	1.36	0.19	0.178
n	OVM-high	1.45	0.19	0.187
o	FY-agricultural	1.69	0.45	0.229
p	GD-agricultural	2.44	0.75	0.322
q	OVD-valley floor	2.17	0.84	0.366
r	OVM-valley floor	3.72	0.89	0.692

TABLE 2. The area descriptions are as follows: GD - Gran Desierto, OVD - Owens Valley December 1982, and OVM - Owens Valley May 1985, FY - Fayyum, Egypt. For both the Fayyum and Gran Desierto high and low estimates were taken in areas with no vegetation cover where minimum and maximum vegetation estimates were observed. For the east valley of Owens Valley, vegetation is less than 15% cover and is all dead biomass. The Owens Valley estimates low and high were taken at the same location in the image for each estimate. The Owens Valley low vegetation estimates were taken at the bottom of the alluvial fans and the high estimates were made near the top of the alluvial fans. The first seven areas (a - g) have no green vegetation cover. For each image a relatively high vegetation cover is also measured in the agricultural areas. Each estimate is the average of 400 pixels (e.g. 20 x 20 pixel box). The data were ordered by the magnitude of the vegetation abundance obtained using SMA.

The SMA estimates of vegetative cover utilized reference spectra from study areas where a significant amount of effort was directed to determine the spectral identity of the endmembers. However, we desire to map sparse vegetation regionally or globally without knowing exactly what the soils are in any one area. Spectral endmembers in table 1 provide an estimate of the soil variability. Different soils mask as mixtures of other soils given the TM bandpasses. The set of soils in Table I exceed the spectral input dimensionality of TM. Given these conditions, is there a single set of endmembers that is applicable to all three areas and what is the uncertainty in measuring sparse vegetation using a single set of endmembers?

Using the soil endmembers in table 1 we compute the minimum σ_v and σ_m from all possible combinations of two, three, and four soil endmembers listed in table 1 on a single green (e.g. poplar) and dry vegetation endmember (e.g. dry grass). For each number of endmembers and bands the soil endmember combination giving the minimum uncertainty is listed in table 3. From table 3 we find that using the minimal number of bands for either green vegetation or NPV does not provide the mimimal level of abundance uncertainty. The uncertainties for two bands and three endmembers are applicable to two band indices. While these uncertainties are not the highest they are also not minimal. TM bands 3 and 4 provide the minimal uncertainty for detecting green vegetation if we are limited to three endmembers for the entire image. However, bands 5 and 6 provide the mimimal uncertainty if we are attempting to detect NPV using three endmembers.

The uncertainty in NPV estimates using three bands is about 4 times higher than that of green vegetation (table 3). The high NPV uncertainty, due to lack of spectral contrast with soils, is a likely reason that no indices exist to map NPV. From table 3 it is evident that there is a greater advantage to using more bands and more endmembers for green vegetation than for NPV (e.g. a factor of two reduction in mapping uncertainty for green vegetation compared to a factor of only 0.3 reduction for NPV). The uncertainties for NPV are, however, also significantly reduced by using more than two bands and three endmembers.

The uncertainties of green vegetation (table 3) determined from the soil endmembers are similar to those inferred from the image data (figure 4) for band ratios and NDVI. An abundance resolution of 1 part in 5 is approximately correct for a 99% confidence level for three endmembers. By using more endmembers it is possible to achieve mapping accuracies which are a factor of two better (e.g., 0.070 for three endmembers and two bands versus 0.036 for seven endmembers and six bands). If we continue to refine the problem such that we select specific areas where we know the soil types then these uncertainties converge to that produced by the instrument which is nominally < 0.01. However, it is usually not known in regional mappings exactly what spectral endmembers apply to what areas without performing some spectral measurements of field samples.

Using the uncertainties of table 3 as indicative of mapping uncertainties inherent in indices such as NDVI and ratios it is important to realize that these uncertainties are obtained by selecting the best endmember set. The endmembers are fixed for NDVI and ratios and thus they may contain uncertainties which are higher than that expressed by two bands and three endmembers in table 3. Thus, the uncertainties in table 3 are best interpreted as the optimal uncertainty possible for two band indices.

From table 3, we find σ_v uncertainties are highest at low numbers of endmembers regardless of vegetation type (e.g. NPV or green). When modeling with more than three endmembers using TM images most of the uncertainty in vegetation estimates is contained in σ_m because the mixture dimensionality is nearly equivalent to the measurement dimensionality. In cases where σ_v is high it is possible to use more than one set of endmembers to model the spectral variation in the image because we can detect pixels which do not fit as mixtures of endmembers. In the case of σ_m, it is not typically possible to detect spatially where the uncertainties are located without field verification and spectral measurements. Cases where σ_m uncertainties may be analytically detected are when the fractions are negative and / or are greater than one.

Inclusion of a band in table 3 is an indication that the included band contains useful information in the context of spectral mixtures. In contrast, a band which is excluded is an indication that the band adds more noise to the SMA estimates than it resolves given the soils we used in the analysis. For green vegetation and NPV, we find a distinct reduction in uncertainty by not including all TM bands. In addition, we find that models with fewer endmembers are equal or better than models containing as many soil endmembers as can be utilized with six TM bands. For green vegetation, TM bands 2 and 5 are typically less useful in detection of vegetation given the variability of soils. In contrast for NPV, we find that the minimum uncertainties are achieved by omitting TM bands 3 and 4.

Another way to determine the uncertainty of vegetation estimates is to exclude it from the analysis and determine at what point it could be detected as a residual. To maximize detection as a residual requires that the endmember set be a minimal set (e.g., we wish to minimize the mixture volume to optimize detection as a residual). Figure 5 summarizes the results of 1095 two-endmember-combinations between the 14 soils of table 1. For each two endmember combinations we determine the amount of vegetation to be detectable as a residual when the residual may be due to another soil in table 1 or to photometric shade (e.g. zero reflectance for all TM bandpasses). Also figure 5 illustrates the effect of different vegetation spectra on the distribution of abundance uncertainties for the 14 soils in table 1.

For the same green vegetation spectrum (*Populus*) used in table 3 we find that when vegetation is treated as a residual, the mean uncertainty level is reduced by a factor of two compared to including it as an endmember. Very few endmember pairs exceed an uncertainty level of 0.03. In the case of green vegetation it does not make much difference if we invest the time to know what soil endmembers are mixing over the entire scene. Certainly improvements are made but perhaps not worth the effort.

Knowing the soil endmembers is much more important in the case of mapping NPV. The decrease in spectral contrast of NPV to the soils result in many soil mixture vectors looking much like NPV. In those cases we cannot separate many soil residuals from residuals produced by NPV. Previous studies in Owens Valley of AVIRIS images find NPV to be indistinguishable from mixtures of soils. It is important to realize that there are many different spectral types of NPV. The litter spectra has a much higher contrast to the soils than does dry grass. However, unlike green vegetation the wide distribution of uncertainties for some NPV indicate it much more advantageous to know specifically what soils are mixing.

4. Discussion

The conceptual models used to determine uncertainty greatly simplify many of the complexities in the data. For the residual model one can argue that the soil vectors do not actually express all directions possible for soil endmembers as there are mixtures of mixtures. In addition, most of the soil mixing lines do not even exist for uncertainties illustrated in figure 5. None of the mapping techniques applied was able to completely separate the soils from vegetation with the six TM bandpasses. Although none of the mapping techniques can explicitly handle the spectral complexity, we find there are significant strategies to maximize detection of vegetation using

Landsat TM data sets. The more quantitative one can be regarding variability and spatial distribution in vegetation and background spectra the more realistic is the spectral estimates of sparse vegetation.

	Green Vegetation			NPV		
Em's	σ_v	σ_m	TM bands	σ_v	σ_m	TM bands
3	0	0.070	3,4	0	0.300	5,7
3	0.023	0.046	3,4,7	0.088	0.270	4,5,7
3	0.027	0.043	3,4,5,7	0.233	0.191	1,2,5,7
3	0.032	0.074	1,2,3,4,5	0.010	0.457	1,2,3,4,5
3	0.034	0.076	1,2,3,4,5,7	0.264	0.206	1,2,3,4,5,7
4	0	0.043	3,4,7	0	0.234	1,5,7
4	0.015	0.037	3,4,5,7	0.012	0.212	1,2,5,7
4	0.005	0.056	2,3,4,5,7	0.007	0.232	1,2,3,5,7
4	0.005	0.069	1,2,3,4,5,7	0.008	0.282	1,2,3,4,5,7
5	0	0.037	1,3,4,7	0	0.234	1,2,5,7
5	0.006	0.031	1,3,4,5,7	0.003	0.234	1,2,4,5,7
5	0.007	0.035	1,2,3,4,5,7	0.009	0.241	1,2,3,4,5,7
6	0	0.036	1,2,3,4,7	0	0.242	1,2,4,5,7
6	0.009	0.044	1,2,3,4,5,7	0.009	0.0521	1,2,3,4,5,7

TABLE 3. The uncertainties for spectral abundances of green and nonphotosynthetic vegetation for different numbers of soil endmembers and TM bandpasses. The estimates for green vegetation utilize the spectrum of a stack of poplar leaves while for NPV we utilize the spectrum of drygrass. The uncertainties are the minimum for all possible band and soil endmember (table 1) combinations in each group. For the three endmember case there is a single soil, shade, and vegetation. For the six endmember case there are four soils, shade, and vegetation.

The uncertainties observed in the image data are consistent with the analytical framework for computing uncertainties. It is apparent by the uncertainties listed in Table III that there is not a fixed set of bands or endmembers for minimizing uncertainties. The variability of soils in some bands do not fit as linear mixtures of soil and vegetation and are best left out of the analysis of green vegetation. This effect is even more noticeable with NPV which has a reduced spectral contrast to the soils compared to green vegetation. In measuring NPV we minimize abundance uncertainty by using fewer bands and less endmembers. We interpret the lower uncertainties corresponding to a band set that does not include all TM bands as an indication of an intrinsic dimensionality in the scene that cannot be separated using the six TM bandpasses.

In mapping sparse vegetation using TM data sets we find it relevant to 1) access the variability of soil and vegetation endmembers. A significant increase in vegetation detection is possible by optimizing band selection and endmembers using multiple scenarios. Because of the difficulty in assessing the different uncertainties in vegetation estimates we have utilized laboratory spectral measurements of soils and vegetative samples. The uncertainties obtained from these samples help in focusing field verification efforts aimed at reducing uncertainties of vegetative abundances.

In interpreting the uncertainties of vegetation estimates using indices and SMA we must examine the assumptions involved. In the case of detecting vegetation as a residual we assume that the soil endmember pairs are representative of the spectral directions of soils in the three areas. Greater soil spectral variability is possible in areas such as the Australian deserts yielding somewhat higher variability in NDVI and ratio indices than observed in the images analyzed. Using mixtures of mixtures does not significantly change the histograms illustrated by figure 6. The 14 soils listed in table 1 include a range of spectral diversity representative of arid and semi arid regions. This does not imply there are no exceptions but rather presents a basic framework useful to find exceptions. The uncertainty framework presented is not fixed but can be modified to maximize the detection of specific endmembers. In the case of including vegetation as an endmember we assume that the soils in table 1 are equally important in spatial distribution and that the full mixture volume is realized over the region. Also it is assumed that the mixture volume represented by the three images can be extrapolated to much larger regions.

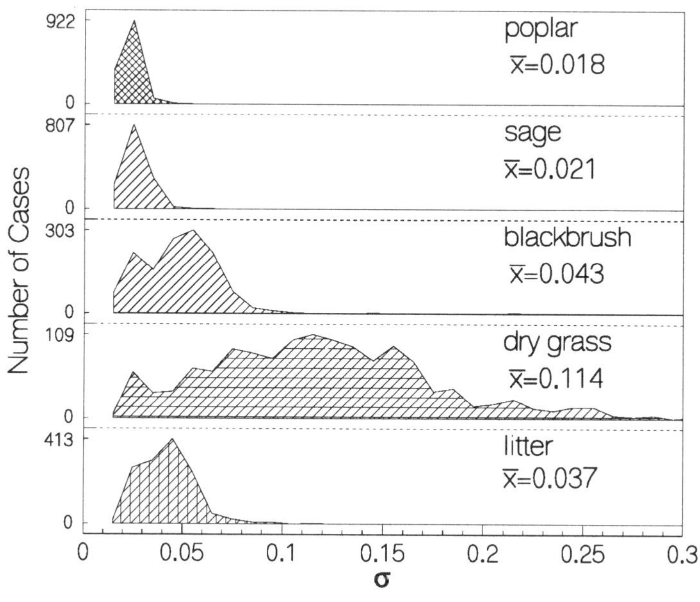

FIGURE 5. Uncertainty of vegetation abundances computed by determining at what point a given vegetation spectrum is detectable as a residual. The histograms summarize the detectability of vegetation for 1095 combinations of two soils or a soil and shade listed in table 1.

Indices are often used as quick estimates of vegetation. The potential risks of quick estimates derived from indices are significant without consideration of endmember uncertainties. Images acquired under similar conditions (moisture, season, solar angle, etc.) typically have similar ratios. Spatial comparisons will incorporate spectral variation from the soils. The largest bias of two band indices of vegetation is likely to be in the spatial distribution not in temporal estimates because of the high σ_m embedded in indices.

The east valley areas (e.g., areas c and f of table 1 and figure 3) of the Owens Valley images did contain vegetative cover which did not show up in any of the abundance estimates. This vegetation is NPV consisting mainly of dead shrubs, stems and litter. Detecting many types of NPV is difficult due to the lack of contrast between NPV and soils. Sparsely vegetated areas of arid and semi arid regions are more likely to contain NPV than green vegetative communities.

The interpretation of changes in vegetation must be made in the context of mapping techniques that optimize its presence. A greening of the vegetation on the east side of the valley does not necessarily indicate a change in climate but perhaps more likely environmental conditions which are optimal for new growth. Areas with sparse cover often green-up rapidly and for short periods to local and seasonal weather patterns.

5. Conclusions

The meaningful detection of sparse vegetation is closely tied to the uncertainties of techniques that map vegetation. A focus on new techniques that explore conceptual models of abundance uncertainty is most likely to result in future mapping improvements. Although the estimates made with SMA are consistent due to the ability to change endmembers for each area, it is possible to select a single group of endmembers for all three areas which does not significantly affect the detection of vegetation. To find the endmember model applicable for all areas requires knowledge of spectral variability of the background endmembers.

Independent of the type of analysis applied (e.g. classification, discriminant analysis, spectral matching, etc.) it is not possible to escape the effects of spectral contrast of the background. If spectral contrast is not an explicit part of the data analyses, it is not possible to determine the extendibility of a mapping technique. Green vegetation is significantly more detectable than some NPV's due to the differences in spectral contrast. The lack of contrast between NPV and soil in Owens Valley, for example, was not detectable by any of the techniques. Detection of some NPV's requires near 100% ground cover.

In cases where detection is critical, considering vegetation as a residual offers the greatest potential for mapping sparse cover at the expense of increased complexity in analysis. Unlike making vegetation an endmember, the application of residual analysis permits focusing efforts to access the likelihood of endmember combinations with low contrast. With a refined understanding of the changes in vegetation contrast to soils in sparsely vegetated areas, it is possible define the confidence level which can be applied to different mapping techniques.

6. Acknowledgments

We gratefully acknowledge the W.M. Keck Foundation for computer equipment and support and also the Joint Research Centre, Ispra, Italy for their financial support in developing these concepts.

7. References

Blount, G., M.O. Smith, J.B. Adams, R. Greeley, and P.R. Christensen (1990) 'Regional aeolian dynamics and sand mixing in the Gran Desierto: evidence from Landsat thematic mapper images', *J. Geophys. Res.*, 95,15463-15482.

Buck, P.E., S.C. Willis, and M.O. Smith (1986) 'A mixture modeling approach to remote sensing of archaeological sites in Egypt using Landsat imagery', in *GSA annual meeting, Archaeological Geology Symposium, San Antonio.*

Gillespie, A.R., M.O. Smith, J.B. Adams, S.C. Willis, A.F. Fischer, and D.E. Sabol (1990) 'Interpretation of residual images: spectral mixture analysis of AVIRIS images, Owens Valley, California', *Proceedings of the Airborne Science Workshop, JPL Publ.,*90-54, 243-270.

Huete, A.R., R.D. Jackson, and D.F. Post (1985) 'Spectral response of a plant canopy with different soil backgrounds', *Remote Sensing of Environment*, 17, 37-53.

Roberts, D.A., M.O. Smith, J.B. Adams, D.E. Sabol, A.R. Gillespie, and S.C. Willis (1991) 'Isolating woody plant material and senescent vegetation from green vegetation in AVIRIS data', *Proceedings of the Airborne Science Workshop: AVIRIS*, JPL Publ.,91-28, 43- 50.

Roberts, D.A., M.O. Smith, D.E. Sabol, J.B. Adams, and S.L. Ustin (1992) 'Mapping the spectral variability in photosynthetic and non-photosynthetic vegetation, soils and shade using AVIRIS', *in* R.O. Green (ed.) *'Summaries of the Third Annual JPL Airborne Geoscience Workshop, Vol I., AVIRIS Workshop'*, June 1-2, 38-40.

Sabol, D.E., J.B. Adams, and M.O. Smith (1992) 'Quantitative subpixel spectral detection of targets in multispectral images', *J. Geophys. Res.*, 97, 2659-2672.

Satterwhite, M.B. and P.J. Henley (1987) 'Spectral characteristics of selected soils and vegetation in northern Nevada and their discrimination using band ratio techniques', *Remote Sensing of Environment,* 23, 155-175.

Smith, M.O., S.L. Ustin, J.B. Adams, and A.R. Gilliespie (1990a) 'Vegetation in deserts I: A regional measure of abundance from multispectral images', *Remote Sensing of Enivronment,* 31,1-26.

Smith, M.O., J.B. Adams, and A.R. Gillespie (1990b) 'Reference endmembers for spectral mixture analysis', *Proc. 5th Australian Remote Sensing Conference, Perth, Western Australia,8-12 October,* vol. 1, 331-340.

Tucker, C.J. (1979) 'Red and photographic infrared linear combinations for monitoring vegetation', *Remote Sensing of Environment,* 8, 127-150.

Tucker, C.J. and L.D. Miller (1977) 'Soil spectra contributions to grass canopy spectral reflectance', *Photogramm. Eng. Remote Sensing,* 43, 721-726.

Ustin, S. L., J.B. Adams, C.D. Elvidge, M. Rejmanek, B.N. Rock, M.O. Smith, R.W. Thomas, and R.A. Woodward (1986) 'Thematic mapper studies of semi-arid shrub communities', *BioScience*,36, 446 - 452.

LAND DEGRADATION AND SOIL EROSION MAPPING IN A MEDITERRANEAN ECOSYSTEM

JOACHIM HILL, WOLFGANG MEHL, and MICHAEL ALTHERR
Institute for Remote Sensing Applications
Commission of the European Communities
Joint Research Centre
I-21020 Ispra (Va), Italy

ABSTRACT. The degradation of the permanent semi-natural vegetation and the resulting acceleration of soil degradation and erosion constitute important elements of land degradation processes in the Mediterranean basin. Under the European Commission's DGXII "Research and Development Programme in the Field of the Environment", the need for identifying, mapping and controlling such desertification phenomena is expressed, and the Joint Research Centre has initiated the development remote sensing-based of methods for detection and repeated monitoring of soil and vegetation characteristics.

In this paper we present an approach for mapping soil conditions and erosion features from hyperspectral images. It requires radiometric rectification of the multi-spectral data and the availability of spectral libraries, before linear spectral mixture modelling is used to decompose image spectra into spectrally distinct mixing components. The resulting abundance estimates (fractions) then permit to identify soil conditions and erosion features, and to obtain an improved measure of vegetation cover. Our results suggest that this approach holds some potential for operational applications, including monitoring of erosion processes and changes in vegetation cover which are important elements for desertification monitoring.

1. Introduction

Until recently, the monitoring of natural resources and environmental conditions has been based mainly on traditional techniques, such as direct field observations or the analysis of aerial photographs. With the advent of operational earth resources satellites, however, a repetitive and regular observation of Mediterranean environments from space has become possible. However, routine interpretation of optical sensor data requires the design and development of efficient and reliable data analysis methods, including pre-processing tools to remove contamination of the sensor signal through combined sensor-atmosphere effects as well as advanced mapping algorithms or suitable inversion models.

Imaging Spectrometers provide reflectance spectra in the complete solar spectrum through a sequence of contiguous spectral bands. The analysis of such data sets is therefore of equal importance for studying basic mechanisms of energy/target interactions and for developing methods that may be also applied for the interpretation of data from operational earth observation satellites.

J. Hill and J. Mégier (eds.), Imaging Spectrometry - a Tool for Environmental Observations, 237–260.

2. Land degradation in the Mediterranean basin

As a transition zone between arid tropics and the more humid (temperate) climates to the north, the Mediterranean basin includes an enormous variety of topographic, lithologic and edaphic conditions. It is primarily the arido-humid climate with an alternation of hot, dry summers and more humid winter periods that provides the major criterion for delineating Mediterranean ecosystems (Nahal, 1981). Accordingly, the natural vegetation has adapted to growth conditions which are mainly constrained by the coincidence of high temperatures and a severe shortage of water during summer. But the dry summer periods also coincide with occasional, but nevertheless violent rainstorms. Soil erosion preferably develops in areas where the vegetation cover has already been seriously damaged (e.g., through overgrazing, wood collection or wildfires), and it causes average yearly soil losses above 15 tons/ha in more than one third of the Mediterranean basin (Grenon and Batisse, 1989). Such an excessive loss of soil, nutrients and seeds from the ecosystem also degrades the regeneration capacity of the vegetation, and might cause irreversible environmental damage. Inadequate land use practises (overgrazing, overtrampling, wood collection, and accidental or repeated burning) further contribute to the acceleration of these degradation processes. In particular the southern and eastern parts of the Mediterranean basin suffer from overstocking and illegal timber extraction, while other regions are more heavily affected by the excessive use of water resources through modern agriculture, tourism and urban growth.

Since increased human pressure upon the environmental resources appears to coincide with a phase of accelerated climatic change it has become an important task to identify regions at risk of severe degradation. Mapping and repeated monitoring of degradational processes forms the basis for drafting and implementing rational development plans for a sustained use of mediterranean land resources, and there is a need for spatial information with adequate resolution for integration with existing topographic maps at local level (i.e. scales 1:100,000 to 1:200,000). Such scales are not only required to precisely locate already degraded areas, but form an essential prerequisite for monitoring dynamic processes on local level before they can also be observed in a regional context. Conventional mapping approaches (i.e. field surveys, air photo interpretation) are usually not suited to provide such solutions, mainly due to high costs, deficient mapping quality in difficult or inaccessible terrain and insufficient standardization and repeatability. Remote sensing offers alternative information sources, but the collected radiation data do not yet correspond to the information we need and must first be interpreted in terms of the desired information.

3. Reflective Properties of Soils

The upper part of the unconsolidated materials overlying the parent rock is exposed to physical, chemical, and biological weathering processes, which, in time, lead to the development of soils with horizontal layers that are distinguished by variations in composition and physical properties. Unlike pedologists in the field, who may refer to exposed soil profiles permitting the observation of genetic soil horizons and the extraction of samples for laboratory analysis, soil mapping through optical remote sensing is restricted to surface reflectance that can be directly observed from the radiation measurements of the sensor system (Escadafal, 1994). Since the characteristics of radiation from a material are a function of material properties, observations of soil reflectance carries information on the properties and the state of the topsoil. This means, in turn, that only such degradation effects can be mapped that have caused significant changes of the soil surface characteristics.

3.1 TYPICAL SOIL SPECTRA

The spectral reflectance of soils is a cumulative property which derives from the inherent spectral behaviour of heterogeneous combinations of minerals and organic matter and soil water (Baumgardner et al., 1985; Irons et al., 1989). Spectral absorption and reflectance effects in the solar spectrum (0.4 - 2.5 μm) constitute diagnostic features that can be used to directly identify important constituents of soils, such as iron oxides (i.e. hematite, limonite, goethite), clay minerals (i.e. kaolinite, montmorillonite) or carbonates. These are caused by electronic transitions and vibrations in the crystal lattice of, for example, OH-groups (i.e. Al-OH bearing minerals such as kaolinite at 2.2 μm) or carbonates ($CaCO_3$ at 2.35 μm) (Huntington et al., 1989; Goetz, 1992). Soil mineralogy, in turn, is often interrelated with soil texture and organic matter content

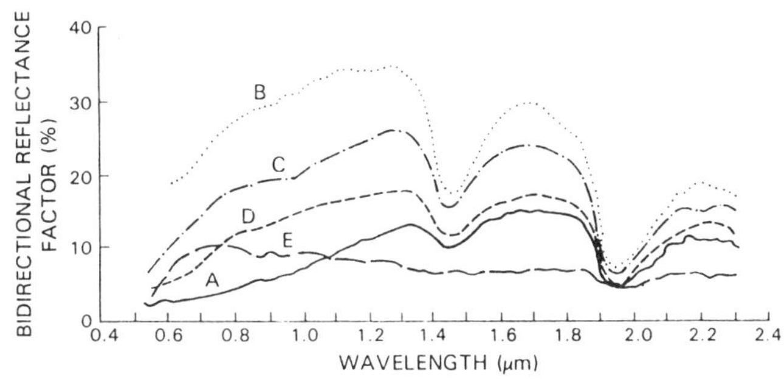

Curve A: developed, fine textured soils with high (> 2%) organic matter content; B: undeveloped soils with low (< 2%) organic matter and low (< 1%) iron oxide content; C: developed soil with low (< 2%) organic matter and medium (1-4 %) iron oxide content; D: moderately course textured soils with high (> 2%) organic matter content and low (< 1%) iron-oxide content; E: fine textured soils with high (> 4%) iron oxide content.

FIGURE 1. Characteristic bi-directional reflectance spectra of dominant soil categoeries (from Stoner and Baumgardner, 1981).

But in addition to specific absorption features, soil reflectance is largely characterised by the spectral reflectance continuum, i.e. the overall shape and albedo. Numerous studies describe the relative contributions of various soil parameters, such as organic matter, soil moisture, particle size distribution, soil structure, iron oxides, soil mineralogy, and parent material to this reflectance continuum (Escadafal, 1994). Stoner and Baumgardner (1981) have defined five distinct soil reflectance types which can be identified by curve shape, the presence or absence of spectral absorption bands caused by organic matter content, iron oxides and soil minerals (figure 1). It is believed that any observed soil spectrum resembles one of these spectral curves.

Eroded soils can often be recognised through typical soil colour changes which are due to the removed topsoil (Weismiller et al., 1984). It is, however, rather difficult to develop generally applicable interpretation principles. Our approach refers to basic concepts which consider soil development to be either *progressive* or *regressive* with time (Birkeland, 1990). Under progressive development, soils become better differentiated by horizons, and horizon contrasts become stronger. Pedogenetic processes involve the formation of clay-size particles by weathering of larger grains, the alteration of clay minerals to other clay-mineral species, and the release and accumulation of iron by weathering. Some solids (silt, clay and $CaCO_3$) and ions (Ca^{2+}, Na^+, etc.) dissolved in rainwater are added from the atmosphere, and topsoil organic matter contents increase with the decomposition of plant and animal residues. Transfers within the soil profiles result in the accumulation of silt and clay, Fe, Al, $CaCO_3$, gypsum or halite in the B horizon, or, due to bioturbation processes, to the soil surface. In contrast, regressive pedogenesis refers to the addition of material to the surface at a rate that suppresses soil formation (i.e. eolian dunes, glacial moraines, distal fans, etc.), or suppression of pedogenesis by surface erosion.

3.2 Remote Mapping of Eroded Soils

Both progressive and regressive processes cause alterations of the soil surface that, to a certain extent, are spectrally detectable. For a majority of Mediterranean soils (e.g., cambisols, fluvisols, luvisols, vertisols, regosols, rendzinas) brunification or rubification, and the organic matter content of the topsoil material provide the most important diagnostic features for a spectral identification of developed soil substrates. In the typology of Baumgardner et al. (1985), reflectance spectra from such soils generally correspond to their spectral types C and D, while spectral measurements of undeveloped or disturbed soils (leptosols) frequently resemble type B (figure 1) and exhibit spectral absorption features which are more characteristic for the parent material.

These principles seem to provide a widely applicable, general framework for relating spectrally detectable surface phenomena to soil conditions, thereby satisfying an important requirement for the successful application of remote sensing techniques. However, transitional situations (e.g., when erosion occurs in thick colluvial accumulations, or excavates paleosols) may still be misinterpreted, and other soil categories may, under identical conditions, even exhibit different phenomena related to soil colour and/or mineralogy. While this makes it already difficult to define standard approaches for the remote mapping of soil reflectance, additional problems result from intrinsic limitations of multi- or hyperspectral systems. Although, under ideal conditions, even narrow spectral absorption features can be identified with high spectral resolution spectroradiometers or imaging instruments, the major obstacle for using remote sensing data in soil observation is the difficulty to consistently interprete soil spectral characteristics under a wide range of environmental conditions (e.g. variable soil moisture, organic matter content, surface roughness). Additionally, most soils are covered with varying amounts of vegetation, so that remotely sensed spectra usually represent a mixture of soil and vegetated components.

4. Study Site

Our test site is located in the southern Ardèche area (44° 20' N, 4° 15' E) in Mediterranean France, and had been established for several experimental campaigns with airborne imaging spectrometers (Hill and Mégier, 1991; Hill, 1991). Its climate is sub-mediterranean humid (Bornand et al., 1977), and the site belongs to that part of France where the most erosive rainfalls occur (figure 2). It was also selected because its permanent vegetation includes a wide range of mediterranean oak woods, shrub- and rangelands the species composition of which closely resembles that of mediterranean woodlands in Corsica, Italy, Sardinia and north-east Spain. The dominant tree- and shrub species (*quercus pubescens, quercus ilex, quercus coccifera, juniperus oxycedrus and buxus sempervirens*) are associated with abundant herbaceous perennials, biennials and annuals. Most of the site exhibits a moderately variable to flat topography with an average altitude of about 200 m, with a single mountain range rising up to 450 m.

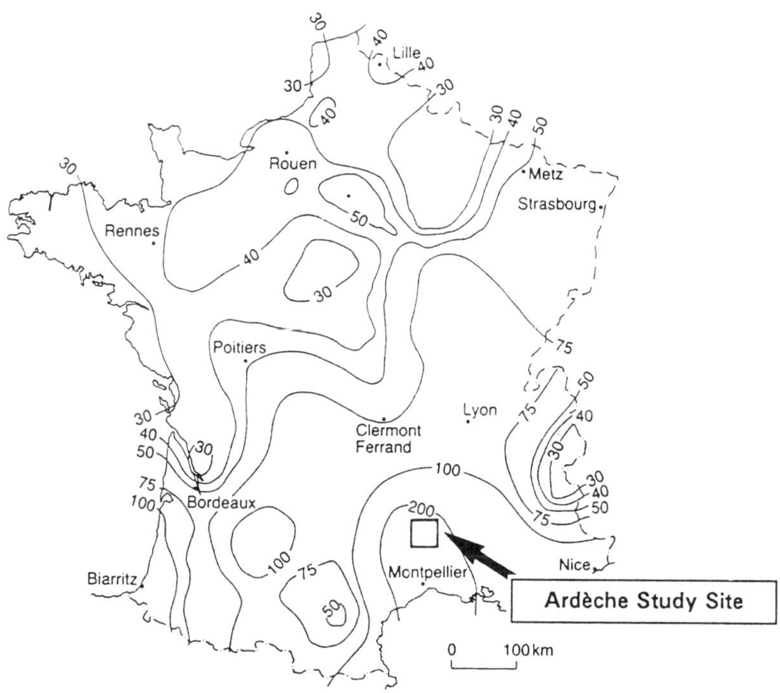

FIGURE 2. Mean annual erosivity of rainfalls in France (from Morgan, 1986)

Limestone and marls dominate the lithology of the study site. Undisturbed soils (fluvisols, orthic rendzinas and cambisols) mainly occur on alluvial and fluvial deposits, while most soils on the limestone and marl areas fall into the category of leptosols, being limited in depth by continuous hard rock or highly calcareous material (Bornand, et al., 1977; Riezebos et al., 1990). Calcite and quartz are the dominant mineral constituents in most soils, the main accessories being feldspars, kaolinite and illite (Negendank et al., 1990). In particular the marl areas include extended locations which exhibit results of frequently occurring, highly erosive rainfall events (badlands).

5. Regional Soil Conditions

In the Ardèche site, the undisturbed soils on fluvial accumulations are mainly calcaric fluvisols with weakly differentiated horizontation. Alluvial sediments, limestone and marl areas with low erosion risk are dominated by soils with rather deep A/C profiles (rendzinas, regosols) which, under favourable conditions, may have developed to cambisols. Brown loamy soils (chromic luvisols, vertic cambisols) are paleosols and can hardly be found "in situ". They only occur as isolated patches on locations which have been protected against erosion processes.

These soil types constitute a regional optimum (climax). A comparison of representative field spectra from the test site and Stoner and Baumgardner's (1981) spectral curves shows that these soils fall into the transition between categories C and D (figures 1 and 3). There is evidence for the presence of iron oxides (spectral absorptions around 0.65 and 0.9 μm), and additional spectral absorptions at 2.2 and 2.35 μm relate to clay minerals (illite, kaolinite) and various amounts of carbonates (Negendank et al., 1990).

Under less favourable conditions the same types of parent rock have produced only weakly developed soils (various leptosols), or existing soils have, as a consequence of changing environmental conditions, been removed through erosion processes (Riezebos et al., 1990). Due to a deficiency of organic components, the corresponding field measurements reach higher reflectance levels and resemble the spectral curve type D from figure 1. The spectra are characterised by absorption features of parent rock material (carbonates and clays at 2.35 and 2.2 μm respectively), but there is only weak evidence of iron oxides which are considered a major indicator for pedogenetic processes (figure 4).

It follows that the brunification (rubification) and organic carbon content of the topsoil provide the most important diagnostic features for the spectral characterisation of developed soils, while eroded or weakly developed soils are dominated by spectral characteristics of the parent material. Fischer (1989) has been able to refer to similar characteristics for mapping soils of different age in an area of young glacial deposits with data from an airborne imaging spectrometer (AVIRIS). In our experiment, we have tried to use this correspondence for the spectral identification of degraded and eroded soils.

6. The Airborne Visible/Infrared Spectrometer (AVIRIS)

AVIRIS contains four spectrometers (A - D) that view the ground through a whiskbroom-like scanner that is coupled via fibre optics to array detectors. Silicon and antimonide line array detectors cover the spectral range of 0.4 to 2.45 μm with a band width of approximately 10 nanometers. Operating from NASA's ER-2 aircraft in a flight altitude of 20,000 meters, image dimensions are 10.5 km wide by 10 to 100 km in length, with approximately a 20 by 20 m spatial resolution.

Date	Time (GMT)	Lat 1	Lon 1	Lat 2	Lon 2
24 June 91	10:51	44-27.2	04-11.1	44-19.8	04-20.9
16 July 91	11:30	44-19.8	04-20.9	44-27.2	04-28.4

TABLE 1. AVIRIS scenes acquired over the study site during NASA MAC Europe 1991

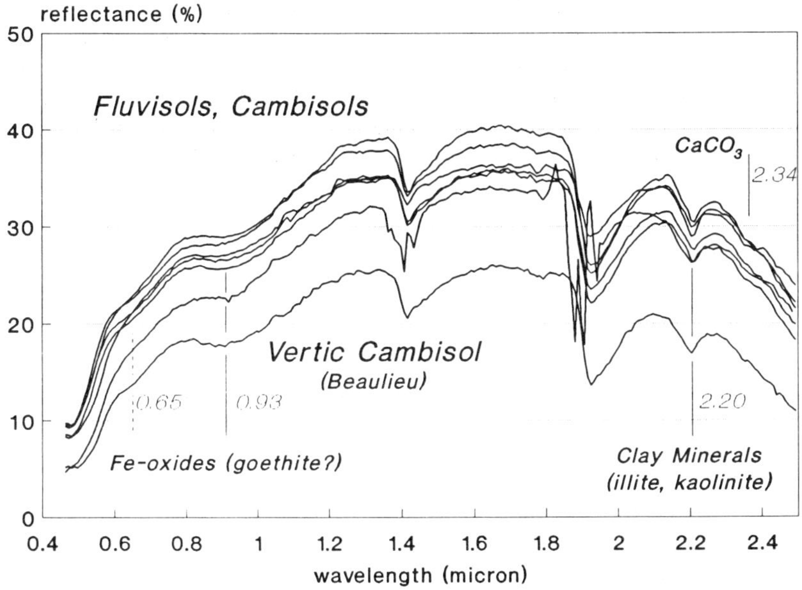

FIGURE 3. Field- and laboratory-measured reflectance spectra of fluvisols and cambisols from the Ardèche site.

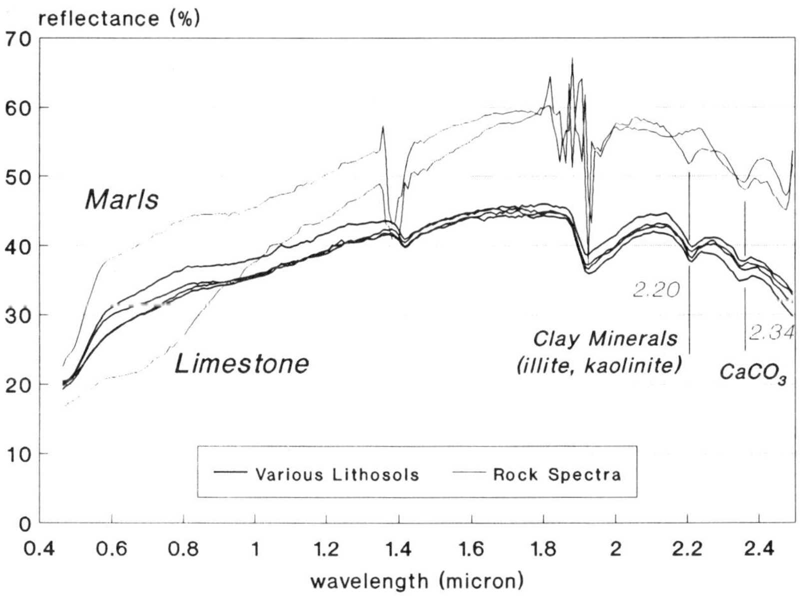

FIGURE 4. Field- and laboratory-measured reflectance spectra of typical leptosols, marls and limestone rocks from the Ardèche site. The spectra of figures 3 and 4 were measured with a GER-SIRIS spectroradiometer of 2.5 nm spectral sampling intervals (Altherr et al., 1991).

Our primary data set includes two AVIRIS scenes which were acquired in summer 1991 during NASA's Multiple Airborne Sensor Campaign (MAC) in Europe (table 1). We have concentrated our analysis on the scene from July 16, which was acquired under slightly better atmospheric conditions. Unfortunately, the optical fibres of the fourth AVIRIS (D-) spectrometer were defect during the European ER-2 deployment, reducing signal throughput to about 15 % of nominal. The resulting data quality in the 2.0-2.5 μm range was therefore not sufficient to use this part of the spectral information.

7. Soil condition and erosion mapping

The thematic analysis of our airborne imaging spectrometry data involved three separate stages. Fundamental pre-processing steps (radiometric corrections) were followed by the spectral decomposition of the original image spectra (spectral mixture modelling). After disturbing effects of terrain illumination and partial vegetation were compensated through renormalization techniques, various soil condition classes were mapped with a euclidian minimum distance classifier.

7.1 Radiometric Pre-Processing

The integration of spectroradiometric field/laboratory measurements into the analysis of optical data requires a conversion of the airborne images to physical quantities, such as reflectance factors. Radiometric corrections of imaging spectrometry data have been performed with different techniques such as the residual or scene average method, flat-field correction, single-spectrum method, empirical line (or regression) method and modelistic radiative transfer calculations.

All methods have advantages and disadvantages. Scene-based approaches (residual method, flat-field corrections) are very sensitive to image-derived quantities or reference target characteristics, and they bear the risk of introducing artifacts or increasing noise features. Single-spectrum and empirical line method heavily depend upon the availability of spectrally well-characterized ground targets. In particular the empirical line (or regression) method requires at least two targets of contrasting and equivalent spectral resolution for which both surface reflectance and airborne spectrometer responses are known.

Atmospheric corrections through radiative transfer calculations have proven quite efficient for converting airborne spectrometry data to reflectance factors. It was particularly encouraging that reasonable agreement was found between the results of different radiative transfer models that employ multiple scattering calculations (Conel et al., 1988). However, it must be noted that the successful application of radiative transfer codes heavily depends on the availability of valid instrument calibration coefficients and atmospheric parameters to adequately characterize the conditions during the flight (i.e. atmospheric optical depth). These parameters are not readily available, in particular for airborne imaging spectrometers which also cover those parts of the solar spectrum that are heavily affected by absorption effects due to spatially variable atmospheric gases (i.e. water vapour).

We have adopted an approach that combines a radiative transfer code with the empirical line method. The atmospheric correction method is originally based on the formulation of radiative transfer as developed by Tanré et al. (1990), and it was modified to account for atmospheric extinction processes as a function of sensor and terrain altitude (Hill, 1991). The modified '5S' code makes extensive use of analytical expressions and preselected atmospheric models, resulting in a rather short execution time. It provides corrections for atmospheric absorption, scattering and pixel adjacency effects, where diffusion and absorption processes are assumed to be independent. Upward and downward transmission coefficients are derived by introducing the auxiliary quantity of optical thickness τ which measures the total extinction of a light beam due to molecular and aerosol scattering when passing through an airmass. Multiple scattering is accounted for according to Sobolev's approximate solution (Sobolev, 1963), and the absorbing atmospheric gases (H_2O, O_3, CO_2, O_2) are assumed to condense at the top of the atmosphere and at the top of the layer between the earth surface and the sensor altitude.

Atmospheric correction of the AVIRIS data involved two steps. Firstly, reflectance measurements from selected calibration targets of bare soils in the study site (Altherr et al., 1991) were compared to atmospherically corrected AVIRIS spectra. For this comparison the atmospheric model was constrained with standard atmospheric data (aerosol optical depth for a horizontal visibility of 40 kilometers, gaseous absorption optical depth for a LOWTRAN mid-latitude summer atmosphere). Outside atmospheric absorption windows the instrument calibration gains were modified until AVIRIS reflectance matched the field-measured reflectance factors. The required modifications rarely deviated by more than 5 per cent from JPL's pre-flight calibration gains (figure 5).

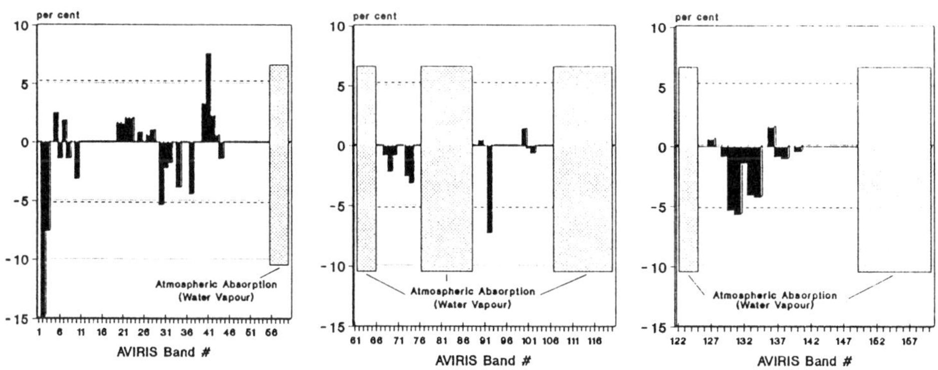

FIGURE 5. In-flight adjustments of calibration gains for the A-, B- and C- spectrometer of AVIRIS

In-flight calibration thus being confirmed, laboratory reflectance measurements of soil samples from the calibration sites were used to update also the water vapour optical depth in the various absorption windows by correspondingly inverting the radiative transfer code.

After having thus calibrated the sensor-atmosphere system with reference to controlled field locations, additional image spectra with a wide range of intensity levels were extracted from homogeneous targets in the AVIRIS scene, and then converted to reflectance factors by applying the fully constrained radiative transfer code. To reduce computational requirements, count-to-reflectance conversion for the complete AVIRIS scene was then performed with the empirical line (regression) method. For each spectral band i, a linear regression between AVIRIS digital counts (DN_i) and modeled AVIRIS reflectance (ρ_i) from the selected targets was applied to develop equations of the form

$$\rho_i = a_o + a_1\, DN_i \,.$$

The regression parameters indicate a partial failure of the method in spectral regions that are strongly affected by atmospheric absorption, but otherwise confirm the validity of this radiometric rectification (figure 6). As a difference to the original 5S code this approach also implies that the environmental reflectance effects for each pixel are no longer considered. These, however, should be minimum in case of exceptionally clear atmospheres like in the present case.

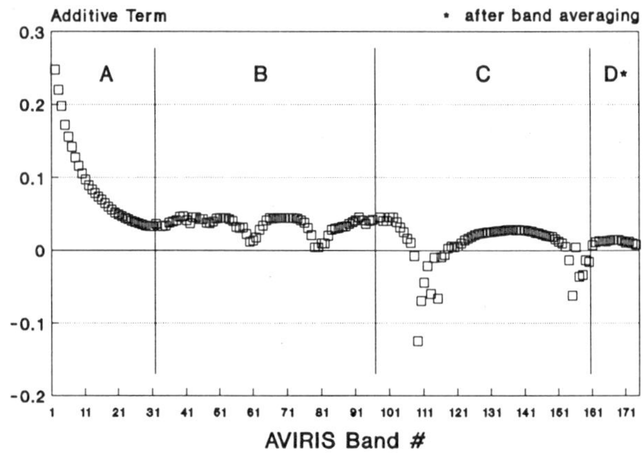

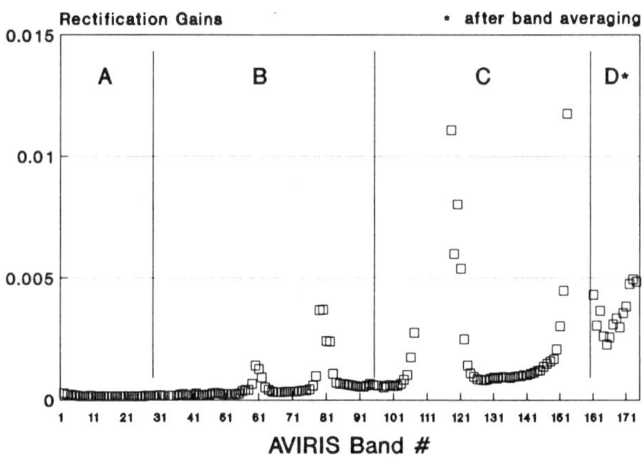

FIGURE 6. Rectification gains and complemented additive terms for the AVIRIS radiometric correction (empirical line method)

Due to the absence of strong horizontal and vertical variations in atmospheric water vapour concentration, the correspondence between field and corrected AVIRIS spectra is good within the complete image frame. However, problems still arise in the 2.0-2.5 μm region because the signal in the recomputed spectral bands of the D-spectrometer appears still very noisy. We have therefore excluded these bands from the further analysis. After eliminating additional noisy channels from the atmospheric absorption regions, totally 124 bands between 0.4 and 1.8 μm were maintained for further data analysis.

7.2 SPECTRAL MIXTURE ANALYSIS

The method of computationally decomposing spectra into their proportions of spectral prototypes ("endmembers") has become known under the term "spectral mixture analysis" (Adams et al., 1986; Boardman, 1989; Smith et al., 1990).

Spectral mixture analysis (SMA) assumes that most of the spectral variation in multi-spectral images is caused by mixtures of a limited number of surface materials (i.e. vegetation, soil, shade) with different reflectance spectra (Smith et al., 1990). They commonly mix at the sub-pixel scale, producing mixed-pixel spectra. As a first approximation, spectral mixing is modeled as a linear combination of pure component ("endmember") spectra, such that

$$R_i = \sum_{j=1}^{n} F_j \cdot RE_{ij} + \varepsilon_i \quad \text{and} \quad \sum_{j=1}^{n} F_j = 1 \tag{1}$$

where $\boldsymbol{R_i}$ is the reflectance of the mixed spectrum in band $\boldsymbol{i}$, $\boldsymbol{RE_{ij}}$ the reflectance of the endmember spectrum $\boldsymbol{j}$ in band $\boldsymbol{i}$; $\boldsymbol{F_j}$ denotes the fraction of endmember $\boldsymbol{j}$, and $\boldsymbol{\varepsilon_i}$ the residual error in band $\boldsymbol{i}$. A unique solution is possible as long as the number of spectral components does not exceed the number of bands plus one.

Linear mixing assumes that the surface components are large and/or opaque enough to allow photons to interact with only one component. Assuming the radiative transfer processes to be additive, spectra can then be unmixed by inverting the linear mixing equation (equation 1) using a least squares regression, while constraining the sum of the fractions to one. The objective is to isolate the spectral contributions of important surface materials ("endmember abundance") before these are edited and recombined to produce thematic maps (Adams et al., 1989). Spectral endmembers are chosen from spectral libraries, but can also be retrieved from the image itself.

Controlling the quality of mixing models is possible by computing the average root-mean-squared (RMS) error

$$RMSE = \sum_{k=1}^{m} \left(\sqrt{\left(\sum_{j=1}^{n} \left(R_{jk} - R_{jk}' \right)^2 \right) / n} \right) / m \tag{2}$$

where $\boldsymbol{R_{jk}}$ refers to the modeled and $\boldsymbol{R_{jk}'}$ to the measured reflectance of a pixel, $\boldsymbol{n}$ denotes the number of spectral bands and $\boldsymbol{m}$ the number of pixels within the image, or by using the wavelength-dependent band residuals

$$\varepsilon_i = R_i - \sum_{j=1}^{n} F_j \cdot RE_{ij} \tag{3}$$

which are calculated by subtracting the modeled reflectance in each spectral band from the sensor-measured reflectance.

The average RMS error for the image provides a measure how much of the spectral variability was explained by the selected endmembers, and an image of the RMS error for each pixel will highlight such objects which could not be adequately modeled. Positive band residuals $\boldsymbol{\varepsilon_i}$ (equ. 3) occur when the *measured* spectrum has higher reflectance at a specific wavelength than the *modeled* spectrum, indicating that the *modeled* spectrum contains absorption features which were lacking in the *measured* spectrum. Negative band residuals indicate the presence of absorption features in the *measured* spectrum which are absent or less pronounced in the *modeled* spectrum. Both, overall RMS error and band residuals provide important diagnostics for handling uncertainties of the mixing model and for its optimization.

7.3 REFERENCE SPECTRA

Initially, it was attempted to model AVIRIS reflectance as a linear combination of green vegetation, soil and shade spectra. In this combination of spectral endmembers, the green vegetation spectrum accounts for varying amounts of photosynthetically active vegetation, while the soil spectrum represents the spectral contribution of the background. An additional spectral endmember ("shade") is required to isolate the influence of shading and shadows which relate to vegetation and soil/rock roughness elements, topography and solar elevation. "Shade" can mix with each of the other endmembers or with their mixtures, thereby modelling the spectrum of the endmember material when it is not fully illuminated (Adams et al., 1989).

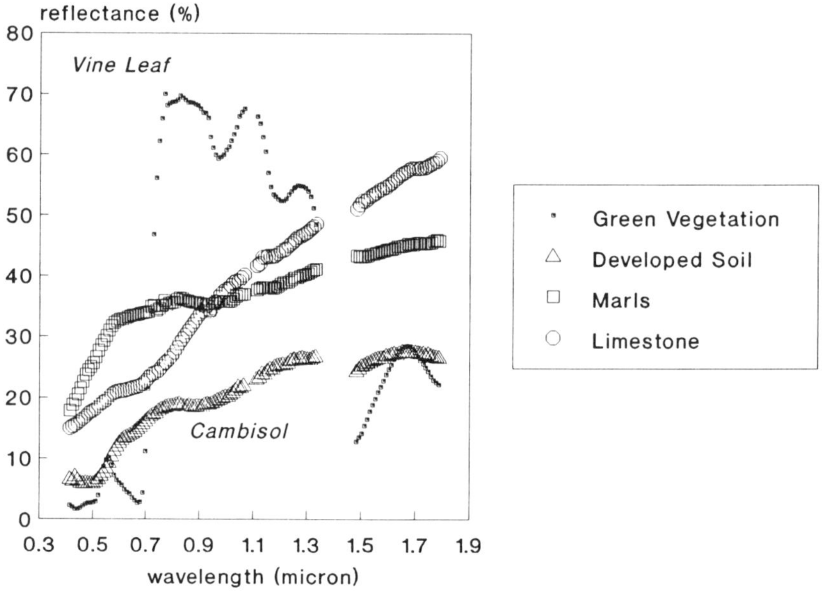

FIGURE 7. Spectral endmembers of the expanded linear mixing model

All reference spectra were chosen from a library of high spectral resolution spectroradiometric measurements from the study site that were collected with the GER "Single-field-of-View Infra-Red Intelligent Spectrometer" (SIRIS) (Altherr et al., 1991). These spectra were convolved to match the retained AVIRIS bandset (figure 7). In our initial approach, green vegetation was modeled with a reflectance spectrum of vine leafs, a cambisol spectrum represented the spectral contribution of soil. Shade was approximated with a spectrum of continuous zero reflectance, assuming that additive atmospheric scattering was successfully removed through the radiometric rectification.

Although the overall root-mean-squared error (r_{rmse} = 0.023) was close to the S/N specification of AVIRIS, substantial modeling errors occured in such areas were the selected soil endmember was different from the actual substrate type. Substituting the "background" endmember through either marl or limestone spectra caused similar modeling errors which of course appeared at different locations (figure 8). The fact that none of the three-endmember models was fully adequate to model the scene confirmed that even the reduced AVIRIS bandset was sensitive to variable vegetation-background configurations.

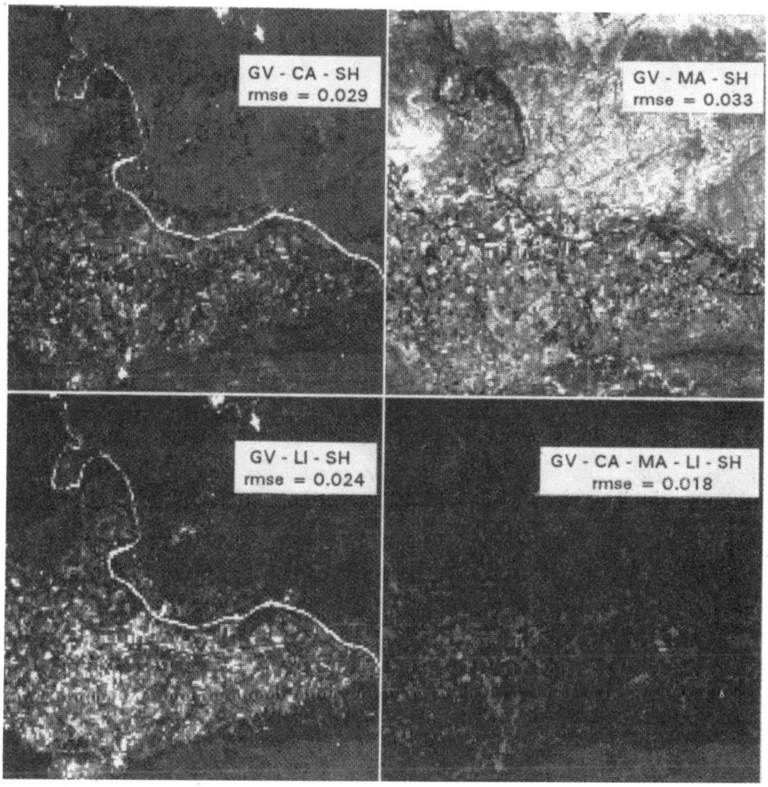

FIGURE 8. RMS error images of spectral mixture models with different endmember sets (GV: green vegetation; CA: cambisol; MA: marl; LI: limestone; SH: shade)

There exist two basic strategies to solve this problem. Firstly, it can be attempted to analyze band residuals of the mixing model for mapping those background surface characteristics which were not accounted for by the selected endmember configuration. Secondly, the mixing model may be expanded by directly including reference spectra for the non-modeled surface components, in this case marls and limestone. Adopting the latter approach a five-endmember mixing model was defined (figure 7), assuming that the surface reflectance of well-developed soils will be dominantly modeled by the cambisol spectrum (EM2), while degraded soils or erosion features would respond with increasing fractions of corresponding parent rock spectra (EM3 and EM4). The rationale behind this strategy is that soil substrates are considered mixtures where, according to soil conditions, either components of the parent lithology or material produced by pedogenetic processes have become predominant. The expanded mixing model in fact provided the lowest overall RMS error ($r_{rmse} = 0.018$) with a minimum of recognizable spatial features (figure 8).

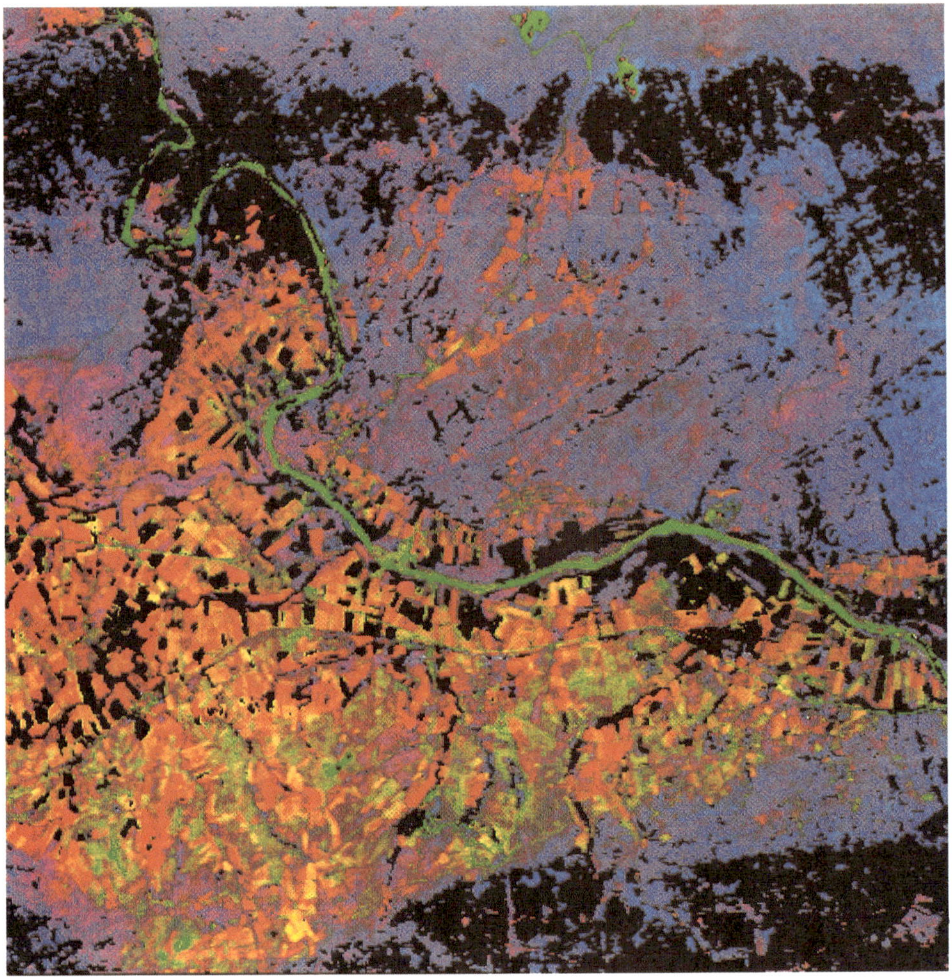

FIGURE 9. Colour composite of renormalised fractions of substrate endmembers (EM2 - EM4) used to spectrally unmix the AVIRIS scene of the Ardèche study site. Areas of permanent vegetation cover ($Fr_{(EM1)} > 0.5$) are masked in black

Further analysis involved selective editing of the resulting fraction images. For better identification of substrate-related spectral components the effects of vegetation and macroscopic shade were removed by renormalizing the three substrate fractions (soil, marl and limestone) with respect to the vegetation and shade fractions. Initially, the fraction of shade ($\boldsymbol{F}_{shade}$) was deleted by rescaling the remaining fractions with the normalisation factor

$$f = 1/\left(1 - F_{shade}\right) \tag{4}$$

so that they sum again to unity

$$\sum_{j=1}^{(n-1)} F_j \, f = 1 \, . \tag{5}$$

For example, a pixel having 0.33 vegetation, 0.33 substrates (i.e. EM2+EM3+EM4) and 0.33 shade would convert to 0.5 vegetation and 0.5 substrates (Adams et al., 1989). We can thus eliminate part of the disturbing influence of illumination and variable soil moisture, retaining that part of the signal which purely relates to vegetation and soil reflectance. Subsequently, also the fraction of green vegetation was eliminated for such pixels that had a vegetation fraction of less than 0.5. All other pixels have been masked, assuming that a more dense coverage with green vegetation would increasingly devaluate the enhancement of soil-related spectral information. For all remaining pixels, however, the rescaled substrate fractions (EM2 to EM4) strongly emphasise relative abundances of developed soils and parent rock material, allowing an interpretation in terms of substrate endmembers only (figure 9).

7.4 AUTOMATIZED MAPPING OF SOIL CONDITIONS AND EROSION FEATURES

A map of soil conditions and erosion features was then obtained through a combination of cluster analysis and automatic classification of the renormalised fraction images. Initially, characteristic spectral mixing patterns were identified by applying the "k-means" unsupervised clustering method (Anderberg, 1973) to representative areas of the averaged fraction image. Eleven substrate classes were identified which, according to their relative proportions of endmembers EM2, EM3 and EM4, were associated to four soil degradation levels. The abundance estimate for endmember 2 (cambisol) appears strongly related to the intensity of pedogenetic processes, while relative proportions of endmembers 3 and 4 reflect variations of the parent lithology (table 2).

Class	EM2 (Cambisol)	EM3 (Marls)	EM4 (Limestone)	Soil Condition	Soil Taxonomy (acc. to FAO, 1988)	Level
1	78.5	34.0	-12.5	fair to satisfactory	Cambisols, Fluvisols	I
2	71.1	18.0	9.4		Orthic Rendzinas & Calcaric Regosols	
3	57.3	6.4	36.3	degraded	Calcaric Regosols	II
4	56.2	-13.2	57.0		& Rendzic Leptosols	
5	46.4	14.3	62.1			
6	44.3	48.6	7.1	severely degraded	Rendzic, Eutric	III
7	38.4	29.1	31.1		& Lithic	
8	39.7	-10.9	71.2		Leptosols	
9	4.0	91.7	4.3	denuded	Marls	IV
10	20.3	-6.4	86.1		Limestone	
11	5.4	-1.5	96.1		Limestone	

TABLE 2. Correspondence of mixing patterns (in percent) , soil condition levels and soil types

It should be recalled that it is impossible to identify a direct correspondence between spectral mixing classes and individual soil types, unless pedogenetic characteristics are significantly correlated with spectral surface features. What appears feasible, however, is to establish a relation between soil development levels (I to III) and grouped pedogenetic types (table 2). The known characteristics of these soil types (i.e. Commission of the European Communities, 1985) can then serve to associate quantitative descriptions with the spectral mixing classes, which in turn may help to apply the mapping results in the context of land management programmes. Thematic maps can be straightforwardly obtained by the applying simple classification methods (e.g., euclidian minimum distance) to the substrate fraction image, thereby assigning each pixel to that soil development class which most closely resembles its vector of mixing fractions (table 2).

FIGURE 10. AVIRIS-derived soil condition classes (I to IV), mapped from the renormalized substrate fractions of the expanded mixing model with 5 spectral endmembers (brown: I = well-developed; ochre: II = degraded/eroded; yellow: III = severely degraded/eroded; red: IV = denuded marls; blue: IV = denuded limestone; black: not mapped, because green vegetation < ~50 %)

Comparisons to a set of 95 control sites in the field (75.4 % overall accuracy) and to available air photos confirm that the mapping results are in very good agreement to the spatial patterns of well-developed and degraded soils in the study site, even in areas where different soil associations are found within the same lithological unit (figure 10).

7.5 RESIDUAL ANALYSIS

The RMSE-image of the expanded (5-endmember) mixing model shows that some spatial features in the AVIRIS scene are still deficiently modeled (figure 8). A thorough analysis of band residual can help to explain these modeling errors.

Our approach to the analysis of modeling residuals concentrates on the importance of identifying the 'correct' substrate endmember (here also termed "background") for unmixing vegetation-soil spectra. The problem becomes clear when we tentatively model a synthetic spectral mixture, such as typically produced by row crops (i.e. vineyards), by using endmember sets with different substrates (marls, limestone and a vertic cambisol). The resulting abundance estimates differ from the original fractions, and only the mixing model with the most similar soil type (vertic cambisol) is capable to reproduce a green vegetation fraction which is close to the original mixture (figure 11).

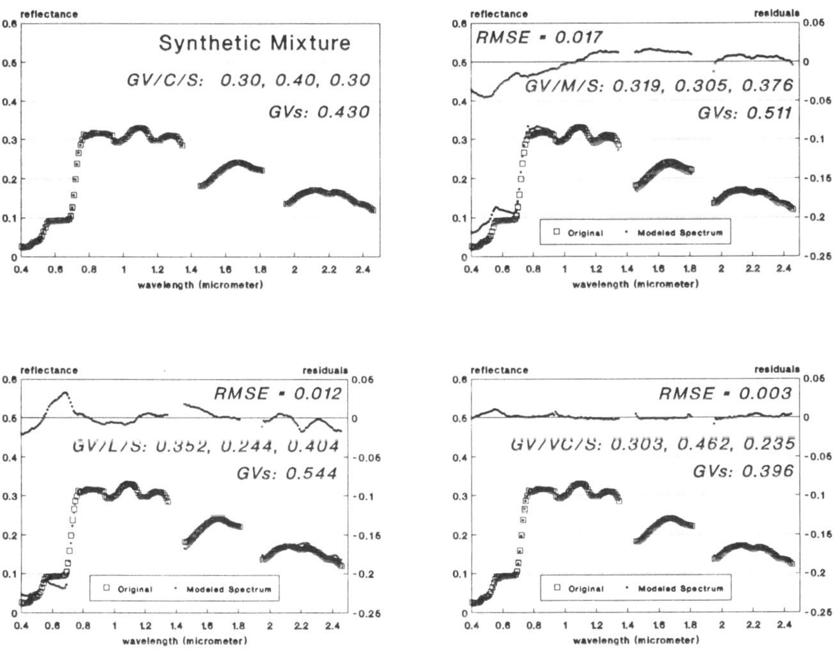

FIGURE 11. Spectral unmixing of a synthetic spectrum, representative for a row crop (e.g. vineyard), using different endmember sets. Fractions are: GV (green vegetation), S (shade), C (cambisol), M (marls), L (limestone), and VC (vertic cambisol); the subscript *s* denotes shade-normalized abundance estimates.

However, since cambic vertisols are darker than the cambisol which was used in the synthetic spectral mixture a larger soil fraction is required to model the mixed spectrum, which is compensated through a lesser amount of shade. It s also seen that the overall RMS error varies according to the aptness of the mixture model, and that the band residuals indicate spectral regions where the particular model is deficient. Using, for example, a limestone spectrum as "background" endmember creates negative band residuals at 2.2 μm, indicating the presence of clay minerals (mainly kaolinite) in the original cambisol (figure 11).

Similar effects also occur when a spectral mixture of senescent vegetation, soil and shade (i.e. complete absence of green, photosynthetically active vegetation) is tentatively unmixed with endmember combinations that involve no dry vegetation component (figure 12). While the "green vegetation - soil (i.e. cambisol) - shade" model is still capable to approximate the overall shape of the original spectrum, it produces band residuals which indicate the absence of combined lignin/cellulose absorptions around 1.75, 2.15 and 2.3 μm, being diagnostic for the presence of senescent plant material. As soon as the dry vegetation spectrum is included into the mixing model, for example as part of three- or four-endmember configurations, band residuals and RMS error diminish, thereby providing quite accurate estimates of the originally used component fractions (figure 12).

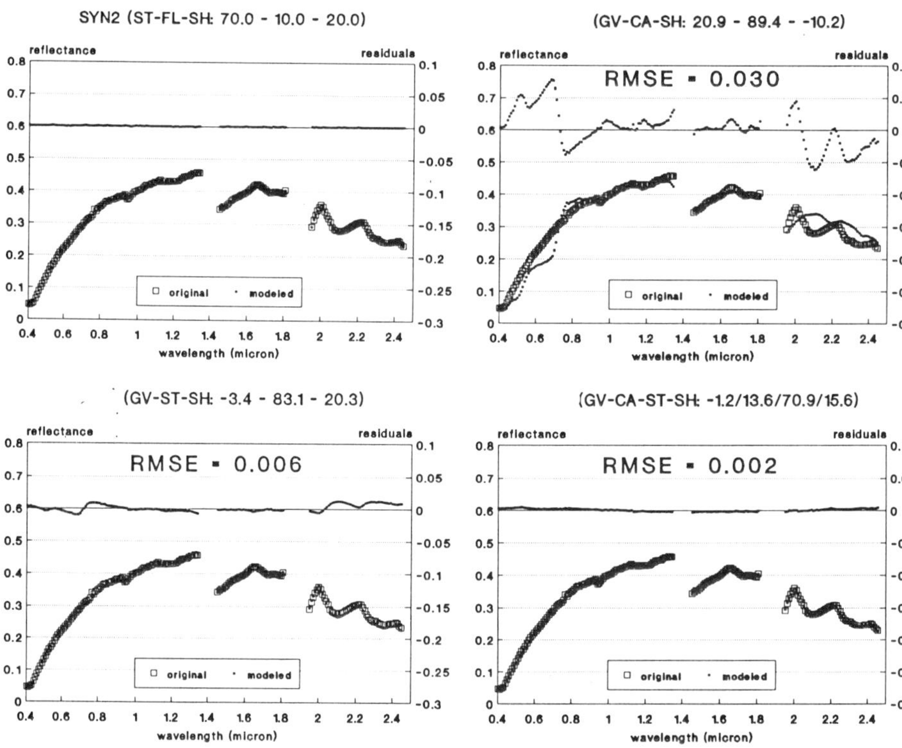

FIGURE 12. Spectral unmixing of a synthetic mixed spectrum describing senescent vegetation in the study site (SYN2: 0.70 straw [ST], 0.10 soil (fluvisol) [FL] + 0.20 shade [SH]), using various endmember sets (GV = green vegetation, CA = cambisol)

The analysis of AVIRIS spectra from, according to the RMSE-image, deficiently modeled parts of the scene reveals very similar characteristics. It is recognized that "senescent vegetation" spectra are erroneously interpreted as being mainly composed by the soil (i.e. cambisol) spectrum, finally producing thematic mapping errors. Since the local modeling errors (band residuals) appear particularly sensitive to such mismatches, it is suggested to use this parameter to adaptively optimize the spectral unmixing process for each pixel.

Based upon these results, a strategy has been developed which considers each AVIRIS pixel to be composed from three major spectral components, namely "foreground"- and "background"-materials, and "shade". Assuming that the foreground material constantly consists of green vegetation (which may be present or not), the background components are considered to be spatially variable. Based on the smallest modeling error it is then determined through the unmixing process itself which set of spectral endmembers provide the optimal solution for a specific pixel. Taking into account that the modeling error in a single spectral band is too noise-sensitive a search window (with a width of ***n*** bands, and ***n*** depending on the spectral bandwidth of the sensor) is used to scan the vector of band residuals, thereby finding the maximum *local RMS error* for a particular endmember set. After testing all available configurations, the algorithm finally applies the endmember combination which produces the smallest local difference between measured and modeled spectrum, assuming that the presence or absence of spectral absorptions is indicative for specific background materials. The selected endmember set is documented in an index map.

Selective editing of index map and resulting fraction images (combining, for example, pixels with low green vegetation abundance and 'senescent vegetation' as the dominant background fraction) was then used to successfully flag those pixels which were incorrectly mapped as well-developed soils.

7.6 Estimating green vegetation cover

The identification and quantification of green plant material is of great interest since leafs and needles are the site of photosynthesis and the prime link between the biosphere and the atmosphere (Pinty & Verstraete, 1992). Considering vegetation the functional, tangible equivalent of terrestrial ecosystems (Graetz, 1990), it follows that changes in vegetation structure and structural dynamics provide important indications for land degradation (desertification) processes. Their precise characterisation, together with the identification of soil conditions (i.e. detection of erosion features) therefore constitute core elements for monitoring the dynamic behaviour of Mediterranean ecosystems.

Since plants have a distinct spectral signature with low reflectance in the visible and short-wave infrared part of the spectrum and high reflectance in the near-infrared region, attempts have been made to use the spectral information for estimating green vegetation cover and biomass through various *vegetation indices* (Perry and Lautenschlager, 1984). The 'Normalized Difference Vegetation Index' (NDVI) is one of the most common indices for exploiting spectral signatures of plant materials. Although images of vegetation indices exhibit significant structure, much concern has been expressed about their sensitivity to the state of the atmosphere, illumination and observation geometry and to the background reflectance of soils and rocks. Pinty and Verstraete (1992) have therefore proposed an alternative index ('Global Environment Monitoring Index', GEMI) which compensates most of atmospheric and illumination conditions, and Huete (1988) and Clevers (1989) have suggested vegetation indices with an improved ability to account for reflectance contributions from background substrates. However, as these indices implicitly or explicitly employ "average" soil spectra, it still appears difficult to compensate effects that result from *spatially variable* substrate types.

Given the sensitivity of spectral mixture analysis to varying background spectra, as it has been demonstrated in the previous section, it is believed that mixing models with multiple endmember sets may also provide better solutions for biomass surveys in complex ecosystems. We have applied these methods (NDVI, GEMI, fixed and multiple mixing models) to three control sites, for which the fractional vegetation cover had been measured during the AVIRIS overflight (table 3). The first site is a large agricultural field of wheat stubble (fractional green vegetation cover 0 %), the second includes an area of severely eroded marls with scattered dwarf shrubs (fractional cover 5-10 %), and the third is covered by very dense vine plants (fractional cover virtually amounting to 100 %).

Estimation Method	Site 1 (0 %)	Site 2 (5-10 %)	Site 3 (100 %)
NDVI (range 0 to 1)	0.175	0.076	0.919
GEMI (range 0 to >1)	0.319	0.206	1.091
SMA fixed (fraction · 100) 1)	27.4	6.7	118.8
SMA variable (fraction · 100) 2)	-1.8 (d)	7.0 (b)	117.2 (a)

2) fixed endmember set: green vegegation (GV), cambisol (CA), marls (MA), limestone (LI), shade (SH)
3) variable endmember sets: (a) GV-CA-SH, (b) GV-MA-SH, (c) GV-LI-SH, (d) GV-DV (dry vegetation)-SH

TABLE 3. Green vegetation estimates through vegetation indices (NDVI, GEMI) and spectral mixture analysis (SMA) for selected control sites.

All methods, except spectral unmixing with variable endmember sets, overestimate green vegetation abundance when the target is characterised by large amounts of senescent plant material, while the use of variable endmember sets, being automatically selected according to the smallest modelling error (see section 7.5), seems to provide a good direct estimate of fractional green vegetation cover (table 3). Although green vegetation cover is somewhat overestimated for the closed canopy (where non-linear multiple scattering processes may have an important function), it is concluded that the adaption to local background/substrate variations is an important requirement for obtaining less biased estimates of green vegetation cover. But further studies and method developments might be needed to more efficiently control the use of multiple spectral endmember configurations.

8. Conclusions

During summer 1991, AVIRIS data over a mediterranean test site in the Ardèche province were acquired during the NASA Multisensor Airborne Campaign (MAC) in Europe. Although spectral data from the D-spectrometer (2.0-2.5 μm) could not be used for our analysis, the spatial extension of degraded and eroded soils could be reliably mapped. This is based on the fact that soil degradation was significantly correlated with spectrally detectable surface characteristics (influenced by the relative amounts of developed soil substrates and material of the parent lithology). The approach, which is primarily based on linear spectral mixture analysis, provided results at a level of precision and spatial differentiation which, due to methodological and financial constraints, are difficult to obtain with conventional mapping approaches.

It must be emphasised that spectral mixture analysis interprets optical images in the context of physically validated constraints, since the data analysis is directly controlled by spectroradiometric field and laboratory measurements. The method requires radiometric corrections of the images prior to the thematic analysis of the data sets but, once these corrections are applied, also holds the potential to be largely standardised in terms of the required processing parameters (i.e. spectral endmembers).

The use of multiple endmember sets for the spectral unmixing of imaging spectrometer data provided promising results. It was possible to identify areas with large amounts of senescent vegetation which, with a single set of spectral endmembers, had been erroneously identified as soils. While the background reflectance of soils may significantly affect estimates from conventional vegetation indices, it is believed that spectral unmixing with multiple endmember sets may also provide better solutions for biomass surveys in complex ecosystems.

Both, mapping results for soil conditions and the estimates of green vegetatiuon abundance suggest that high spectral resolution images may provide important advances in the remote observation of Mediterranean ecosystems currently at risk of further degradation. It is believed that more accurate abundance estimates and increased mapping accuracy may be achieved when the 2.0-2.5 μm spectral range is available, too. Parallel investigations are undertaken for testing possibilities of transfering the proposed method to data from operational earth observation satellites.

9. Acknowledgements

The authors wish to thank the Exploratory Research Programme of the CEC Joint Research Centre for supporting this research, and we wish to thank NASA for the opportunity to work with data from an outstanding instrument. We are also grateful to John B. Adams, Milton O. Smith and their colleagues from the Department of Geological Sciences, University of Washington, Seattle, U.S.A., for many discussions and critical comments on our work. The support of J. Bodechtel (Munich) in providing the GER-SIRIS spectroradiometer during MAC Europe is gratefully acknowledged. We also thank G. Maracci and his team for the assistence during the field campaign.

10. References

Adams, J.B., M.O. Smith and A.R. Gillespie (1989) 'Simple models for complex natural surfaces: a strategy for the hyperspectral era of remote sensing', *Proc. of the IGARSS '89 Symposium,* July 10-14, Vancouver, Canada, 16-21.

Adams, J.B., M.O. Smith and P.E. Johnson (1986) 'Spectral mixture modeling: a new analysis of rock and soil types at the Viking Lander I site', *J. Geophysical Research*, vol. 91, 8098-8112.

Altherr, M., J. Hill, W. Mehl and P. Pollicini (1991) *NASA multi-sensor airborne campaihn: reflectance spectroscopy Ardèche June-July 1991*, Technical Note No. I.91.120: JRC, Ispra Establishment.

Anderberg, (1973) *Cluster analysis for applications*, Academic Press, New York.

Baumgardner, M.F., E.R. Stoner, L.F. Silva and L.L. Biehl (1985) 'Reflective properties of soils', *in* N. Brady (ed.) *Advances in Agronomy*, 38, Academic Press, New York, 1-44.

Birkeland, P.W. (1990) 'Soil-geomorphic research - a selective overview', *Geomorphology*, 3, 207-224.

Boardman, J. (1989) 'Inversion of imaging spectrometry data using singular value decomposition', *Proc. of the IGARSS '89 Symposium*, July 10-14, Vancouver, Canada, 2096-2072.

Bornand, M., J.P. Legros and J. Moinereau (1977) *Carte Pédologique de France a Moyenne Echelle, Privas N-19, 1:100000*, CNRA, Versailles.

Commission of the European Communities (1985) *Soil map of the European Communities, scale 1:1,000,000*, Directorate General for Agriculture, Office for Official Publications of the European Communities, Luxembourg.

Clevers, J.P.G.W. (1989) 'The application of a weighted infra-red vegetation index for estimating leaf area index by correcting for soil moisture', *Remote Sensing of Environment*, 29, 25-37.

Conel, J.E., R.O. Green, R.E. Alley, C.J. Bruegge, V. Carrere, J.S. Margoslis, G. Vane, T.G. Chrien, P.N. Slater, S.F. Biggar, P.M. Teillet, R.D. Jackson and M.S. Moran (1988) 'In-flight radiometric calibration of the Airborne Visible/Infrared Imaging Spectrometr (AVIRIS)', *in* P.N. Slater (ed.) *Recent advantages in sensors, radiometry, and data processing for remote sensing*, Proc. SPIE, vol. 924, 179-195.

Escadafal, R. (1994) 'Soil spectral properties and their relationships with environmental parameters - examples from arid regions', *in* J. Hill and J. Mégier (eds.) *Imaging spectrometry - a tool for environmental observations*, Kluwer Academic Publishers, Dordrecht, (this volume).

Fischer, A.W. (1991) 'Mapping and correlating desert soils and surfaces with imaging spectroscopy', *Proc. of the Third Airborne Visible/Infrared Imaging Spectrometer (AVIRIS) Workshop*, JPL Publication 91-28, 23-32.

Goetz, A.H.F. (1992) 'Imaging spectromerty for earth remote sensing', *in* F. Toselli and J. Bodechtel (eds.) *Imaging spectrometry: fundamentals and prospective applications*, Kluwer Academic Publishers, Dordrecht, 1-19.

Graetz, R.D. (1990) 'Remote sensing of terrestrial ecosystem structure: an ecologist's pragmatic view', *in* R.J. Hobbs and H.A. Mooney (eds.) *Remote sensing of biosphere functioning*, Springer, New York, 5-30.

Grenon, M. and M. Batisse (eds.) (1989) *Futures for the Mediterranean basin, the Blue Plan*, Oxford University Press: Oxford.

Hill, J. (1991) 'Analysis of GER imaging spectrometer data acquired during the European Imaging Spectrometry Aircraft Campaign (EISAC) '89. Quality assessments and first results', *EARSeL Advances in Remote Sensing*, vol. 1, 64-77.

Hill, J. and J. Mégier (1991) 'The use of imaging spectrometry in mediterranean land degradation and soil erosion hazard assessments', *Proc. 5th Int. Colloquium on "Physical Measurements and Signatures in Remote Sensing"*, Courchevel, France, 14-18 January, ESA SP-319, 185-188.

Huete, A.R. (1988) 'A soil-adjusted vegetation index (SAVI)', *Remote Sensing of Environment*, 25, 409-421.

Huntington, J., A.A. Green and M.D. Craig (1989) 'Identification - the goal beyond discrimination. The status of mineral and lithological identification from high resolution spectrometer data: examples and challenges', *Proc. of the IGARSS '89 Symposium*, July 10-14, Vancouver, Canada, 6-11.

Irons, J.R., Weismiller, R. and G.W. Petersen (1989) 'Soil reflectance', *in* G. Asrar (ed.) *Theory and applications of optical remote sensing*, Wiley Interscience, New York.

Morgan, R.P.C. (1986) *Soil erosion and conservation*, Longman Group, Burnt Mill, Harlow, UK.

Nahal, I. (1981) 'The mediterranean climate from a biological viewpoint', *in* F. Di Castri, D.H. Mooney and R.L. Specht (eds.) *Mediterranean-type shrublands*, Ecosystems of the World, 11, Elsevier Scientific Publishing Company, Amsterdam, Oxford, New York, 63-86.

Negendank, J.F.W., H. Baumann, H. Siemann and H.B. Brückner (1990) *Geological mapping and mineralogic analysis of surface samples for soil erosion assessment with high resolution spectroscopy data (Ardèche)*, Final report, study contract no. 3790-89-08 ED ISP D, Dep. of Geology, University of Trier, Germany.

Perry, C.R. and L.F. Lautenschlager (1984) 'Functional equivalence of spectral vegetation indices', *Remote Sensing of Environment*, 14, 169-182.

Pinty, B. and M.M. Verstraete (1992) 'GEMI: a non-linear index to monitor global vegetation from satellites', *Vegetatio*, 101, 15-20.

Riezebos, H.Th., S.M. de Jong, J.C. van Hees, and P.B.M. Haemers (1990) 'Physiographic and pedological mapping for soil erosion hazard assessment (Ardèche test site)', Final report study contract no. 3787-89-08 ED ISP NL, Inst. of Geographical Research, University of Utrecht, The Netherlands.

Smith, M.O., S.L. Ustin, J.B. Adams and A.R. Gillespie (1990) 'Vegetation in deserts: I. A regional measure of abundance from multispectral images', *Remote Sensing of Environment*, vol. 31, 1-26.

Sobolev, V.V. (1963) *A treatise on radiative transfer*, Van Nostrand, Princeton.

Stoner, E.R. and M.F. Baumgardner (1981) 'Characteristic variations in reflectance of surface soils', *J. American Soc. Soil Science*, vol. 45, 1161-1165.

Tanré, D., C. Deroo, P. Duhaut, M. Herman, J.J. Morcrette, J. Perbos and P.Y. Deschamps (1990) 'Description of a computer code to simulate the signal in the solar spectrum: the 5S code', *Int. Journal of Remote Sensing*, vol. 11, no. 4, 659-668.

Weismiller, R.A., G.E. Van Scoyoc, S.E. Pazar, K. Latz and M.F. Baumgardner (1984) 'Use of soil spectral properties for monitoring soil erosion', *in* S.A. El-Swaify, W.C. Moldenhauer and A. Lo (eds.) *Soil erosion and conservations*, Soil Cons. Soc. of America, Ankeny, Iowa.

IMAGING SPECTROSCOPY IN HYDROLOGY AND AGRICULTURE - DETERMINATION OF MODEL PARAMETERS

WOLFRAM MAUSER and HEIKE BACH
Institute for Geography
University of Munich
Luisenstr.37
D-8000 Munich 2, Germany

ABSTRACT. In the framework of hydrologic research a better knowledge of the characterisation of the land surface is essential for improved modelling of the global hydrological cycle. The upcoming Imaging Spectroscopy space missions will provide a more exact view of the land surface. To prepare this, the potential of airborne Imaging Spectroscopy data of the GER-IS-scanner and the CASI-sensor for the assessment of agricultural and hydrological parameters is analysed. The data of the aircraft campaigns are combined with a set of agricultural ground truth and field spectrometry to give an insight into methods of integrated data analysis.

After atmospheric correction and calibration, the resulting reflectance values of the wavelength region, which contains the most characteristic spectral features of vegetation, the red edge, are parameterised. Two methods (inverted Gaussian fit, second derivative) are used and their advantages and limitations are demonstrated. A strong correlation between the vegetation height of corn and the inflection wavelength of the red edge is found. This correlation exists independent of sensors, different methods for extracting the inflection wavelength, different times and different soil backgrounds. The possibility to use data of the future MERIS-sensor for quantitative red edge analysis is tested.

1. Introduction

The key question in future hydrological research is the precise quantification of the elements of the hydrological cycle on a global basis. This is becoming increasingly important in the frame of modelling of the global and regional climate and its change. In the context of understanding of the global energy budget the hydrological parameters (rainfall, evapotranspiration, runoff, soil-moisture) are the ones, which are yet most uncertain and which on the other hand are responsible for app. 80% of the global atmospheric energy transport through water vapour. This situation is further complicated through the action of man, who constantly changes the land surface of the Earth through increased and intensified agriculture, irrigation, the destruction of forests and the growth of urban areas. In order to provide food for an increasing population through agriculture and to quantify the effects of changes of the land surface on the hydrologic cycle a better parameterisation of the Earth's land surfaces is needed. This enables an improved modelling capability of global and regional energy fluxes.

The key parameter in most climates for both agricultural production and energy fluxes on the land surfaces is evapotranspiration. The quantity of evaporated and transpired water defines the

J. Hill and J. Mégier (eds.), Imaging Spectrometry - a Tool for Environmental Observations, 261–283.

agricultural yield and the amount of water vapour, which is released to the atmosphere and can then be deposited as rain at another point on the globe. Evapotranspiration to a great extent is determined by the transpiration of the vegetation cover and can be modelled using soil-vegetation-atmosphere transfer models (Lynn & Carlson, 1990; Sellers et al., 1986). For proper operation these models are in addition to meteorological parameters (radiation, temperature, humidity, wind speed) mainly depending on a set of plant specific seasonal parameters (plant species, plant height, leaf area index) which have to be provided as spatially distributed data sets in order to determine the spatial distribution of evapotranspiration. Since it is not possible to determine the spatial distribution of these parameters on the ground, remote sensing methods have to be found to accurately assess these parameters from space. Existing remote sensing data sources do not provide enough spectral resolution to do the job. This is why Imaging Spectroscopy is looked upon as being a promising tool for this kind of studies.

The plant species as one land surface parameter can quite easily be determined through statistical classification. Quantitative, dynamic plant parameters, such as the LAI, are commonly extracted from spectral information by means of ratios (Refl. (NIR) / Refl.(red), NDVI) or simple grey values. In recent studies the 'red edge' of reflectance spectra and especially the wavelength position of the red edge is investigated in greater detail (red edge = strong increase of the reflectance of vegetation spectra towards the near infrared). So far, a number of studies were carried out with field spectrometers (Boochs et al., 1990; Miller et al., 1991; Plummer et al., 1991). Their results show that the position and shape of the red edge contains information about biomass, chlorophyll content and physiological stress of vegetation. Apart from these field studies, only few analyses applied to airborne data.

The actual availability of imaging spectrometry data of agricultural units leads to the question, which pre-processing steps have to be conducted to allow the interpretation of the spectral data, and which agricultural information can be extracted from them. To answer these questions airborne imaging spectrometry data of the GER-IS-sensor and the CASI-sensor are evaluated together with agricultural ground truth. For combinations with IS-data it is preferable to use a plant parameter, which can easily, fast and accurately be measured with a complete aerial coverage. Therefore, the following analysis concentrates on the vegetation height of corn.

The airborne IS-sensors which are used can be understood as prototypes for future satellite based systems. The better spatial and spectral resolution of the airborne sensors permits simulations of future satellite systems, like MERIS or MODIS, which will allow global monitoring. These simulations are undertaken to determine the possible applicability of the developed procedures to extract plant parameters by means of future satellite missions.

2. Data Base

The data base consists of airborne imaging spectrometry (IS) data, field spectrometry data and agricultural ground truth acquired in the field during two aircraft campaigns, which took place in the summers of 1989 and 1990 respectively.

2.1 EISAC '89 CAMPAIGN (GER-IS-SENSOR)

In May and June 1989 the first European Imaging Spectroscopy Aircraft Campaign (EISAC) was conducted by JRC and ESA (Bodechtel and Sommer, 1991). The Freiburg test site was selected as one of the agricultural test areas. On May 13th at 8:25 UTC a flight with the GER-IS-sensor was conducted in the Upper Rhine valley. In the centre of the flight strip the Freiburg test site was covered with an extension of 4 by 6 km. The flight altitude was 2700 m above ground. At this altitude the IFOV of 3.3 mrad led to a spatial resolution at ground level of 9 m.

The GER-IS-sensor is an imaging spectrometry scanner with 63 spectral bands from the visible (VIS) to the short-wave infrared (SWIR). Three separate spectrometers allow the measurements in this wide wavelength interval, which covers almost the whole solar spectrum. The nominal bandwidth of the sensor is 12.5 nm between 480 and 850 nm, and 16 nm between 2.0 - 2.44 μm. Four 120 nm wide bands are situated between 1.44 and 1.8 μm.

2.2 CASI'90 CAMPAIGN

As another part of EISAC, JRC conducted a second imaging spectrometry flight over the Freiburg test site in 1990. On July 20th, 1990 the Compact Airborne Spectrographic Imager CASI (ECOSCAN, Herrenberg) was flown over the Upper Rhine Valley. The technical specifications of the CASI are summarised in table 1.

Spectral Range	380 nm - 890 nm
Number of Bands	up to 288
Spectral Sampling Interval	1.8 nm
True Bandwidth	2.9 nm
Radiometric Resolution	12 bit
Radiometric Sensitivity	0.08 μW/cm² sr nm, 50 lines/sec Noise Equiv. Radiance at 635nm
Swath (FOV)	15 - 60 degrees
Spatial Resolution	up to 578 pixels
Spatial Mode	8-15 bands, 578 pixels across track
Spectral Mode	288 bands, 39 pixels across track

TABLE 1. Technical specification of the CASI

The CASI is a pushbroom scanner, which uses a 578 x 288 CCD (Charge Coupled Device) to measure the radiance between 380 and 890 nm. The CASI is characterised by two different operating modes, which either sample data with a high spectral or high spatial resolution. In the so called 'spectral mode' all 288 spectral elements are collected for a limited number of look directions. The number and spacing of the look directions are programmable. The 'spatial mode' provides the ability to obtain maximum spatial resolution for a limited number (up to 15) of spectral bands, which can be chosen out of 288.

The CASI-flight was performed on July 20th between 8:00 and 9:30 UTC. Data were acquired in several flight strips in spectral mode. No compensation or registration of the aircraft movements was carried out. Figure 1 shows the spatial coverage of spectral data. The illustration of the band at 768 nm shows that the spectral data were acquired in 5 flight strips. The width of each flight strip is about 700 to 800 m.

The fact, that the CASI only covers a small part of the solar spectrum, however, with a continuous coverage and a very high spectral resolution, restricts its use to the analysis of waterbodies, atmosphere and vegetation. The influence of the water content as well as soil constituents can only be investigated with the GER-IS-sensor, which is sensitive also in the SWIR.

FIGURE 1. Spatial coverage of the Freiburg test-site through the CASI-data in spectral mode (the spectral band at 768 nm is shown)

2.3 FIELD SPECTROMETRY

Field spectrometry measurements of selected surfaces, which were performed synchronously with the overflights, aimed at the verification of the calibration of the airborne data and of the calculations, which transform the radiance values of the imaging spectrometry data into reflectance values. The measurements were conducted from a cherry picker at a height of 10 m with a SIRIS field spectrometers of the GER-Corporation. The SIRIS is a single field of view instrument, which covers the wavelength region between 400 and 2500 nm with a spectral resolution of 2 nm in the VIS and near infrared (400 - 1000 nm), 4 nm between 1000 - 1900 nm and 6 nm in the SWIR (1900 -2500 nm).

2.4 AGRICULTURAL GROUND TRUTH

In 1989, an intensive ground data collection was carried out in the area covered by the GER. This test site was mapped and for each field the following main agricultural features were determined:

1. vegetation type
2. vegetation height
3. row distance
4. proportion of soil covered by vegetation
5. plant damage

The field boundaries of the study area were digitised as one component of a Geographical Information System (GIS) and the vectors transformed into digital raster maps of the parameters listed above. The results of this procedure are 5 digital maps of the main agricultural parameters with a spatial resolution of 10 m (Bach and Mauser, 1989).

In 1990, the agricultural data collection was repeated. For the whole test area the land use was mapped and transformed into a digital map. During the day of the overflight of the CASI the main agricultural features of 74 fields of corn, wheat, barley, soybeans, sugar beets and potatoes were collected additionally .

3. Data Pre-processing

The imaging spectrometry data were delivered with radiance calibration. The radiance reflected from a target is not only depending on the properties of the target, but also on the solar irradiance, on atmospheric, sensor and flight conditions. For the comparison and combination of remote sensing data from different sensors, different dates and different flight conditions, the radiance values must be converted into reflectance values, which are mainly depending on the scattering and absorption characteristics of the target. The conversion into reflectance values is possible with the LOWTRAN-7 model (Kneizys et al., 1988).

Apart from the radiometric pre-processing, which will be described below, a geometric rectification is applied to the GER-data. The geometric corrections permits the combination of the GER-IS data with the digital ground truth maps. The CASI data are left in their original geometry, as they were not delivered roll-corrected and heavy aircraft movement occurred during data acquisition.

3.1 Conversion of the IS-Data into Reflectance Values

LOWTRAN-7 is a radiative transfer code, which includes the option to calculate the solar radiance reaching a sensor. Flight conditions and atmospheric conditions are taken into consideration for this calculations. The radiance reaching a sensor, which is emitted by the sun and influenced by the absorption, scattering and reflection processes in the atmosphere and on the ground, is modelled as a function of the reflectance of the surface. This allows the calculation of the unknown spectral reflectance of a surface by means of its reflected radiance, which in turn is measured by a sensor.

This principle is used to convert the GER and CASI data into reflectance values. For these calculations the actual atmospheric conditions are considered using data available through the German Weather Service. The visibility at ground level near Freiburg and radiosonde data of Stuttgart are used as an input for the LOWTRAN-7 calculations. For each overflight four LOWTRAN calculations are carried out. The radiance reaching the sensor is calculated for the whole solar spectrum with the maximum spectral resolution assuming a ground albedo of 0 %, 10 %, 30 % and 60 % respectively. The resulting 4 radiance spectra are convolved with the according sensor response function. That leads to radiance values for each band for given reflectance values of 0 %, 10 %, 30 % and 60 %. The values between these four pivots are interpolated linearly. The error introduced through the linear interpolation was determined to be less than 2 % in the blue spectral range and less than 1 % for the rest.

To verify the results of the atmospheric model, a reflectance calibrated spectrum of the CASI-sensor is compared with reflectance measurements conducted with the field spectrometer during the overflight. Figure 2 shows the result for the spectra of a soybean field.

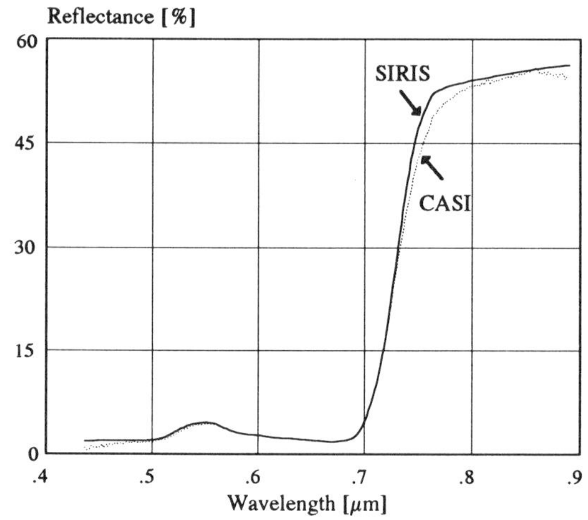

FIGURE 2. Comparison between an airborne CASI-spectrum and a field measurement conducted with a SIRIS spectrometer of the same field of soybeans

This representative result confirms the use of the LOWTRAN-7 model for converting imaging spectrometry data into reflectance values. Nevertheless, the radiometric pre-processing of the imaging spectrometry data includes problems that should not be neglected. These difficulties are sensor-specific and will therefore be summarised for each sensor separately.

3.2 Sensor-specific Data Quality Aspects of the GER

Data quality analysis shows that the signal to noise ratio (SNR) of the GER-scanner is rather low. It was determined to be 5-10 for low reflectance targets at 550 nm and 20-50 for high reflectance targets at 800 nm (Bodechtel and Sommer, 1991). This SNR does not permit a pixel-by-pixel analysis of the data set, but restricts spectral analysis to averages over larger areas.

In addition to this low spectral sensitivity, sensor defects during this campaign caused a total failure of band 28. This failure also deteriorated the neighbouring bands 27 and 29, which led to a poor data quality in the spectral range between 810 and 840 nm. The spectral resolution in the VIS and near infrared (NIR) did not confirm the nominal 12.5 nm, but was determined to be 50 nm.

Furthermore, the GER-image shows sensor dependent off-nadir-effects, which are typically caused by scanners with a large scan angle (GER: 90°), and which overshoot the LOWTRAN simulations of the angle dependency of atmospheric off-nadir effects. To eliminate these effects a statistical approach can be used. First the column-statistics (mean, standard-deviation) are calculated over a large portion of the image (1000 rows). The subsequent fitting of a 2nd-order polynomial to the mean-values of the columns leads to one correction-factor for each column and band. This correction technique allows the radiometric homogenisation of the GER-image.

3.3 SENSOR-SPECIFIC DATA QUALITY ASPECTS OF THE CASI

The raw CASI images show distinct striping along the flight direction. These variations between the 39 look directions were caused by a freezing of the sensor during the overflight, which produced ice patterns on the CCD. Based on the assumption that the effect of the ice crystals on the signal is mainly absorption and scattering, the spectra of one flight strip (512 rows) are averaged for each look direction. The ratio is calculated between the mean spectrum of the whole image and these averaged spectra of each look direction to quantify the effect of the ice patterns. The result is an array of multiplicative coefficients, one value for each look direction in each band. Multiplying the CASI raw data with these coefficients corrects the ice irregularities.

The improvement of the spectral quality can be quantified by calculating the signal to noise ratios of the CASI data before and after the pattern removal. The SNR is approximately assessed by computing the ratio of the mean and the standard deviation of several homogeneous targets. For low reflectance targets (pine trees) the SNR increases from 15 to 19 after the correction. That is an improvement of more than 25 %. The SNR for high reflectance targets (bare soil) is determined to be 55 after applying the pattern removal.

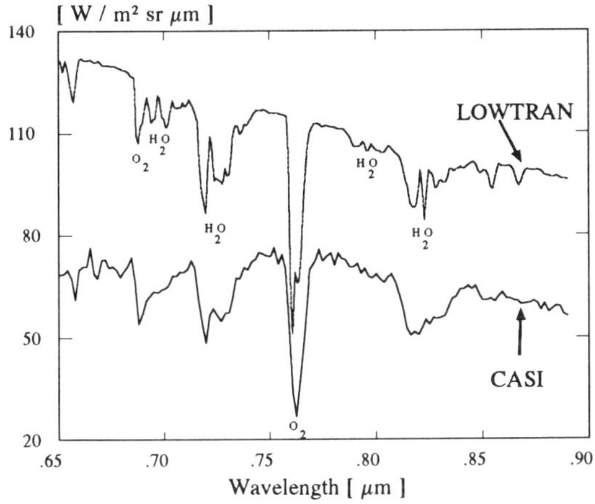

FIGURE 3. Comparison of the spectral resolution of a radiance spectrum measured with CASI and a spectrum calculated with LOWTRAN-7 (For better comparison the CASI spectrum was shifted downward by 40 W/m^2 sr μm).

In order to convert the radiance data of the CASI into reflectance values, the LOWTRAN calculations were carried out with the highest spectral resolution possible (20 cm^{-1}). The very high spectral resolution of the CASI data in spectral mode can be illustrated most clearly by comparing a CASI-radiance spectrum with a simulated LOWTRAN spectrum (figure 3). The simulated spectrum is calculated for a target with 30 % reflectance. It has a spectral resolution of 1 nm. The measured spectrum shows the radiance reflected from gravel, which has a reflectance of app. 30 % in this spectral range. The true bandwidth of the CASI nominally is 2.9 nm. Note, that for better comparison the LOWTRAN spectrum is shifted by 40 W /m^2 sr μm. The atmospheric absorptions, which are marked in figure 3, show distinct features in both spectra.

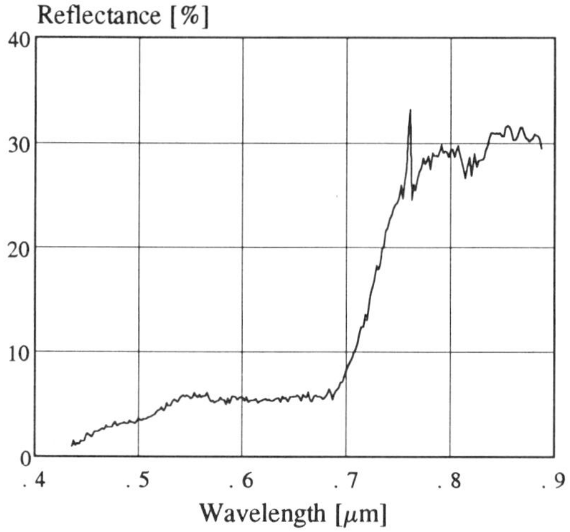

FIGURE 4. CASI reflectance spectrum derived from a radiance spectrum with nominal radiance calibration applied.

The strongest absorption in figure 3 is caused by oxygen at 760 nm. The modelled O_2-absorption is bimodal, whereas in the CASI spectrum the two minima are blurred. The modelled, very narrow oxygen absorption feature becomes broader in the CASI spectrum, and the absorption depth is stronger in the simulated than in the measured spectrum. These characteristics, which depend on the higher spectral resolution of the LOWTRAN calculations, can also be seen in the water absorptions at 700, 725 and 820 nm.

For the conversion of the CASI-data into reflectance values a precise knowledge of the spectral characteristics of the CASI is required. A reflectance spectrum converted by using the calibration data from the CASI-company is shown in figure 4. This reflectance spectrum was derived from a radiance spectrum, which was averaged over all pixels of one flight strip (app. 20000 pixels). The reflectance spectrum in figure 4 looks noisy in the visible and shows a peak at the oxygen absorption at 760 nm and water absorption features at 820 nm, which should not occur in a reflectance spectrum. These irregularities in the reflectance spectrum are not caused by random variations in the scene, but are introduced by incorrect calibration information.

The radiance calibration was conducted with an integrating sphere by the CASI-company with an accuracy of 2 %. This accuracy was diminished by the freezing on the CCD to about 4% (ITRES-Research Canada, 1991). Therefore, errors in the radiance calibration cause the systematic irregularities and 'noise' in the reflectance spectrum in figure 4.

To improve the spectral information the following procedure was developed. The procedure assumes that contrary to the radiance spectrum, which is characterised by distinct absorption features, the averaged reflectance spectrum has a smooth shape with no peaks and atmospheric features. Therefore, the reflectance spectrum (figure 4) was smoothed by fitting a spline function to the reflectance values. The resulting reflectance spectrum, which is shown in figure 5, is re-transformed into radiance values. This derived radiance spectrum is much smoother than the original radiance spectrum, but the atmospheric absorption features are still distinct. The ratio between the two radiance spectra is calculated and a smoothed radiance calibration file derived, which allows the conversion of the raw CASI radiance values into the corrected values.

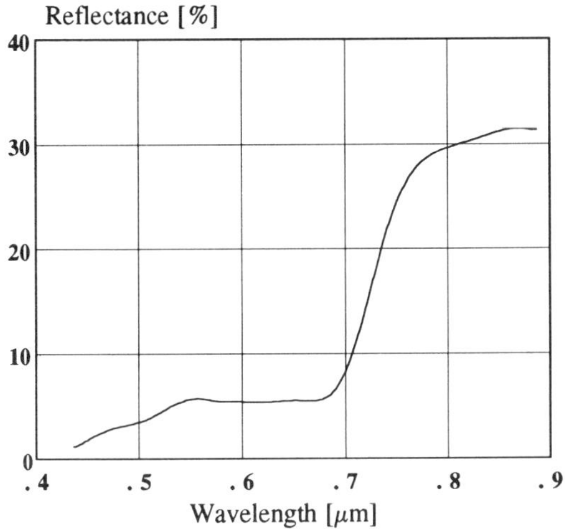

FIGURE 5. Smoothed CASI reflectance spectrum, used for deriving a smoothed radiance calibration file

The smoothed radiance calibration file is then applied to all CASI data in spectral mode. This global application results in an optimised elimination of system inaccuracies in the calibration of the data set. Because of the elimination of calibration inaccuracies this procedure also leads to a considerable smoothing of single-pixel spectra. Contrary to a pixel-by-pixel smoothing, the global smoothing of the radiance calibration file preserves comparability of spectra between pixels. It is therefore preferable to a pixel-by-pixel smoothing of spectra using spline operations.

4. Spectral Signatures of Agricultural Units

As the GER-data are very noisy, the extraction of reflectance spectra on a pixel-by-pixel basis is not adequate for the GER-data. The easiest way to reduce the noise is by averaging the pixels of single fields or selectable frames in the images. A result of this procedure is shown in figure 6 for CASI-data, which were averaged over boxes of 3 x 3 pixels.

In figure 6 reflectance spectra of corn, soybeans, mature wheat and a field of wheat, which was already harvested, are shown as they were measured by CASI. The reflectance of corn and soybeans is very similar in the visible, differing only in a slightly stronger chlorophyll absorption from the soybeans at 680 nm. The difference between these two land uses becomes evident in the near infrared, where the reflectance of the soybeans is much higher than the one of corn. The two spectra of wheat show distinct differences. The mature wheat has a much lower reflectance over the whole spectral range than wheat after harvesting.

It is the intention of this paper to provide objective analyses of the IS-data with regard to their information content about agricultural units. Because of the poor SNR of the GER, a method for extracting spectra is used, which combines the IS-data with the ground truth data of 1989. The GER-data are overlaid with the digital ground truth maps (land use, vegetation height, etc.) in order to provide a set of surface parameters for each pixel. Using this information, all pixels of a certain plant parameter can be averaged over the whole GER-image.

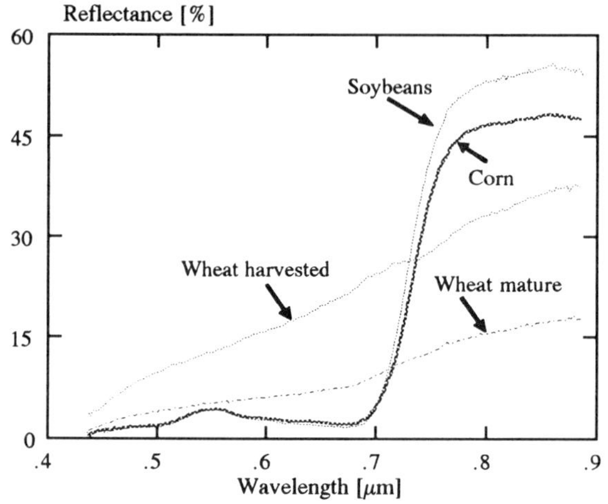

FIGURE 6. CASI reflectance spectra of corn, soybeans, mature and harvested wheat.

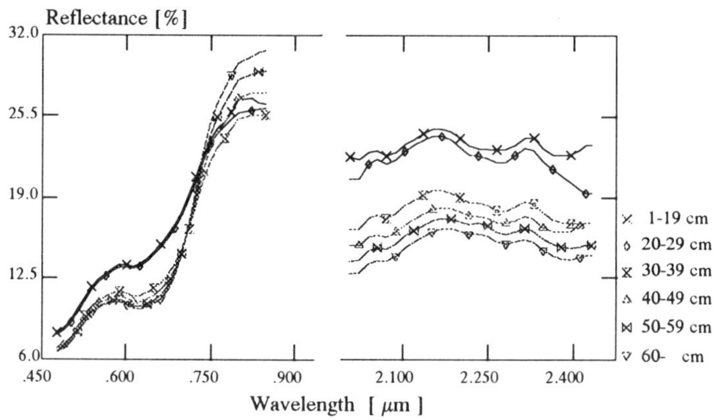

FIGURE 7. GER reflectance spectra of corn with varying plant height averaged by means of the digital maps.

An example, how spectra can be extracted by means of the digital maps, is shown in figure 7 for corn with varying vegetation height. During the overflight of the GER in 1989, the plant height of corn varied between 10 cm and 1 m. First, for the whole test site the pixels of corn are extracted (app. 32000 pixels) and then grouped into 6 classes of different plant height. For these classes the mean spectra are then calculated.

The resulting spectra, which are shown in figure 7, display obvious changes with varying vegetation height. The higher the plants, the stronger is the chlorophyll absorption at 675 nm, and the higher is the reflectance in the NIR. The largest spectral differences can be seen in the SWIR, where a continuous decrease of the reflectance with increasing plant height can be noted. This is caused by the decreasing influence of the soil signal on the signal of the mixel with increasing vegetation height.

5. Red Edge Modelling

CCD imaging spectrometry sensors, as CASI, are still restricted to the wavelength region, where silicon is sensitive, that is the visible and the near infrared. In this spectral range the typical reflectance characteristics of vegetation can be summarised by the chlorophyll-absorptions and the increase of the reflectance towards the NIR. As the blue spectral range is difficult to calibrate and strongly affected by atmospheric influences, the red edge is the characteristic feature for vegetation in the visible and near infrared.

5.1 Parameterisation of the Red Edge by Means of a Gaussian Fit

One attempt to extract the information of the red edge can be made by parameterising the course of the reflectance in the wavelength interval from 670 to 830 nm with an inverted Gaussian model (Bonham-Carter, 1988). Fitting a Gaussian function to the spectral data in this wavelength region leads to four parameters, which represent the red edge characteristics (Bach and Mauser, 1991). The red edge parameters are listed in figure 8.

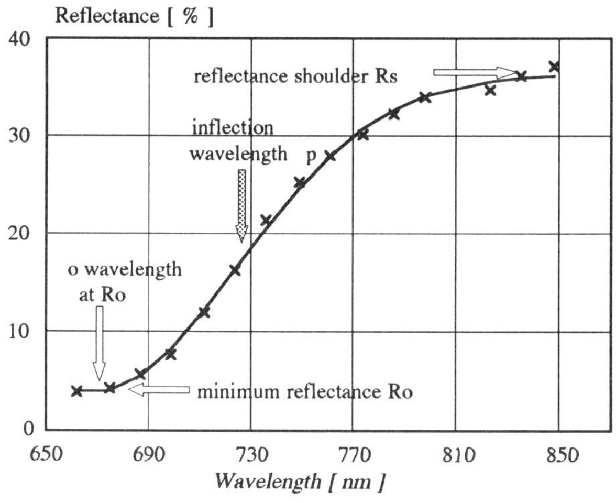

FIGURE 8. Inverted Gaussian fit of a GER-spectrum of forest.

5.2 Calculation of the Inflection Wavelength by Means of Derivatives

A more direct way to determine the inflection wavelength of the red edge, is through the calculation of the main inflection point of the red edge by means of derivatives. Figure 9 illustrates this procedure. The upper curve shows a reflectance spectrum of corn measured by the CASI sensor. To reduce noise, the reflectance spectrum is smoothed by approximating a spline function to the spectral data with an allowed average deviation of 0.1 % (absolute). The first derivative of the smoothed reflectance spectrum is shown in the centre of figure 9. It shows a distinct maximum at 730 nm, which represents the maximum slope of the reflectance spectrum. The wavelength position of this maximum slope, the inflection point of the red edge, is calculated as the second derivative intersection with the x-axis.

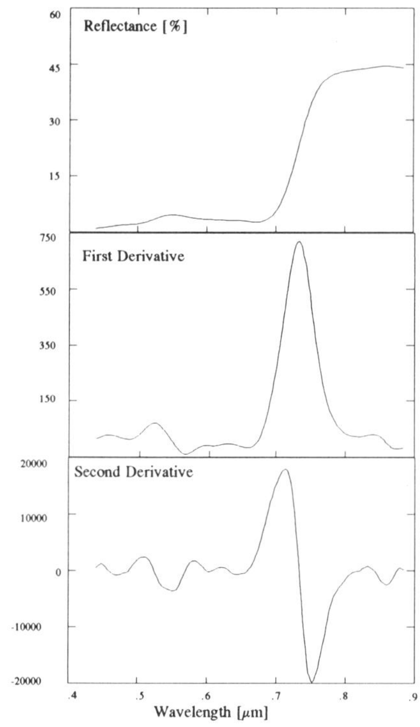

FIGURE 9. CASI-reflectance spectrum and its first and second derivatives.

5.3 Comparison of Both Methods to Parameterise the Red Edge

Of the two methods described to parameterise the red edge, the Gaussian fit has the advantage of 4 resulting parameters for the red edge. Not only the inflection wavelength, but also values of the minimum and maximum reflectance and the wavelength position of the minimum reflectance can be extracted by means of the Gaussian fit. Which of the two methods is the best to parameterise the red edge, still is a question of research. Is there a limitation for the use of the Gaussian fit and the derivative method respectively? Data of which instruments should be parameterised by means of which method? These questions will be analysed considering different bandwidths of the spectral data.

The goodness of the Gaussian fit can easily be inquired through the RMS error (RMS = root of mean square error), which quantifies the difference between measured and calculated reflectance values. In order to analyse the validity of the Gaussian fit for different spectral data CASI spectra of corn with different vegetation height are selected. From these original spectra, which are characterised by their very high spectral resolution of about 2 nm, spectra with broader bandwidth and a continuous spectral coverage are then simulated. The Gaussian fit is calculated for each simulated spectrum with varying bandwidth and the RMS error is determined. Figure 10 shows the results for two CASI spectra of corn with a vegetation height of 230 and 270 cm respectively. The RMS error is high for spectra with bandwidth of 2 to 20 nm. The broader the bands are the better the goodness of the Gaussian fit becomes. Spectra with a bandwidth of more than 20 nm are well suited with an RMS of less than 0.2 % (absolute). The higher the plants are, the stronger is the decrease of the RMS with increasing bandwidth. This illustrates that the chlorophyll absorption of the reflectance spectrum of dense vegetation becomes so strong and the red edge so steep that the shape is no longer a smooth Gaussian curve. Spectral data with a high spectral resolution preserve these edges, whereas broad bands smooth them. Therefore, the Gaussian fit should not be applied to spectra with bandwidths of less than 20 nm.

The simulated corn spectra with varying bandwidth are also used to analyse the goodness of the second derivative to calculate the inflection wavelength. To quantify the goodness of the derivative method through simulated spectra, the following steps are conducted. The decreasing part of the second derivative around the inflection wavelength (figure 9) is extracted and a regression line calculated between the second derivative and the wavelength. It is assumed that the tighter the regression is, the better the goodness of fit of the derivative method is. As the regressions for derivative spectra of different bandwidths are calculated for various numbers of data points, the resulting correlation coefficients cannot be compared directly. Therefore, the T-values of the different regressions are calculated and related with varying bandwidths of the spectral data.

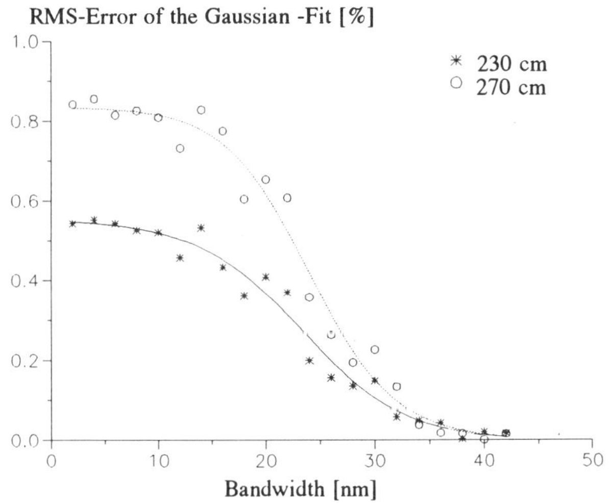

FIGURE 10. RMS-error of the Gaussian fit for calculating the inflection wavelength of spectra with varying bandwidth

The results for the corn spectra with a plant height of 230 and 270 cm, which are also used for figure 10, are shown in figure 11. A decrease of the T-value with increasing bandwidth is obvious. That means that the derivative method becomes less safe with increasing bandwidth. This time there is no distinct difference between the T-values of the two spectra of different vegetation height, contrary to the results in figure 10. Together with the T-values, also the 95% and 99% significance levels are illustrated in figure 11. The regressions are very significant up to a bandwidth of 12 nm on a 99%-level. Spectral data with a bandwidth of more than 16 nm have no significant correlations (95 % -level). One can summarise that the derivative method is very suitable for spectral data with a resolution of better than 10 nm, and is still suitable for spectral resolutions of up to 16 nm.

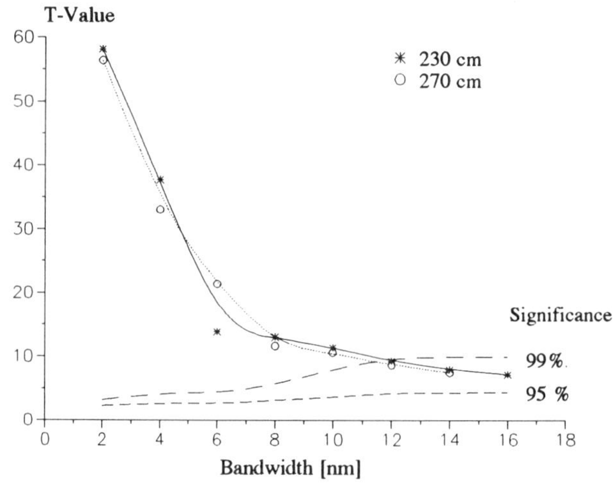

FIGURE 11. T-value of the regression for calculating the inflection wavelength by means of the second derivative of spectra with varying bandwidth

Therefore, one can conclude

- that the red edge of the GER-like-spectra should be parameterised with the Gaussian-Fit
- that the red edge of the CASI-like-spectra and the spectra measured with the field spectrometer (SIRIS) should be parameterized with the derivative method

5.4 Determination of the Vegetation Height of Corn by Means of the Red Edge Parameters

Collins (1978) showed that the wavelength position of the red edge is a good parameter to describe the growth of vegetation because of the increasing chlorophyll-a-concentration. Clevers and Büker (1991) showed through model calculations that the inflection wavelength is strongly influenced by the LAI (leaf area index), which is a key parameter quantifying the development of vegetation and the amount of transpiring leaf area. Based on these results, the expected relation between the growth and development of corn and the red edge of the corn spectra are analysed by correlations between the red edge parameters and the vegetation height of corn.

First, the results for the GER data are summarised. The comparison of the two methods to parameterise the red edge shows that the Gaussian fit is adequate for the GER-data. Therefore, the Gaussian fit is applied to the averaged GER-spectra of corn for different vegetation height (shown in figure 7). Correlations between the resulting red edge parameters and the plant height are determined. They lead to significant correlation coefficients as listed in table 2. The 'minimum reflectance' correlates negatively with the vegetation height of corn. The higher the plants are, the stronger is the chlorophyll absorption and therefore the minimum reflectance decreases. The other Gaussian parameters show a positive correlation, which is strongest for the inflection wavelength.

Min. Reflectance Ro	Reflectance Shoulder Rs	Wavelength Minimum o	Inflection Wavelength p
-0.93	0.85	0.94	0.97

TABLE 2. Correlation coefficients of the regressions between the parameters of the Gaussian fit derived from the GER-spectra of corn with varying plant height and the vegetation height of corn.

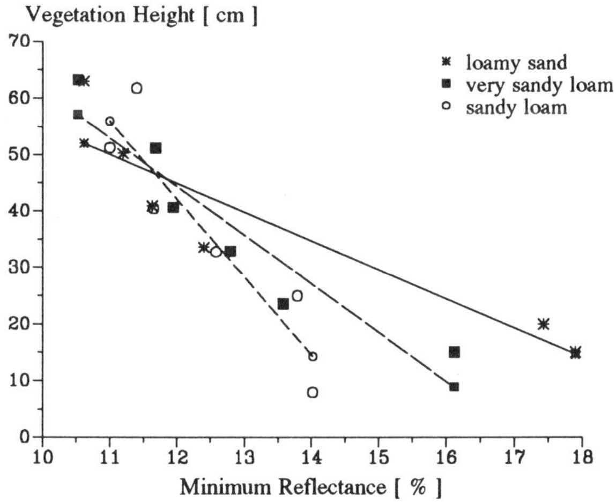

FIGURE 12. Regressions between the minimum reflectance and the vegetation height of corn for different soil types.

Although there is a correlation between the Gaussian parameters and the vegetation height of corn, it is still not clear, how strongly the underlying soils, which vary from sandy loam to loamy sand in the test area, influence the results. To analyse this, a digital soil map of the test area is included in the calculations. This soil map is based on the 50 m sample grid of the German Special Soil Estimation and was digitised from the 1:5000 topo map. It contains information on the soil type, fertility and water balance.

The averaged spectra of corn with different plant heights and for different soil types are then extracted from the data set and the Gaussian parameters are calculated. Figure 12 shows the correlation of the derived minimum reflectances and the vegetation height of corn. The different style of the data points symbolises the different soil types. For each soil type linear regressions can be applied and are shown as well. It is obvious that the influence of the underlying soil is not negligible and of specific importance for corn fields with plants smaller than 40 cm.

The correlation between the inflection wavelength and the vegetation height of corn is shown in figure 13. For each soil type the data points are close together and one single regression line can be calculated for all soil types. That shows that the inflection wavelength is a soil independent parameter for the description of the plant height of corn. The correlation coefficient of 0.99 is very high. With increasing plant height the inflection point of the red edge shifts towards the higher wavelengths.

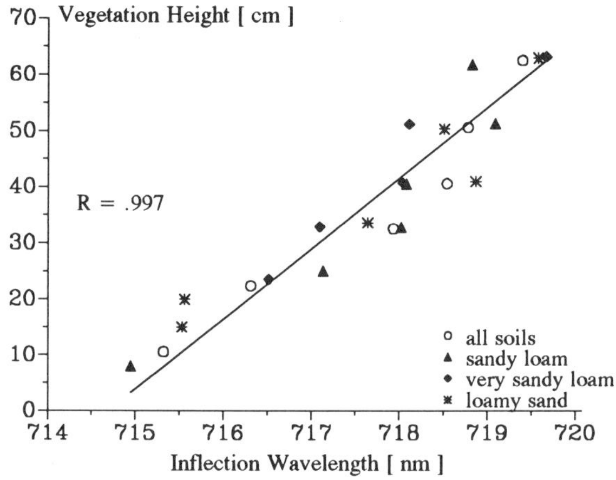

FIGURE 13. Regression between the inflection wavelength and the vegetation height of corn for different soil types.

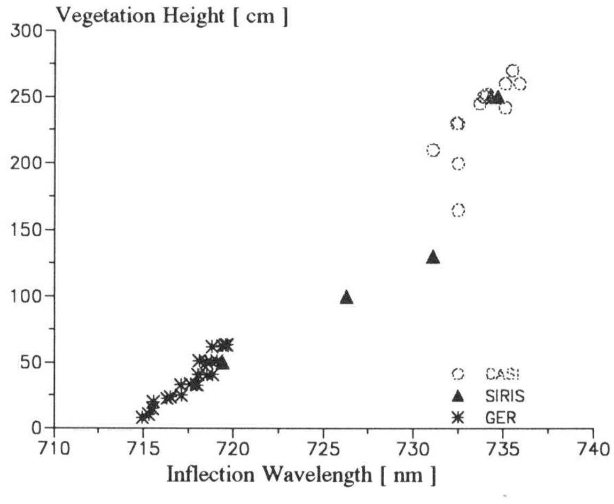

FIGURE 14. Relation between the inflection wavelength of corn spectra and the vegetation height of corn (circle = data of the CASI airborne IS-sensor; triangle = SIRIS field spectrometer; cross = GER airborne IS-sensor).

These investigations confirm that the inflection wavelength is a very good descriptor for plant characteristics, because it is independent of the underlying soil. This was also shown by Clevers and Büker (1991) in a theoretical approach with the SAIL model. They modelled that the inflection wavelength is independent from soil background and irradiation conditions, and only moderately influenced by the solar zenith angle. This emphasises the extraction of the inflection wavelength also for the CASI data and the spectra of the SIRIS-field spectrometer.

For 12 corn fields, whose vegetation height was mapped during the CASI-overflight, CASI-reflectance spectra are extracted and evaluated. The plant height of these corn fields varies from app. 2 to 3 m. In addition to these CASI-data also field measurements of 4 different samples conducted with the SIRIS are taken into consideration. The inflection wavelengths of the CASI-spectra and the SIRIS-spectra are determined by means of the second derivatives (because of the very high spectral resolution of the data) and related to the vegetation height. Figure 14 shows the relation between the plant height of corn and the derived inflection wavelengths of spectra of corn, measured with different airborne sensors and field instruments. The GER-data set displays very small plant (0 - 80 cm), the CASI-data set displays very large plants (200 - 280 cm). Therefore, it is important that the field measurements, which were conducted with the SIRIS at different phenological stages, agree both with the GER and CASI data and show a connection between both data collectives.

To summarise the results: A shift of the inflection wavelength of spectra of corn from 715 nm to 740 nm is evident during growth. This shift can be determined by determining the inflection wavelength of the red edge of the reflectance spectra. The correlation between the vegetation height of corn and the inflection wavelength of corn spectra is strong. This correlation was confirmed for different sensors, different soil backgrounds and different years.

6. MERIS Simulation

It shall be inquired, whether the results derived from airborne data can be extrapolated for future satellite missions. One future, satellite based imaging spectroscopy system will be MERIS, Medium Resolution Imaging Spectrometer. MERIS is designed to perform measurements with a high spectral and moderate spatial resolution in the visible and near infrared. This sensor is a passive optical pushbroom instrument with 15 spectral bands between 400 and 1050 nm. MERIS will possibly be installed on the first Polar Platform. The swath width is proposed to be 1500 km with a spatial resolution of 250 m over land. The technical specifications of the MERIS-sensor are summarised in table 3.

Spectral Range	400 - 1050 nm
Number of Bands	15
Spectral Bandwidth	10 nm at 410 to 755 nm 2.5 nm at 765 nm 25 nm at 1025 nm
Absolute Calibration	< 2 %
$NE\Delta\rho$	5×10^{-4} at sun angle 60°
Polarisation	<= 1 %
Swath Width	1500 km
IFOV	250 m

TABLE 3. Technical specifications of the MERIS sensor (Rast, 1991).

A detailed listing of the proposed band setting is shown in table 4. Although MERIS is mainly dedicated to oceanographic and atmospheric studies, an extension to land applications is possible. To improve the band setting for land applications 3 modified bands are proposed by the ESA MERIS land applications panel. Band 7 should be shifted from 665 to 670 nm, band 13 from 880 to 860 nm, and band 14 from 900 to 960 nm (central wavelengths).

Band 1	405 - 415 nm	
Band 2	440 - 450 nm	
Band 3	485 - 495 nm	
Band 4	515 - 525 nm	
Band 5	560 - 570 nm	
Band 6	615 - 625 nm	
Band 7	660 - 670 nm	
Band 8	680 - 685 nm	
Band 9	705 - 715 nm	
Band 10	750 - 760 nm	
Band 11	761.25 - 763.75 nm	(2 out of 3)
Band 11/12	763.75 - 766.25 nm	(2 out of 3)
Band 12	766.25 - 768.75 nm	(2 out of 3)
Band 13	875 - 885 nm	
Band 14	895 - 905 nm	
Band 15	1010 - 1035 nm	

TABLE 4. MERIS - band setting (Bezy, 1992).

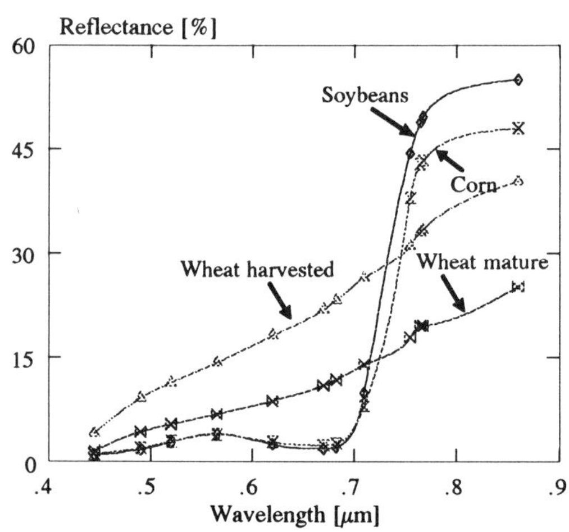

FIGURE 15. Selected reflectance spectra extracted from the simulated MERIS image

The CASI sensor has the same spectral range as the future MERIS sensor, with the exception of the two last MERIS-bands in the NIR, which are not covered by CASI. As the CASI-sensor has a higher spectral resolution, it is possible to simulate the broader MERIS-bands from the CASI data in spectral mode. For this purpose, the radiance values of CASI were integrated to the MERIS-bands and the calibration file was adapted to the new bands. These calculations were carried out for the CASI image to allow the extraction of spectra from the simulated MERIS image.

Reflectance spectra, which are extracted from the simulated MERIS image, are shown in figure 15. Spectra of soybeans, corn, mature and harvested wheat are selected. The bands of the MERIS are marked with symbols. For interpolation between the data points a spline function is used. Besides this more descriptive analysis, the feasibility of the MERIS band setting to observe the red edge was tested. For this purpose, again the CASI spectra of corn with varying vegetation height are considered. For these 12 CASI-spectra a correlation between the inflection wavelength of the red edge and the plant height of corn is determined (figure 16). The question is now, whether this correlation can also be observed with the simulated MERIS spectra. To analyse this, MERIS spectra of corn with varying vegetation height are simulated and, after fitting a spline function to the MERIS-bands, the inflection wavelengths are calculated by means of the second derivative.

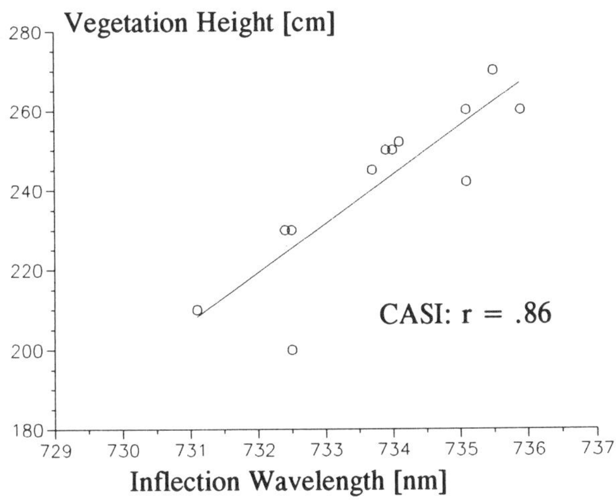

FIGURE 16. Correlation between the inflection wavelength and the vegetation height of corn for CASI-spectra.

The correlation of the resulting values for the inflection wavelength with the plant height was not significant on a 95%-significance level (figure 17). This is not surprising, because there are only 6 MERIS bands in the red edge region, and 3 of the 6 bands are situated very close together in order to observe the oxygen absorption line. Over land surfaces they correlate very strongly (the 3 bands around the oxygen absorption at 760 nm correlate with correlation coefficients of more than .98). Just between 715 and 750 nm, the wavelength region, which is most important for the inflection wavelength, no measurements will be conducted by the proposed MERIS. Therefore an attempt is made to determine, whether the result can be improved if one new band is added in the spectral range of the red edge. This band should be placed outside the water absorption features around 720 nm and positioned optimally within the red edge. The best position is centred around 740 nm (with a bandwidth of 10 nm). Corn spectra of varying vegetation height are then simulated with these new wavelength specifications. The inflection wavelength is again extracted and correlated to the plant height. As a result the correlation increases strongly after adding just one band to the MERIS spectrum (figure 18).

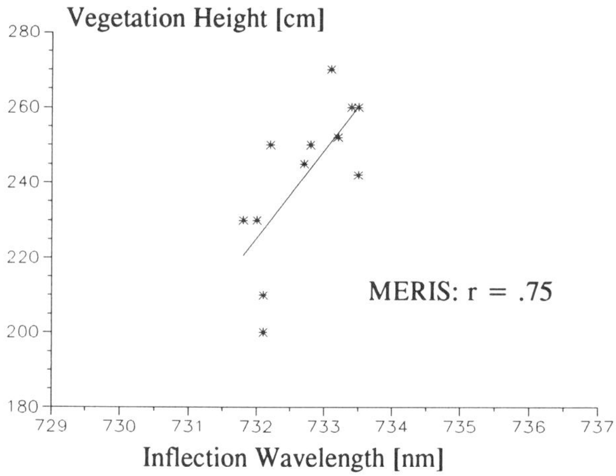

FIGURE 17. Correlation between the inflection wavelength and the vegetation height of corn for MERIS-spectra

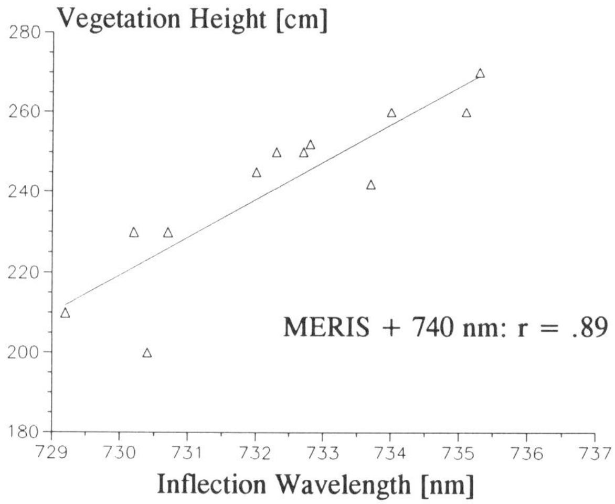

FIGURE 18. Correlation between the inflection wavelength and the vegetation height of corn for MERIS-spectra with one additional band at 740 nm.

The results of the correlations between the inflection wavelength and the vegetation height of corn are summarised in figures 16 - 18 for the different band settings. After adding one band at 740 nm the correlation increases strongly to .89. Although the regression lines vary between the two significant correlations, it is obvious that a MERIS sensor with one additional band in the red edge region would allow to quantitatively determine the red edge of a vegetation spectrum.

Without this additional band it is questionable, whether MERIS can successfully be used for red edge analysis.

7. Conclusions

For the evaluations of imaging spectrometry data several pre-processing steps must be conducted. The radiometric correction is the most important one. An atmospheric model like LOWTRAN-7 allows the conversion of the radiance calibrated data into reflectance values with sufficient accuracy. The derived reflectance values are independent of sensor characteristics, flight conditions, atmospheric circumstances and solar irradiance. However, the quality of available imaging spectrometry data is still characterised by sensor specific problems. The resulting errors, which are caused by scan effects for the GER and ice artefacts for the CASI respectively, can be compensated to a large degree. The necessary corrections are one of the main reasons why the analysis of IS-data is still not operational.

The wavelength region of the red edge has a high information content for vegetation spectra. The red edge can be parameterized by two approaches. The Gaussian fit, which is most suitable for GER data, because of its spectral bandwidths of more than 20 nm. For the CASI and field spectrometer data the inflection wavelength of the red edge has to be determined by means of the second derivative. This method was found to be most feasible for data with spectral bandwidths of less than 16 nm and continuous coverage.

The inflection wavelengths of corn spectra are highly correlated with the plant height of the corn plants. This correlation was found for data gained by different sensors (airborne and ground), different plant height intervals, different years and different soil backgrounds. A shift of the wavelength position of the red edge by 23 nm towards the NIR was determined during the growth of corn. The transfer of these results to other plant parameters like LAI and biomass is expected, because of a high correlation among these geometrical and physiological plant parameters for corn.

Simulations were undertaken to determine the possible applicability of the developed procedures to extract plant parameters by means of future satellite missions, like MERIS. The proposed band setting of MERIS will not allow detailed analyses of the red edge of vegetation spectra. One additional band in the red edge would be necessary for the extraction of plant parameters like the vegetation height.

This paper shows, that a quantitative determination of geometrical and physiological parameters of plants is possible if high resolution spectral in formation is used. The influence of spectral noise originated from the soil background and atmosphere is minimised by analysing the shape of the red-edge. The determined parameter (plant height) can be used as input parameter to soil-plant-atmosphere models. There it determines the aerodynamic roughness of the surface and is a key parameter for the turbulent moisture exchange between surface and atmosphere. Future studies have to clarify, how much further information about the plant physiology and phenology can be drawn from spectral information. Several facts, like the strong correlation between physiological, phenological and geometrical properties within one plant species give reason to an optimistic view. With the application of Imaging Spectroscopy data on a large scale more information on plant species, their growth status and their temporal development will be available to improve the modelling of the hydrologic cycle on the land surface.

8. Acknowledgements

The authors wish to thank JRC and ESA for financing and conducting the EISAC'89 and EISAC'90 campaign and the German Research Association (DFG) for funding the evaluations under the contract Ma 875/2-2. We thank A. Demircan for supporting the agricultural ground data collection and K. Dietz for his critical review of the manuscript.

9. References

Bach, H. and W. Mauser (1989) 'EISAC 1989 - Internal Report on the Ground Data Collection Test Site Freiburg to the JRC, Ispra', Freiburg, Germany.

Bach, H. and W. Mauser (1991) 'The Application of Imaging Spectroscopy data in Agriculture and Hydrology - The EISAC-89-Campaign in the Freiburg Test-Site', *EARSel Advances in Remote Sensing*, vol.1, no.1, 34-42.

Bezy (1992) ESA ESTEC, personal communication.

Bodechtel, J. and S. Sommer (1991) 'The European Imaging Spectroscopy Campaign - EISAC. Review of the First Results and Outlook on Future Aspects of Data Evaluation', *EARSel Advances in Remote Sensing*, vol.1, no.1, 116-120.

Bonham-Carter, G.F. (1988) 'Numerical Procedures and Computer Program for Fitting an Inverted Gaussian Model to Vegetation Reflectance Data', *Computers & Geosciences*, vol. 14, no. 3, 339-356.

Boochs F., G. Kupfer, K. Dockter, and W. Kühbauch (1990), 'Shape of the red edge as vitality indicator for plants', *Int. J. Remote Sensing*, vol.11, no.10, 1741-1753.

Clevers, J.G.P.W. and C. Büker (1991), 'Feasibility of the red edge index for the detection of nitrogen deficiency', *Proc. of the Fifth International Colloquium on 'Physical measurements and Signatures in Remote Sensing'*, 14-18 January, Courchevel,

Collins, W. (1978) 'Remote sensing of crop type and maturity', Photogr. Engineering and Remote Sensing, vol.44, no.1, 43-55.

ITRES-Research Canada (1991) personal communication (contact: Richard Adamson)

Kneizys, F.X., G.P. Anderson, E.P. Shettle, W.O. Gallery, L.W. Abreu, J.E.A. Selby, J.H. Chetwynd, and S.A. Clough (1988) 'Users Guide to LOWTRAN-7', Environmental Research Papers No.: 1010; AFGL-TR-88-0177, Air Force Geophysics Laboratory.

Lynn, B.H. and T.N. Carlson (1990) 'A stomatal resistance model illustrating plant versus external control of transpiration', *Agric. and Forest Met.*, 52, 5-43.

Miller, J.R., Jiyou Wu, M.G. Boyer, M. Belanger, and E.W. Hare (1991) 'Seasonal patterns in leaf reflectance red-edge characteristics', *Int. J. Remote Sensing*, vol.12, no.7, 1509-1523.

Rast, M. (1991) 'Imaging Spectroscopy and its application in spaceborne systems', ESA Publication SP-1144.

Sellers, P.J., Y. Mintz, Y.C. Sud, and A. Dalcher (1986) 'A simple biosphere model for use within general circulation models', *J. Atm. Sciences*, vol.43, no.6, 505-531.

Plummer, S.E., A.K. Wilson, and A. Jones (1991) 'On the Relationship between High Spectral Resolution Canopy Reflectance Data and Plant Biochemistry', *EARSel Advances in Remote Sensing*, vol.1, no.1, 27-33.

ALPINE AND SUBALPINE LANDUSE AND ECOSYSTEMS MAPPING

KLAUS I. ITTEN, PETER MEYER, TOBIAS KELLENBERGER, MICHAEL SCHAEPMAN, STEFAN SANDMEIER, IVO LEISS, and SUSANN ERDÖS
Remote Sensing Laboratories, University of Zurich-Irchel
Winterthurerstr. 190,
8057 Zurich, Switzerland

ABSTRACT. A multisensor multispectral satellite image data set was used to test geometric and radiometric corrections, necessary for landuse and ecosystems mapping in rugged, alpine and subalpine regions. A traditional maximum likelihood classification of ecosystems is discussed versus a clustering after illumination corrections. A forest mapping experiment demonstrates improvements in classification accuracies through slope aspect and atmosphere corrections.

1. Introduction

As a pretest to the application of hyperspectral AVIRIS data over rugged terrain, near simultaneously obtained Landsat TM, and SPOT XS and Pan imagery was used to assess alpine and subalpine landuse. Previously obtained unsatisfactory results in mapping alpine ecosystems, and the first raw signatures obtained from AVIRIS led to suggest and test carefully balanced preprocessing procedures, especially in applications in alpine and subalpine regions.

Alpine and subalpine regions are here defined as high elevation rugged zones with high relief energy. In Switzerland, alpine regions are located between 2000 m and 2400 m/asl and subalpine regions are defined between 1200 m and 2000 m/asl. The nivale zone lies above 2400/2600 m. In other areas of the world these boundaries may vary considerably. In this paper two test regions are used:

- The subalpine site "Rigi" is located in Central Switzerland at the northern border of the Swiss Alps. Elevation extends up to 1700 m, but its major part is subalpine and strongly influenced by agricultaral use such as pasturing. Here we approach the ecological mapping task from the side of botany and the definitions given by botanists. Traditional supervised classification techniques using Landsat TM as well as SPOT XS and Pan data were applied. AVIRIS spectra for some of the classes involved are shown.

- The second subalpine/alpine testsite "Beckenried" lies also in Central Switzerland but southward from "Rigi" with elevations up to 2400 m. Besides tests with forest classifications, unsupervised clustering is applied to thematically masked Landsat TM data.

J. Hill and J. Mégier (eds.), Imaging Spectrometry - a Tool for Environmental Observations, 285–293.

Since digital image data of the AVIRISwiss-91 campaign has only been received recently some of the preliminary remarks could not jet be based on large experience. A rather extensive research programme however has been initiated on the basis of this overflight data, and has started this summer of 1992.

2. Geometric and Radiometric Considerations

If remote sensing is planned to be an operational tool (which in many fields is still some time away), then its results must either be compatible with traditional map data or deliver completely new information that has not been synoptically presented before. It has to be stressed however, that any such attempt of an application fails if we are not able to locate our objects properly, and clearly prove that the measurements show the objects characteristics, - and not some erraneous effects of neighbouring areas. In terms of geometry this means precise geometric corrections, which appears obvious if in an alpine environment we wish to analyze data. Radiometrically we should correct for illumination differences and atmospheric effects, and evaluate the influence of adjacent objects. All together we should understand the physical priniciples, the sensor and platform parameters, and evaluate the major sensor and scene related geometric and radiometric effects before starting a classification.

3. Digital Elevation Models

As a consequence of the before-mentioned it becomes clear, that for any alpine or sub-alpine application with more than coarsely guessed results, a digital elevation model (DEM) is definitely needed. Not only elevation itself is an important factor for our ecological studies, elevation dependent atmospheric effects have to be removed, slope-aspect illumination corrections have to be performed, and above all elevation differences are needed for precise geometric corrections.

After Goodenough et al (1990) for radiometric corrections the resolution of a DEM should be 3 to 4 times finer than the size of the sensors ground resolution. Taken the 20 m AVIRIS pixel or the 30 m TM data a DEM resolution element of betwen 6 and 10 m would be needed. Since no one is in an operational sense today able to work with such high resolution DEM's, research is currently underway to estimate the errors made in our example, working with an 25 m model (x,y). Test with this model and Landsat TM data are showing already encouraging results.

4. Object Reflectance

Rarely an object behaves - in a spectral sense - as an lambertian reflector. This means that similar to the problems working with active microwave signal analysis, where the objects scattering matrix has to be known, its bidirectional reflection distribution function (BRDF) should be known. For non vegetated alpine snow, rock or bare soil this is easly understood, but also for vegetated surfaces especially when viewed by airborne or other rather large swath widths sensors this knowledge is necessary. Only under strict nadir viewing normal reflectance can be assumed. Therefore again, DEM's are needed and the BRDF should be known. Should we as a consequence know then everything about the object before we try to detect it, making remote sensing superfluous? No, but it is easier to solve a complicated equation with less unknowns. Here imaging

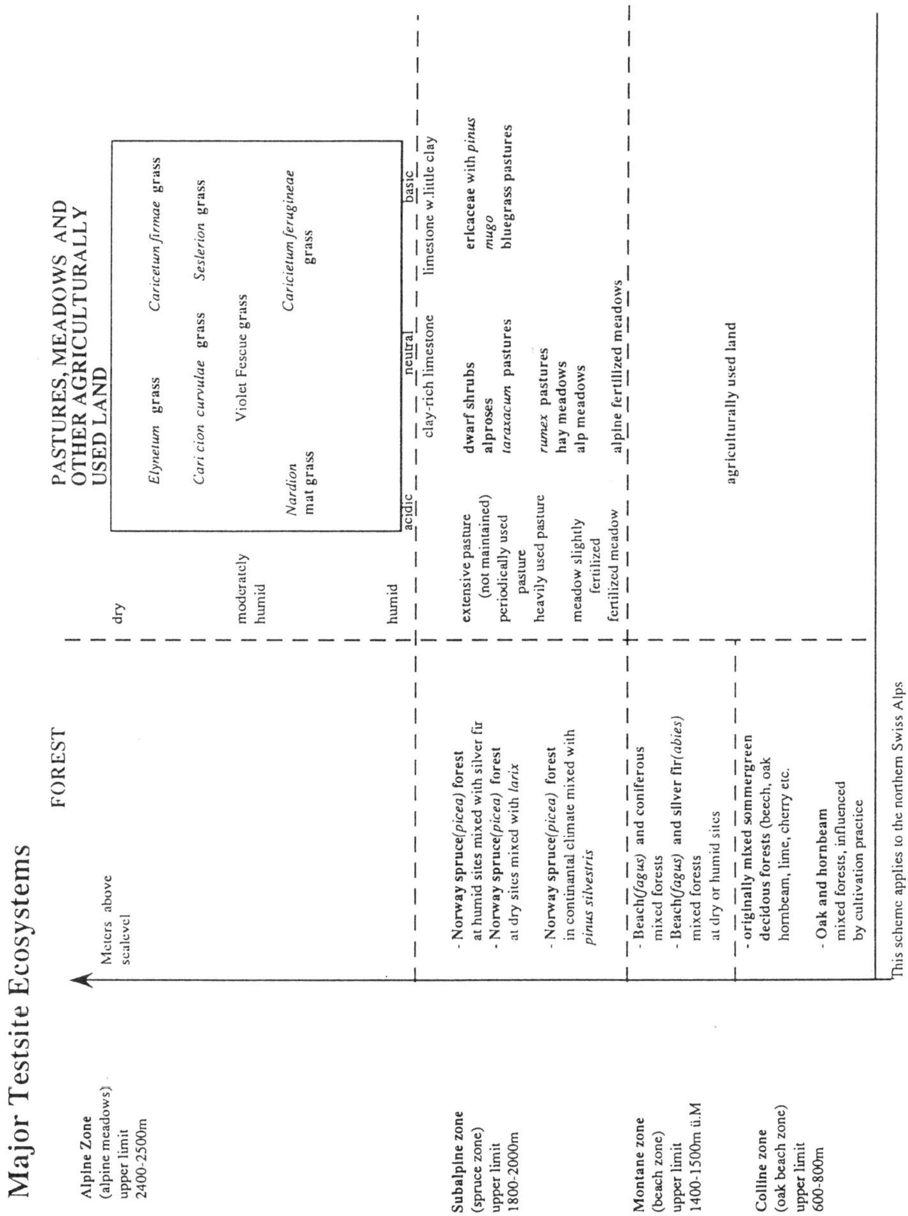

FIGURE 1. Major Test Site Ecosystems.

spectrometry and sensors measuring objects from a variety of look angles may solve many of the open questions.

It is one of the targets of the AVIRISwiss project to analyze the spectral effects of high precision DEM modelling on imaging spectrometry data. Todays experience is based on the rather traditional multispectral satellite imagery.

To circumvent these problems, in the past, more and more classes were needed in a classification attempt. Snow regions in various elevation zones, under varying sun angles etc. were classified individually, and only afterwards merged to one class snow. Unintentionally in clustering exactly this is achieved, and no new elements are detected. DEM modelling of the data should therefore improve possibilities and accuracies in ecological classifications.

5. Thematic Masking

Some sensors bands and therefore wavelenght regions are great in separating land from water but are merely useless in vegetation studies. Some others are good in evaluating biomass etc... With the former sensors it has often been wise to thematically mask out those parts of an image which had been clearly classified, and where an other band did not provide new insights. In imaging spectrometry this may even be more important since the amount of data is too big for traditional procedures. After spectral and spatial masking, only the parts of the spectrum which are necessary are used to solve the remaining problems. Regional masks such as watersheds, elevation zones etc. may also be helpful.

6. Supervised Vegetation Classification, or the Normal Ecological Approach in a Subalpine Region

Based on Landolt (1983) the major Swiss alpine and subalpine ecosystems were compiled, and are shown in figure 1. The Rigi testsite in Central Switzerland belongs to the mid-european flora region. It has however suffered in species richness due to moderate to heavy use of the pastures and meadows. Despite this fact it was tried to search for some of the ecosystems and select training sites for a supervised classification. The used satellite data consisted of near simultaneous Landsat TM, SPOT Pan and XS which were geometrically corrected using a DEM, geocoded and registered, forming (without Landsat Band 6) a 10 channel dataset. In the maximum likelihood approach the three best suited bands were TM5, XS3 and XS2. A scattergram showing the data and training sites are shown in figure 2. In comparison, figure 3 shows raw AVIRIS spectra for some of the classes. AVIRIS, Landsat TM and the SPOT data were taken within 12 days, therefore the landuse is comparable.

7. Unsupervised Vegetation Classification after Illumination Correction

Basing on the results of the previous study and also assuming a reduced atmospheric haze influence at higher elevations the following classification tests were carried out in our alpine-subalpine testarea "Beckenried".

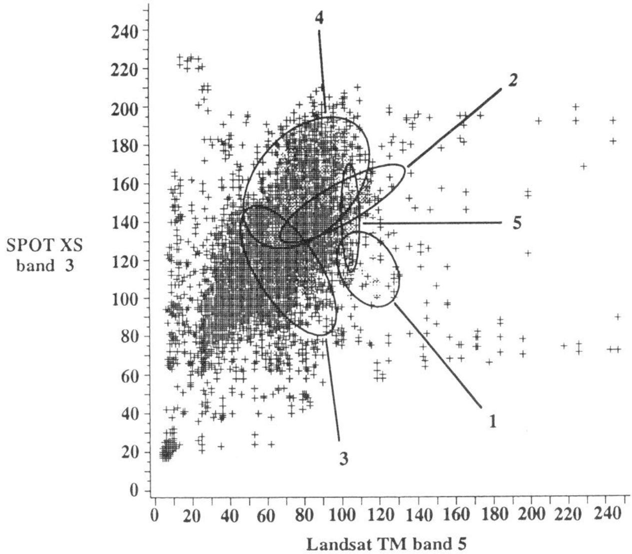

FIGURE 2. Scattergram of data and training sites for classes: dwarf shrubs (5), taraxacum, pastures (3), hay meadows (1), alp meadows (2) and alpine fertilized meadows (4).

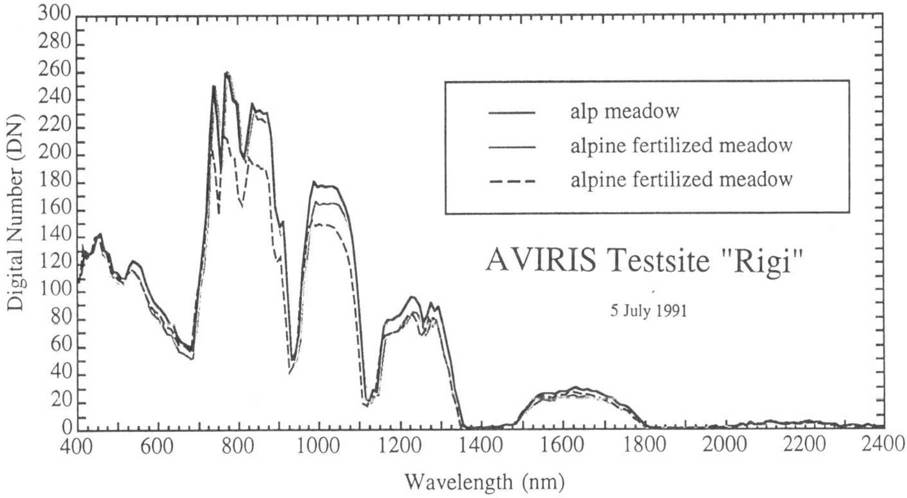

FIGURE 3. AVIRIS raw spectra of two training areas of alpine fertilized meadows under differing illumination and one training area of alpine meadow.

- After the same geometric and geocoding corrections as applied to the previous dataset, slope aspect illumination differences were removed using a semi-empirical C-correction method. Water and forest surfaces were thematically masked out to allow for a concentration on the remaining areas.

- Unsupervised clustering (Sarle, 1983) of the three first principal components of Landsat TM data (-TM6) was then tested on these areas. Class assignment was achieved in comparison with an airphoto interpretation. Figure 4 shows a scattergram of the 6 clusters in the PC2 /PC3 dataspace.

8. The Influence of Illumination and Atmosphere Correction on the Classification of Sub-Alpine Forests

In the Beckenried site tests were carried out within the "Swiss Remote Sensing Forest Mapping Project" to assess the influence of illumination and atmospheric corrections on the accuracy of classifications of forests and forest stands using Landsat TM data. After careful investigations a semi-empirical C-correction algorithm (Teillet, 1986) was used to equalize illumination differences due to slope and aspect variations of the terrain. Atmospheric corrections using the "5S" code (Tanré et al., 1990) revealed interesting results. Tables 1a and 1b show the differences in classification accuracies for forest vs. non forest, and tables 2a and 2b for stand/forest type classifications. In the latter experiment only 3 classes could be distiguished: decidous, mixed, and coniferous stands.

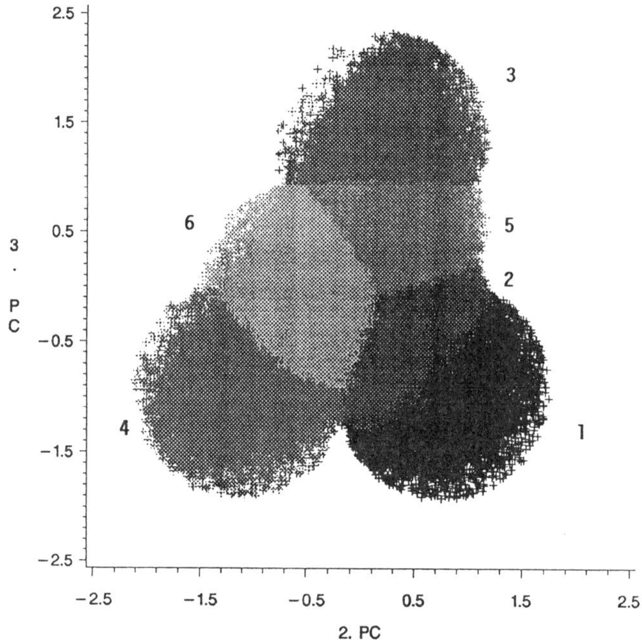

FIGURE 4. Scattergram of the 6 obtained clusters in the PC2 /PC3 dataspace. The following classes could be assigned: Harvested areas in valley bottoms and highest elevation scarse vegetation (1), agriculturally used areas in valley bottom and scarse vegetation on steep slopes (2), fertilized alpine meadows (3), Border areas around lake and forests, and small river (4), pastures (5), low density of trees and shrubs (6).

illumination [cos (i)·100]	in % of the testarea Beckenried	classific. accuracy original TM bands	classific. accuracy C-correction	classification difference
1 to 10	0.56 %	53.15 %	59.18 %	**+ 6.03 %**
11 to 20	0.99 %	67.86 %	68.67 %	+ 0.81 %
21 to 30	1.91 %	76.13 %	76.00 %	- 0.12 %
31 to 40	3.80 %	78.19 %	80.61 %	+ 2.42 %
41 to 50	6.40 %	76.18 %	82.12 %	+ 5.95 %
51 to 60	7.70 %	77.00 %	82.16 %	+ 5.16 %
61 to 70	9.59 %	81.52 %	82.45 %	+ 0.94 %
71 to 80	12.27 %	86.06 %	84.23 %	- 1.83 %
81 to 90	41.01 %	94.28 %	94.10 %	- 0.17 %
91 to 100	15.77 %	82.64 %	84.05 %	+ 1.41 %

TABLE 1a. Accuracies of the *forest vs. non forest classification* in the test area Beckenried in relation to the illumination after slope-aspect correction (100 % = 338,624 pixel)

illumination [cos (i)·100]	in % of the testarea Beckenried	classific. accuracy C-correction	classific. accuracy + atm. correction	classification difference
1 to 10	0.56 %	59.18 %	61.58 %	**+ 2.40 %**
11 to 20	0.99 %	68.67 %	70.89 %	+ 2.22 %
21 to 30	1.91 %	76.00 %	78.02 %	+ 2.02 %
31 to 40	3.80 %	80.61 %	81.84 %	+ 1.23 %
41 to 50	6.40 %	82.12 %	82.96 %	+ 0.83 %
51 to 60	7.70 %	82.16 %	82.31 %	+ 0.15 %
61 to 70	9.59 %	82.45 %	82.65 %	+ 0.19 %
71 to 80	12.27 %	84.23 %	84.65 %	+ 0.43 %
81 to 90	41.01 %	94.10 %	93.91 %	- 0 19 %
91 to 100	15.77 %	84.05 %	83.53 %	- 0.53 %

TABLE 1b. Accuracies of the *forest vs. non forest classification* in the testarea Beckenried in relation to the illumination after slope-aspect <u>and</u> atmospheric correction (100 % = 338,624 pixel).

Through the correction of scene related radiometric effects, especially through slope-aspect illumination corrections and atmospheric corrections, a classification accuracy of forest vs. non forest of almost 90 % could be achieved in our Beckenried test area. The slope-aspect correction alone improved the classification accuracy of faintly illuminated areas by about 5 %, and the subsequent atmospheric correction yielded in the same problem areas another 2 % increase.

In the forest stand / type classification the accuracy improvement with the slope-aspect correction was between impressive 10 % and 30 % for brightly illuminated areas, whereas the correction of the atmospheric effects had a slight worsening effect on the result. Overall forest stand / type classification accuracies for the Beckenried testsite were in the region of 62 %. This accuracy is surely not high. We have to remember however, that the alpine forest ecosystem encountered in the Beckenried area is composed of a large variety of deciduous trees and also heavily mixed with conifers. Therefore we have to contend ourselves with this result under the specific conditions. While interpreting the percentages it has to be kept in mind, that a specific rigorous accuracy measure is applied, which makes it somehow difficult to compare the results with those of other authors. The best accuracy in the forest stand / forest type classification is achieved using a binary hierarchical classification procedure. TM band 4 and a synthetic band (TM 4 - TM 2) are proposed. Also TM band 4 alone offers already useful results.

illumination [cos (i)·100]	in % of the testarea Beckenried	classific. accuracy original TM bands	classific. accuracy C-correction	classification difference
1 to 10	1.79 %	60.51 %	61.03 %	+ 0.51 %
11 to 20	4.80 %	58.13 %	58.13 %	+ 0.00 %
21 to 30	11.10 %	50.91 %	58.18 %	+ 7.27 %
31 to 40	15.95 %	54.89 %	59.72 %	+ 4.83 %
41 to 50	20.32 %	60.86 %	61.81 %	+ 0.95 %
51 to 60	14.26 %	59.91 %	63.00 %	+ 3.09 %
61 to 70	9.65 %	53.23 %	63.21 %	+ 9.98 %
71 to 80	8.58 %	47.38 %	60.21 %	+ 12.83 %
81 to 90	9.25 %	48.91 %	64.29 %	+ 15.38 %
91 to 100	4.29 %	44.44 %	74.57 %	**+ 30.13 %**

TABLE 2a. Accuracies of the *stand / forest type classification* in the testarea Beckenried in relation to the illumination after slope-aspect correction (100 % = 338'624 pixel)

illumination [cos (i)·100]	in % of the testarea Beckenried	classific. accuracy C-correction	classific. accuracy + atm. correction	classification difference
1 to 10	1.79 %	61.03 %	58.46 %	- 2.56 %
11 to 20	4.80 %	58.13 %	54.88 %	- 3.25 %
21 to 30	11.10 %	58.18 %	55.45 %	- 2.73 %
31 to 40	15.95 %	59.72 %	57.31 %	- 2.42 %
41 to 50	20.32 %	61.81 %	60.68 %	- 1.13 %
51 to 60	14.26 %	63.00 %	63.06 %	+ 0.06 %
61 to 70	9.65 %	63.21 %	62.07 %	- 1.14 %
71 to 80	8.58 %	60.21 %	56.26 %	- 3.96 %
81 to 90	9.25 %	64.29 %	63.59 %	- 0.69 %
91 to 100	4.29 %	74.57 %	72.65 %	- 1.92 %

TABLE 2b. Accuracies of the *stand / forest type classification* in the testarea Beckenried in relation to the illumination after slope-aspect and atmospheric correction (100 % = 338,624 pixel)

TM band 2 was best suited for the forest vs. non forest classification. If water surfaces are present in the scene, they have to be masked out in TM band 5 prior to the forest classification.

As a recommendation the following procedures can be suggested:

- geometric correction and geocoding of the TM data with a DEM and nearest neighbour resampling

- evaluation of viewing angle / path length problems
- testing of sensor related radiometric effects

- slope-aspect correction using the semi-empirical C-correction method (good DEM needed)

- evaluation and correction of atmospheric effects using the model "5S" with radio sonde data or standard atmosphere

- if necessary masking out of water surfaces with TM band 5

- classification into forest vs. non forest with TM band 2
- classification into deciduous, coniferous and mixed by using TM4 and a synthetic band (TM 4 - TM 2) in a binary hierarchical procedure, or TM 4 alone.

9. Conclusions

Through careful preprocessing of multispectral data using DEMs, illumination- and atmospheric corrections, major improvements in rugged terrain landuse classifications could be achieved. First analysis of AVIRIS hyperspectral data shows, that the same rigorous corrections have to be applied, and only then it can be hoped to improve information extraction and ecosystems mapping further in alpine and subalpine environments.

10. References

Ellenberg, H. (1982) *Vegetation Mitteleuropas mit den Alpen in ökolgogischer Sicht*, 3.Aufl. Stuttgart, Ulmer, 516-589.

Goodenough, D.G., Deguise, J.C., Robson, M.A. (1990) 'Multiple Expert Systems for Using Digital Terrain Models', *Proc. IGARSS'90*, 961.

Itten, K.I., Meyer P., Kellenberger T., Leu R., Sandmeier St., Bitter P. and Seidel K. (1992) *Correction of the impact of topography and atmosphere on Landsat-TM forest mapping of alpine regions*, Remote Sensing Series, vol.18, RSL, University of Zurich.

Landolt, E. (1984) *Unsere Alpenflora*, Verlag des SAC, 318 pp.

Landolt, E. (1983) 'Probleme der Höhenstufen in den Alpen', *Botanica Helvetica*, 83, 255-268.

Tanré, D., Deroo C., Duhaut, P., Herman,M., Morcrette, J.J., Perbos, J. and Deschamps, P.Y. (1986) *Simulation of the Satellite Signal in the Solar Spectrum (5S)*, Laboratoire d'Optique Atmosphérique, Villeneuve d'Ascq Cédex, France.

Teillet, P.M., Guindon, B., Goodenough, D.G. (1982) 'On the Slope-Aspect Correction of Multispectral Scanner Data', *Canad. J. of Remote Sensing*, vol.8, nr.2, 84-106.

Sarle, W.S. (1983): *The Cubic Criterion*, SAS Technical Report A-108, Cary, NC.

Woodham, R.J. (1989) 'Determining intrinsic surface reflectance in rugged terrain and changing illumination', *Proc.IGARSS'89*, 1-5.

Zumbühl, G. (1983) *Pflanzensoziologisch-ökologische Untersuchungen von gemähten Magerwiesen bei Davos*, Geobot.Inst. ETH Zurich, 83, 101pp.

IMAGING SPECTROMETRY AS A RESEARCH TOOL FOR INLAND WATER RESOURCES ANALYSIS

ARNOLD G. DEKKER* and MARCEL DONZE**
** Institute of Environmental Studies*
Vrije Universiteit
De Boelelaan 1115, 1081 HV Amsterdam, The Netherlands.
*** Faculty of Civil Engineering*
Technical University of Delft,
P.O. Box 5048, NL 2600 GA Delft, The Netherlands.

ABSTRACT. Qualitative and quantitative aspects of optical water quality of inland waters may be determined by remote sensing. The potential of imaging spectrometry for inland waters is discussed following the analytical method where the inherent and apparent optical properties are used to model the reflectance and vice versa. A classification is proposed for describing the contributions to the total absorption spectrum of water and the three main constituents: aquatic humus, photosynthetic pigments and tripton. The relevant literature on airborne spectrometry is discussed. An example is given how photosynthetic algal pigments (i.c. cyanophycocyanin) may be estimated by imaging spectrometry. Absorption spectra of water, aquatic humus and the particulate matter and scattering spectra of 31 inland water samples are presented. Using these inherent optical properties it was possible to compare *in situ* measured subsurface irradiance reflectance *R(0)* with the modelled *R(0)* and with the airborne measured spectra recalculated to *R(0)*.

1. Introduction

Remote sensing of surface water has been mainly used to investigate the oceans where it has contributed greatly to the discovery, mapping and understanding of large scale processes. As compared to applications in research on inland waters this can be understood from an organizational point of view: oceanography as a discipline is much better organized to undertake large scale scientific projects. Most work to develop the optical models to explain and predict the reflected signal from surface water was done in marine waters. Preisendorfer (1976), Jerlov (1976) and Austin (1974) laid a firm physical basis defining the inherent and apparent optical properties of waters. Doerffer (1991) explains how imaging spectrometry may be used for oceans on the basis of these models.

Since inland waters are bordered by land, there is usually no problem to detect their crude structure: it is already given on maps. Most general research in inland water ecology has been concerned in developing zero dimensional models that consider a lake a perfectly mixed system. Different lakes are compared and single lakes are studied by monitoring over long periods. Only when the internal structure of a lake must be known, or when large numbers of lakes must be compared do remote sensing techniques become useful.

J. Hill and J. Mégier (eds.), Imaging Spectrometry - a Tool for Environmental Observations, 295–317.

The desired information from a remotely sensed signal of a pixel of surface water is the upwelling (ir)radiance just under the water surface, the subsurface upwelling irradiance Eu_S. Other contributions to the measured signal are, among others, surface reflectance and backscatter from the atmosphere. Correction methods for these effects exist, but residual systematic effects remain, while their contribution to noise is inevitable. The remote sensing signal arises in the top layer of the water column, so information from below the level where light intensity is below, say, one quarter of surface intensity, cannot be obtained, while detection and quantification of inhomogeneity in a water column probably will not be possible by passive techniques.

From the incident light intensity and subsurface upwelling irradiance information can be extracted on those components in the water that interact with the natural light field. These are mainly the presence or absence of surface layers of oil and algal blooms, concentration of suspended algae with some information on their content of different pigments, suspended matter and aquatic humus. The mathematical deconvolution procedures to do this are usually called algorithms. Also generalized optical water quality parameters such as vertical attenuation, Secchi transparency and optical depth, and in shallow clear water also depth, can be calculated from remotely sensed reflection spectra.

Quantitative images obtained this way can play a role in calibration and validation of two- and three dimensional hydrodynamic and ecological models. Interactions between waterbodies like in- and outflow of polluted water and their mixing processes can be visualised and in part quantified. Resuspension of bottom material in dependence on wind characteristics can now be observed over whole lakes.

Most freshwater systems in the world are affected by man-made eutrophication, leading to undesirable increases in biomass of higher plants an/or planktonic algae. These phenomena often show large local differences and interactions with patterns of water flow. Especially the amount and distribution of nuisance-forming cyanobacteria is a primary concern in water management. Remotely sensed images give indications on the patterns of distribution of water colour and associated physical properties. Often even a few images are useful to design or improve a point sampling monitoring program by careful choice of sampling points.

Large increases in spatial resolution have become available in satellite images during the last decade and much better spectral resolution is now available in instruments flown in airplanes. High spectral resolution is also planned for new satellites in the near future. These developments will make extensive application of passive optical remote sensing to problems in research and management of inland waters attractive in the near future. Here we review current research that anticipates this development.

2. General methodology for extraction of water quality parameters from remote sensed spectral data

The underwater light field is determined by the inherent optical properties which are independent of the ambient light field (i.e. independent of changes in the angular distribution of radiant flux). These properties for light of a certain wavelength are specified by the absorption coefficient a (m^{-1}), the scattering coefficient b (m^{-1}), and the volume scattering function $b(\Theta)$. The last parameter describes the angular distribution of scattered flux resulting from the primary scattering process. The definitions are based on the behaviour of a parallel beam of light incident upon a thin layer of medium (Jerlov, 1976; Kirk, 1981a and b; Kirk, 1983). Direct field measurements of these properties are not yet feasible. Often a particle-dominated normalized $b(\Theta)$ determined by Petzold (1972) for the turbid water of the San Diego Harbour is used in optical models of the underwater light field.

Morel and Gordon (1980) pointed out three different approaches by which measurements of spectral (ir)radiance can be used to estimate concentrations of water constituents in remote sensing. This approach is followed here, it is somewhat extended to include inland waters and methods developed since 1980.

- The empirical method: Statistical relationships are sought between measured spectral values and measured water parameters. The weakest method.
- The semi-empirical method: Spectral characteristics of the compounds sought for are more or less accurately known. This knowledge can be included in the statistical analysis which is focused on well chosen spectral areas and appropriate bands or combinations of bands are used as correlates. Reasonable algorithms can be found by common sense and honed by experience. Quantitatively the coefficients only apply to the dataset at hand so each application must be individually calibrated. This method is commonly used.
- The analytical method: The inherent and apparent optical properties are used to model the reflectance and vice versa. The water constituents are expressed in their specific (per unit measure) absorption- and backscatter coefficients. Subsequently a suite of analysis methods can be used to optimally retrieve the water constituents or parameters from the remotely sensed upwelling radiance or radiance reflectance signal. Such methods are in development.

In Gordon and Morel (1983) a comprehensive discussion of the analytical models available for clear ocean waters through to turbid coastal waters is given. Kirk (1983) extended the discussion to inland waters.

In clear oceanic waters the spectral reflectance is a function of:

1. Absorption by algal pigments and low concentrations of (aquatic) humus at short wavelengths and by pure water at long wavelengths.
2. Scattering by water molecules at short wavelengths and Raman scattering at intermediate wavelengths.
3. Fluorescence caused by algal pigments at longer optical wavelengths.

In turbid coastal water and in almost all inland waters these effects also occur, but the analysis becomes more complicated due to the importance of two additional components:

4. Backscattering from particles is the dominant scattering factor, up to 1000 times the backscattering of the clearest oceanic waters.
5. Absorption at low wavelengths by high concentrations of humus.

This introduces negative relations among the constituents. Increase in silt for example will make algae less visible, and absorption by humus does the same to the molecular backscattering by water. In nature these variables are often highly correlated, while non-linear relations between concentrations and reflection can occur. As a consequence the statistical approach to data analysis is nearly impossible.

3. Optical Classification of Water Types

Jerlov (1976) proposed the first classification of types of **marine water** based on its spectral transmittance of downward irradiance at high solar altitudes. In order of decreasing transmittance three types of oceanic water (I, II and III) and nine types of coastal water (1 to 9) were recognized. Basis for this classification is the observed fact that the shape of the volume scattering function of surface water is relatively constant among different oceanic regions. He concluded that the absorption spectra of particulate material and humus primary determine the spectral signature. Prieur & Sathyendranath (1981) extended this classification scheme by separating the particulate matter absorption into algal pigments and other particulate matter. They also developed a technique based on the spectral form of the absorption curves.

Case I- and case II waters have been studied most intensive being dominant in research on large scale oceanic patterns and processes. The optical characteristics of case III waters are determined mainly by scattering by suspended sediment, but interacting with other constituents.

The optical classification scheme for seawater is in some respects similar to the classification for **inland waters**, as proposed by Kirk (1980). It was based on measurements of absorption spectra of the soluble fraction humus and of the particulate matter from Australian water bodies. Such a classification can be used to compare water bodies on a regional scale. But also processes in time, like a seasonal cycle or heavy rain in the catchment with soil erosion can make a particular water change type.

Here Kirks's classification is extended by adding an absolute reference value for absorption value as well as by discriminating different photosynthetic pigments. This classification is intended as an instrument to indicate the optical characteristics of inland waters. It has been used to decide on the optimal choice of spectral bandsets in some of our remote sensing projects.

The following four main features of absorption are distinguished:

H = aquatic humus absorption
A = photosynthetic pigment absorption of algae
T = tripton absorption
W = water absorption

In addition to the Kirk (1980) method, an absolute ordering is used based on the maximum absorption coefficient in the photosynthetic active radiation range of 400 to 700 nm. The absolute reference point for the pure water absorption coefficient at 700 nm is 0.6225 m^{-1}. Therefore this concept is universally valid and repeatable.

Since H and usually T have a maximum at 400 nm and A has a maximum at one of the pigment absorption peaks, with water absorption fixed, this description of optical water types can also be used as an input preposition in statistical or other spectral signature unmixing algorithms.

The components are ordered according to maximum values of the absorption coefficient of the parameters within the PAR range. As an example a *HTWA* type has H larger than T values at 400 nm, both being > 0.6225 $\cdot$ m^{-1} and A values < 0.6225 at 435 nm or 685 nm or any other pigment absorption peak. For clarity the A type absorptions can be subscripted with the wavelength location(s) of the pigment absorption peaks. A shallow lake in a peat area dominated by cyanobacterial phytoplankon with high phycocyanin content and with considerable resuspension of bottom material could have an optical water type of $A_{435,680}THWA_{630}$. This indicates that the two chlorophyll *a* pigment absorption peaks are > 0.6225 and higher than T and H and the presence of a phycocyanin peak at 630 lower than 0.6225 m^{-1}. By only giving the location and the height of the absorption peak an absolute description of the absorption feature is given instead of a possibly erroneous attribution to a pigment.

Often it is difficult or impossible to discriminate between absorption by algae and tripton, especially outside the pigment absorption peaks. In this case the T and A are placed in brackets *(TA)* or *(AT)* implying the combined signal. The choice between *TA* and *AT* order of notation is made by dividing the maximum absorption value at 670 to 690 nm by the maximum absorption value at 425 to 445 nm (i.e. the two chlorophyll *a* absorption peaks). If this ratio is smaller than 2 it is an *AT* type, otherwise a *TA* type. The underlying hypothesis is that for algae absorption at the first chlorophyll *a* peak at 435 nm is unlikely to be more than twice as high as absorption by the peak at 680 nm. In absence of a clear absorption peak at 435 and 680 nm the water type will be T. Examples of these water types are given in the paragraph on groundbased spectral measurements.

4. Airborne Imaging Spectrometry

To develop applications of imaging spectrometry imaging and non-imaging (line or point) data are of interest. Groundbased surface and subsurface spectral measurements serve as calibration and as a link between the remotely sensed signal and the inherent optical properties.

Line spectrometry measurements over inland waters from the air were reported from Canada by O'Neill et al.(1987), for Ireland by McGarrigle et al. (1990), from eastern Europe and the former Soviet Union by Gitelson (1990, 1991) and Kondratyev and Podzniakov (1990) and from the Netherlands by Dekker et al. (1990a and b; 1992d and c). Imaging spectrometry for inland waters was carried out by Melack and Pilorz (1990) and Dekker et al. (1991, 1992a and b).

4.1 AIRBORNE LINE SPECTROMETRY MEASUREMENTS

Line spectrometry measurements (at 10 nm resolution) by McGarrigle et al. (1990) on 49 lakes in Ireland were analyzed with advanced statistical techniques. No attention was given to the spectral features of the constituents of the water, hampering further interpretation in physical terms. Gitelson (1991) reviewed the results of several hundred airborne spectral measurements in Hungary, Germany and the former USSR together with simultaneous ground data. A large range in water quality parameters occurs in these data. Chlorophyll *a* ranged from 3-350 $\mu g \cdot l^{-1}$, suspended matter from 2-43 mg $\cdot$ l^{-1}, and humus expressed as the absorption at 330 nm from 0.1-10 m^{-1}. The source and composition of suspended matter and humus also varied for the different water bodies. The airborne measurements were made with a 10 channel spectroradiometer with wavelengths centered at 10-12 nm wide bands situated at 480, 500, 520, 560, 630, 650, 675, 685, 700 and 712 nm. Three factors explaining most of the variation were found by factor analysis. They each consist, however, of at least two but sometimes more absorption and scattering features.

4.2 IMAGING SPECTROMETRY MEASUREMENTS

Imaging spectrometer data from the Programmable Multispectral Imager (PMI) was analysed for a lake system in the Netherlands by Dekker et al. (1991) using the semi-empirical model. Secchi disk transparency, vertical attenuation coefficients and surface and subsurface spectroradiometric measurements were made in situ. Seston dry weight, and chlorophyll *a* and phaeopigments were determined from samples. Inherent optical properties were estimated from measured apparent optical properties. The PMI was flown in spectral mode at 1000 m altitude and in spatial mode at 3000 m altitude. In spectral mode 288 spectral channels over 430-805 nm at 1.3 nm intervals were acquired in 32 interspaced lines over the swath. These airborne spectra showed a large similarity with the in situ upwelling irradiance spectra from 500 nm to longer wavelengths. In spatial mode an eight channel spectral bandset, developed by Moniteq for chlorophyll *a* analysis, was used. The spatial mode bandsetting was simulated using 25 subsurface upwelling irradiance measurements and 17 airborne spectral mode measurements. The highest correlation was found for the spectral band ratio of [673-687] \ [708-715] with seston dry weight (r = 0.95; range 3 - 40 mg); chlorophyll *a* (r = 0.96; range 6 - 91 μg); secchi disk transparency (r = 0.98; range 0.3 - 2.25 m) and the vertical attenuation coefficient over 400-700 nm (r = 0.97; range = 0.7 - 5.3). The linear regression equations were applied to log-transformed data. The limited amount of spatial mode data was processed according to these algorithms. The results were a water quality map showing the spatial distribution of these parameters. Simultaneous measurements of three in situ samples indicated the validity of the calculations by fitting close to the remotely sensed results.

The same spectra were used to model the performance of other existing multispectral scanning systems such as the Landsat Thematic Mapper and the SPOT (Dekker et al. 1992c; Dekker and Peters, in press). These spectra were used to develop a dedicated inland water quality spectral bandset for the 9 channel multispectral CAESAR airborne scanner (Dekker et al., 1990a and b). In Dekker et al., (1992a) examples of spectra obtained by the Compact Airborne Spectrographic Imager (CASI) in spectral and spatial mode and spectra and images by CAESAR in the Inland Water Mode configuration are shown to illustrate the application of this type of data in optimising spectral band location for remote sensing of turbid and/or eutrophic waters.

4.3 DETECTION AND QUANTIFICATION OF CYANOBACTERIAL PIGMENTS USING IMAGING SPECTROMETRY

In Dekker et al.(1992b) a method is presented to detect and measure concentrations of cyanobacteria in inland waters with imaging spectrometry. The abundance of cyanobacteria in freshwater systems is a strong indicator for eutrophication. The method is outlined below.

Ten Lakes with cyanobacteria from the Vecht lakes area in the Netherlands were studied with CASI in 1990. The cyanobacterial concentrations were quantified using microscopic algal counts, spectrophotometry, flow cytometry (Dubelaar, 1989) and pigment extraction followed by HPLC. Following the method by Bennett & Bogorad (1973) phycocyanin (CPC) and phycoerythrin (CPE), pigments unique to cyanobacteria, were determined in the lake samples.

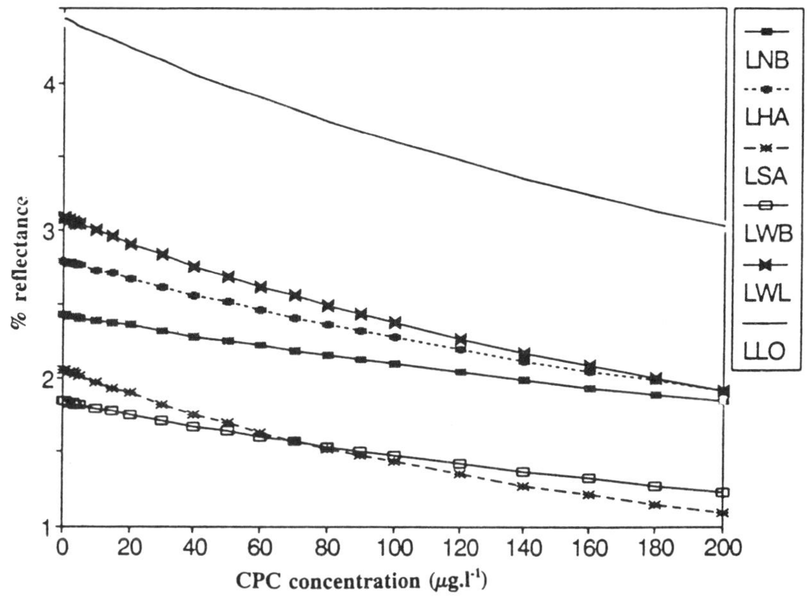

FIGURE 1. Change in reflectance at 624 nm as a function of cyanophycocyanin concentration (LNB = Lake Naardermeer; LHA = Lake Hollands Ankeveen; LSA = Lake Stichts Ankeveen; LWB = Lake Wijde Blik; LWL = Lake western Loenderveen; LLO = Lake Loosdrecht).

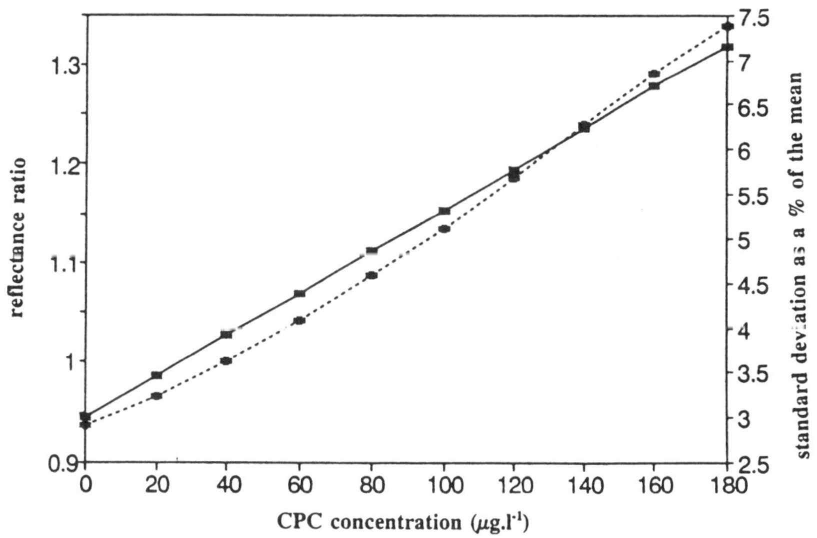

FIGURE 2. Reflectance ratio at 624/648 nm as a function of cyanophycocyanin concentration. Average of 8 lakes (including the lakes in figure 1).

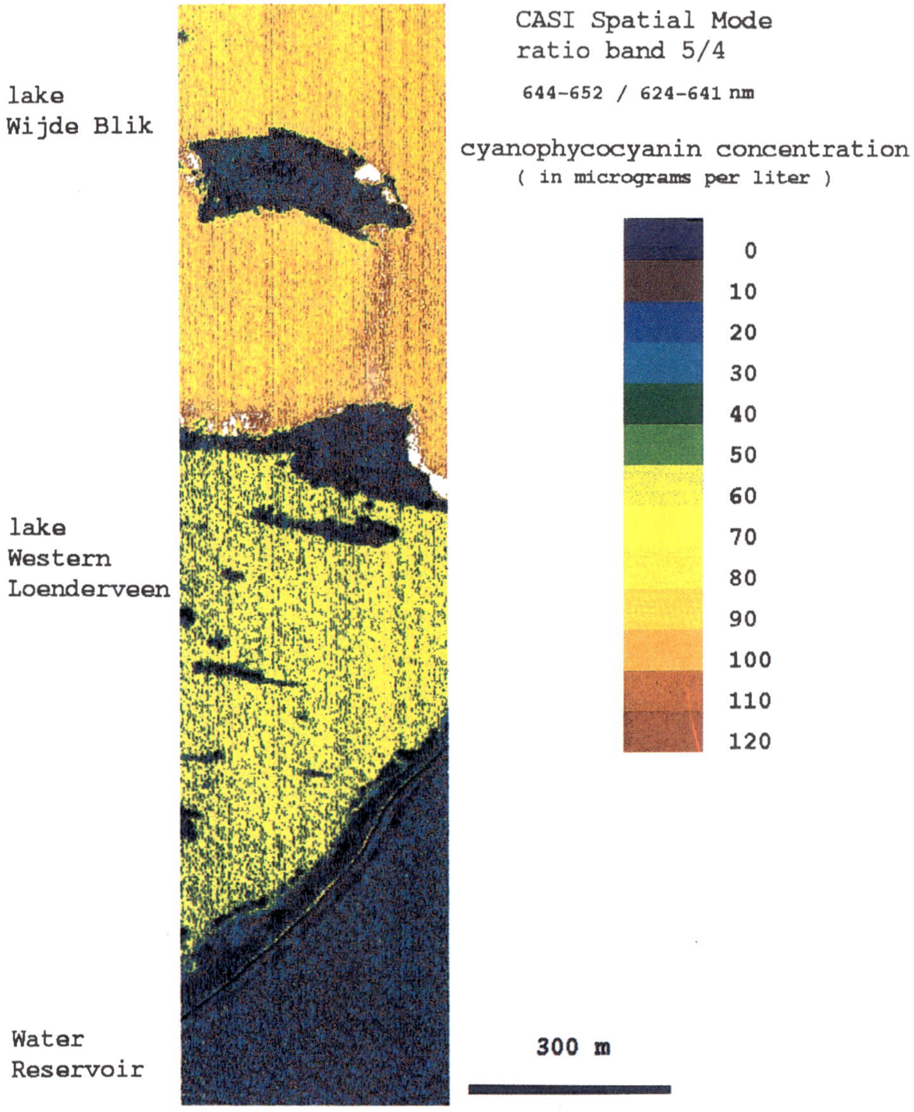

FIGURE 3. CASI spatial mode image of lakes Wijde Blik, Western Loenderveen and the Water Reservoir showing CPC concentrations represented as the ratio of band 5 (644-652nm) / band 4 (624-641nm).

The general species composition varied from green algae, cyanobacteria, diatoms to the prokaryote *Prochlorothrix hollandica*. All shallow hypertrophic lakes were dominated by *Oscillatoria spp.* and *P. hollandica*, dispersed through the water column. Only the deeper Lake Wijde Blik was dominated by *Microcystis aeruginosa*, a colonial species that can form surface scums. The oligotrophic Water Reservoir was dominated by cryptophyceae and *Volvox spp.*, whereas the Amsterdam Rhine Canal contained mainly diatoms.

The presence of cyanobacteria was indicated in the remotely sensed data by the absorption of light by CPC and seen as decreased reflectance around 627 nm. For simplicity it was assumed that increased numbers of cyanobacteria did not influence scattering.

Assuming the model for calculating subsurface reflectance from absorption and backscattering for turbid waters by Gordon & Morel (1983) and Whitlock et al. (1981), where

$$R(0) = \kappa\, b_b / (a + b_b) \tag{1}$$

is valid for these type of waters, the influence on reflectance of 1 $\mu g \cdot l^{-1}$ CPC was calculated. $R(0)$ is subsurface irradiance reflectance, k is a solar angle factor, which was taken to be 0.385 (after Kirk, 1981a), b_b is the backscattering coefficient and a is the total absorption coefficient.

The specific absorption coefficient of CPC at the peak absorption wavelength was estimated from the spectrophotometric absorption measurements to be 0.0035 $m^{-1} \cdot \mu g^{-1} \cdot l^{-1}$. If it is assumed that a(ah) and a(p) will not change with varying absorption by CPC, the effect on $R(0)$ of changing CPC levels was calculated. The backscattering coefficient was assumed to be 0.014 $*$ b (as estimated for plankton-rich water by Kirk, pers.comm.) The effects on $R(0)$ of levels of 1, 2..10; 10, 15..50; 50, 60..100 and 100, 120..200 $\mu g \cdot l^{-1}$ CPC are shown in figure 1. The noise equivalent changes in reflectance for 1 $\mu g \cdot l^{-1}$ CPC were about 1:385 at low concentrations of seston and algae and 1:500 at high concentrations. Other concentration threshold detection levels may be calculated from this value.

By using a spectral band ratio of 648/624 nm a reflectance ratio model of CPC concentration was calculated. The results are presented in figure 2. In 8 lakes the bivariate linear correlation coefficient between the 648/624 nm CASI reflectance ratios and the modelled reflectance ratios was 0.70. The CASI spectra were averaged values per lake, between 3 and 6 spectra per lake, on September 14th, 1990. CPC absorption was determined from samples taken two days before.

Figure 3 is a CASI spatial mode image of lakes Wijde Blik, Western Loenderveen and the Water Reservoir showing CPC concentrations represented as the ratio of band 5 (644-652 nm) / band 4 (624-641 nm). In Lake Wijde Blik there is some spatial structure in the image. Probably this is caused by wind drift effects on floating colonies of Microcystis. Lake Western Loenderveen shows a homogeneous concentration of 60 $\mu g \cdot l^{-1}$ CPC, the Water Reservoir has levels of CPC of 0 to 10 $\mu g \cdot l^{-1}$.

These results show that cyanobacterial pigments can be determined by remote sensing when the spectral bands are sufficiently narrow and are spectrally correctly positioned.

5. Requirements for the Development of Analytical Methods in Imaging Spectroscopy

In this section results obtained with a CASI imaging spectrometer will be used as an illustration of our efforts to develop the analytical method in imaging spectroscopy. As before the work was done in the Vecht Lakes in The Netherlands (Dekker et al.,1990a and b, 1991, 1992 a-d). The water types encountered are:

$GW(TA_{680})$	$(TA_{435,630,680})GW$	$(TA_{435,680})GW$
$G(TA_{435,630,680})W$	$GTWA_{680}$	$TGWA_{680}$

It was found that the spectral reflectance as measured by an imaging spectrometer is directly related to the reflectance signal just above the water surface, which is in turn directly related to the subsurface radiance reflectance. Via the optical model for case III waters of Gordon and Morel (1983) of $R(0) = b_b / (a + b_b)$ the relationship with the main optical components of water will explored. Spectral scattering coefficient are estimated for the first time from this kind of data.

5.1 Ground-based Spectral Measurements

Two types of ground-based measurements were done. Firstly spectrophotometric measurements were carried out on water samples in the laboratory. The spectral extinction and absorption curves of humus and seston were measured. Using these the spectral scattering of the samples was estimated. Secondly *in situ* downwelling and upwelling radiance above and directly under the water surface was measured with a spectroradiometer. These measurements contain the combined spectral effects of absorption, scattering and solar-induced fluorescence from water, humus and the seston. Dekker (in prep.) discusses the groundbased measurements in detail.

Laboratory spectrophotometric measurements were made on water samples taken at sites where radiance and irradiance measurements were made in the field. The spectral absorption of humus and particulate matter was measured. The absorption and scattering spectrum of water is constant, it was taken from literature.

Spectral scattering was determined by subtracting from the attenuation signal the fully corrected absorption signal. This results in the spectral scattering over 5°- 180°; 0°-5° was the acceptance angle of the photomultiplier (see Dekker, in prep., for details).

5.2 Absorption Spectra of Inland Waters

Figure 4a shows the absorption and scattering spectra for pure water. It absorbs light only very weakly in the blue and green regions of the spectrum. Absorption begins to rise as the wavelength increases above 550 nm and it becomes significant in the red region. At 680 nm a 1 meter thick layer of pure water will absorb about 35% of the incident light. There are absorption shoulders at 610-620 and 660-670 nm. The pure water absorption at 700 nm was taken as reference point to classify inland waters. Scattering by water is inversely proportional to wavelength, together with low absorption this causes the blue colour of pure water. In situ Raman scattering can significantly increase scattering by water in nature significantly (Peacock et al. 1990). For these freshwater bodies molecular- and Raman scattering are negligible as compared to scattering by particulate matter.

Figure 4b shows absorption spectra of 31 samples of humus from 16 inland waters in the Netherlands, ranging from shallow hypertrophic lakes to deep mesotrophic lakes to turbid shipping canals.

Humus can efficiently remove blue light in the top few cm's of the water column, imparting a yellow colour to the water. In productive lakes an increase in humus absorption tends to covary with an increase in lake trophic status. Kirk (1983), Visser (1984) and Davies-Colley & Vant (1987) discuss the several origins of humus in detail.

For comparison of humus absorption spectra a reference wavelength of 440 nm was chosen by Kirk (1983), because it corresponds approximately to the mid-point of the first chlorophyll *a* absorption band. For the subject area the range of humus absorption at 440 nm is from 0.83 to 2.80 for the dataset presented here. An extensive review of values for the humus absorption values

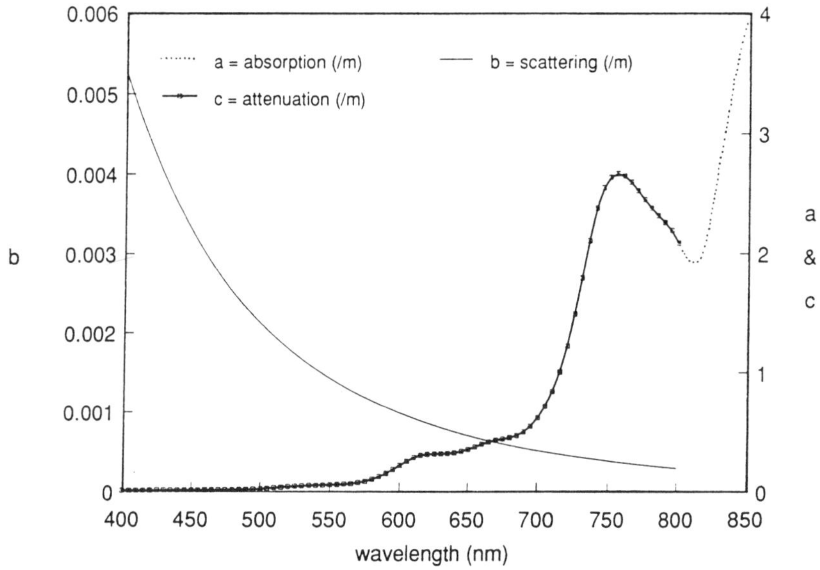

FIGURE 4a. The pure water absorption and scattering spectrum as derived from literature. Buiteveld and Donze (unpubl.) estimated the best values for absorption and scattering spectra of pure water from 300 to 800 nm.

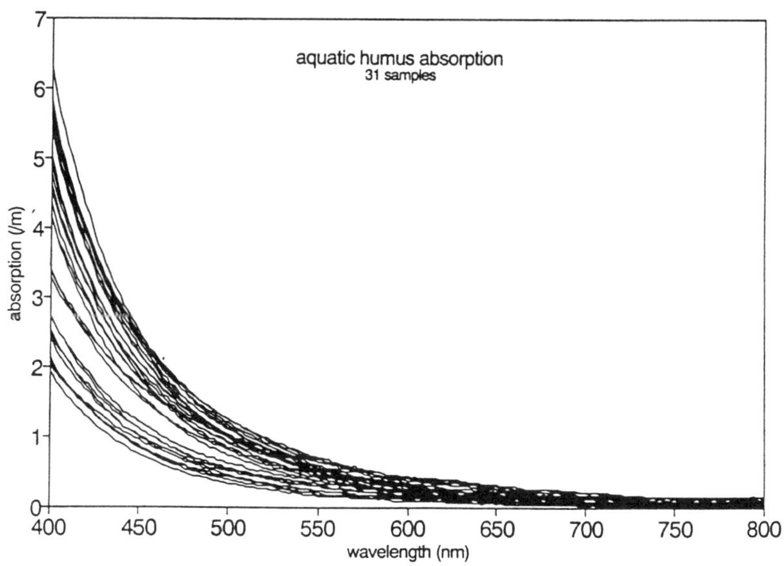

FIGURE 4b. The spectral absorption of aquatic humus. In first approximation it resembles an exponentially declining function of wavelength. The lake water samples were filtered through 0.2 μm (Sartorius membrane) filters. Absorption by the filtrate was measured in 10 cm pathcells using a Perkin-Elmer UV-VIS 551S double beam spectrophotometer at a resolution of 1 nm.

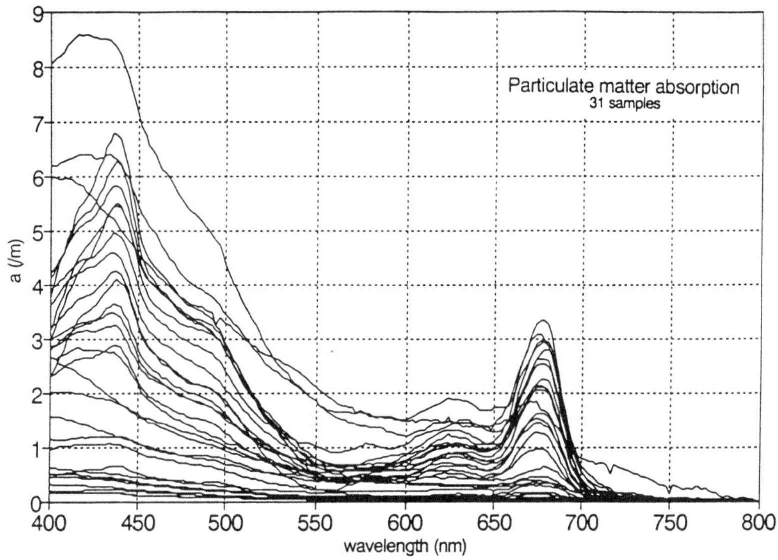

FIGURE 4c. Absorption spectra of the particulate matter for the same waterbodies as figure 4b. Note the light harvesting pigment absorption peaks at 435, 630, and 680 nm. Particulate matter was collected on 0.2 μm membrane filters and resuspended in a smaller quantity of distilled water, following the concentration method of Kirk (1980). The spectral absorption was measured in 1 cm pathlength cuvettes placed next to openings of an integrating sphere, enabling nearly all forward scattered light to be collected. A correction for scattering was made following the method of Davies-Colley (1986), where the value for attenuation using the integrating sphere at 800 nm was assumed to be a measure for the scattering.

at 440 nm is given by Kirk (1983, table 3.2). As maximum range for this absorption approximately 0 for the Sargasso Sea to 19.1 for Lough Napeast in Ireland is given.

Figure 4c shows the absorption spectra of the particulate matter for the same waterbodies as figure 4b. General spectral features are: from 400 to 435 nm an almost flat or slightly decreasing slope in the samples with low absorption. For the samples with a higher absorption this slop increases to 435 nm. The peak at 435 is the blue or 1st chlorophyll *a* absorption maximum (A_{435} in the classification scheme). After the peak at 435 nm a strong decrease occurs till a shoulder at 480 nm becomes visible. This absorption feature is probably due to *b*-carotene, a light-harvesting pigment present in all algae. Beyond 480 nm absorption decreases to a minimum at 550 nm, after which it increases to a relatively small peak at 630 nm. This peak is probably caused by CPC absorption, indicating the presence or the dominance of cyanobacteria in these waters (referred to as A_{630} in the optical classification). After 630 nm a relative low in absorption at 650 nm is followed by a clear increase in absorption to a maximum at 680 nm: this is the red, or second, chlorophyll *a* absorption peak (A_{680}). Beyond 680 nm the absorption decreases till it is almost zero at 720 nm. Although the plant pigments cause the most prominent features in the absorption scans of especially the eutrophic waters, the detritus present in considerable quantities in these lakes also absorbs light strongly. The spectrum of detritus absorption is probably similar to that of humus.

Cases without a significant absorption feature at 435 nm probably are almost pure detritus absorption spectra. The detrital part of a spectrum is given the *T* classification.

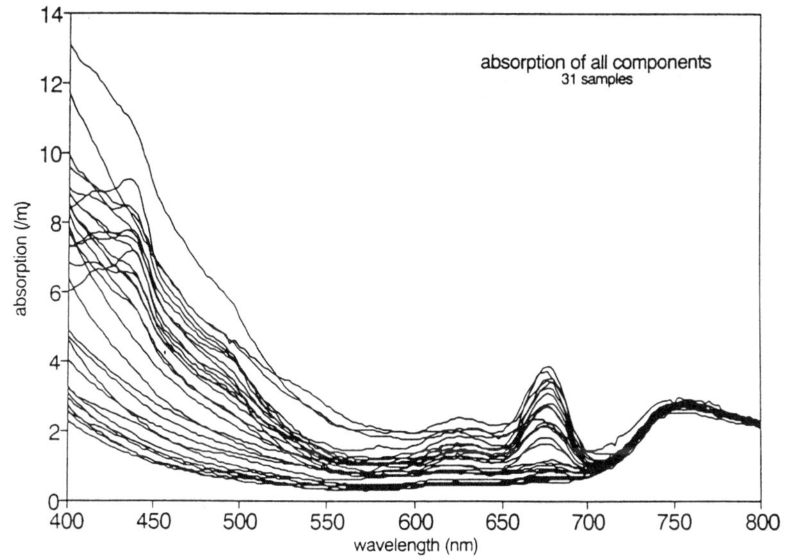

FIGURE 4d. The combined absorption curves of the aquatic humus, the particulate matter and the pure water spectrum. Note the high absorption at short wavelengths caused by aquatic humus, detritus and the first chlorophyll *a* absorption peak. The lowest combined absorption values occur at 550 to 600 nm, at 650 nm and at 705 nm.

The combined absorption curves are shown in figure 4d. It is seen that at short wavelengths humus, detritus and the first chlorophyll *a* absorption peak cause high absorption. This is the reason for the low reflectance observed in this spectral area as will be demonstrated later. Beyond 500 nm the spectral information becomes less ambiguous and reflectance increases allowing better discrimination of spectral features. The lowest total absorption values occur at 550 to 600 nm, at 650 nm and at 705 nm, coinciding with maxima in reflectance.

5.3 SPECTRAL SCATTERING

Before considering spectral reflectance the inherent optical property that causes the upward change in direction of downwelling light, spectral scattering, must be discussed. Published spectral scattering spectra of inland waters are few. Figure 4e shows the spectral scattering for the same samples as used before to measure absorption spectra. It is the spectral scattering over an angle of 5 - 180°. The values are extremely high compared to those found for ocean waters. In general an increase in scattering is associated with an increase in slope of the spectral scattering towards shorter wavelengths. An increase in scattering is closely correlated to an increase in particulate matter as expressed in seston dry weight concentration and in absorption by particulate matter. The shape of the scattering spectrum is a mainly a function of the scattering properties of the particles. In general a steepening of the slope toward shorter wavelengths may be explained as being caused by a decreasing optical cross section of the average particle size in the samples. In inland water samples the particulate matter will always be a mixture of particle types and sizes. Scattering by algae is quite complex: an alga is semitransparent and contains a variety of different materials within its cell structure. Cyanobacteria contain gas vacuoles that cause high scattering at large angles (Donze et al., 1987). Scattering by algae is roughly inversely related to wavelength. Scattering spectra of freshwater algae are given in Klepper (1984) and Donze et al. (1987).

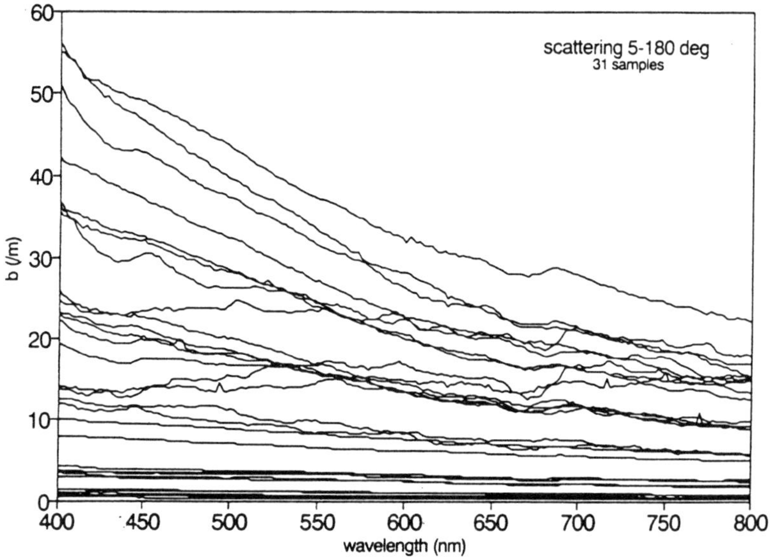

FIGURE 4e. The spectral scattering for the same samples as in the absorption spectra. It is the spectral scattering over an angle of 5-180°. The values are extremely high compared to those found for ocean waters. In general an increase in scattering is associated with an increase in slope of the spectral scattering towards shorter wavelenghts. An increase in scattering is closely correlated to an increase in particulate matter.

The combined effects of humus and detritus plus algal pigment absorption at short wavelengths (< 500 nm) together with the opposing effects of an increased scattering towards shorter wavelengths make this spectral region inappropriate for extracting information from the resultant reflectance (Dekker et al., 1990a and b, 1991, 1992b-d). It is between 500 nm and 720 nm that the absorption of various components can be discriminated more accurately. At 550 nm the algae rich spectra have a minimum absorption; at 630 nm a slight absorption peak due to CPC is visible; at 650 another local minimum in absorption is visible; at 675 nm the red chlorophyll *a* peak is clear; the lowest absorption over the entire spectrum is almost invariably found at 700 -710 nm. it is also evident from the total absorption spectra that beyond 720 nm water absorption is the single most dominant factor. Based on these results it is indeed correct to assume that information extraction from inland waters should take place in the spectral green to nearby infrared areas (Gitelson & Kondratyev, 1991; Dekker et al., 1990 a and b, 1991, 1992 b-d). This is significantly different from ocean colour remote sensing practice where the blue to green spectral area is considered the most appropriate for water quality feature information extraction.

5.4 FIELD SPECTRORADIOMETRIC MEASUREMENTS

Radiance reflectance measurements were made in situ using a portable spectroradiometer (figure 4f). In this experiment, 31 measurements were made under varying conditions of illumination. Irradiance varied from clear skies to complete cloud cover.

The reflectance spectra are similar to the inverted total absorption spectra. Where absorption peaks occur the reflectance is low and vice versa. The increase in spectral scattering towards shorter wavelengths in eutrophic waters is counterbalanced by high absorption at these wavelengths.

Similar subsurface irradiance spectra for eutrophic water bodies are reported by Miyazaki et al. (1987) for Lake Kasumigaura in Japan, by Vos et al. (1976) for Frisian lake water in The Netherlands, Davies-Colley et al. (1988) for New Zealand lakes and by Seyhan et al.(1974) for ditch water in The Netherlands. Gitelson et al. (1986) show normalized spectra for Lake Balaton in Hungary with similar features as did Gitelson & Nikanorov (1988) for the Don and Seversky Donetz Rivers in the Soviet Union.

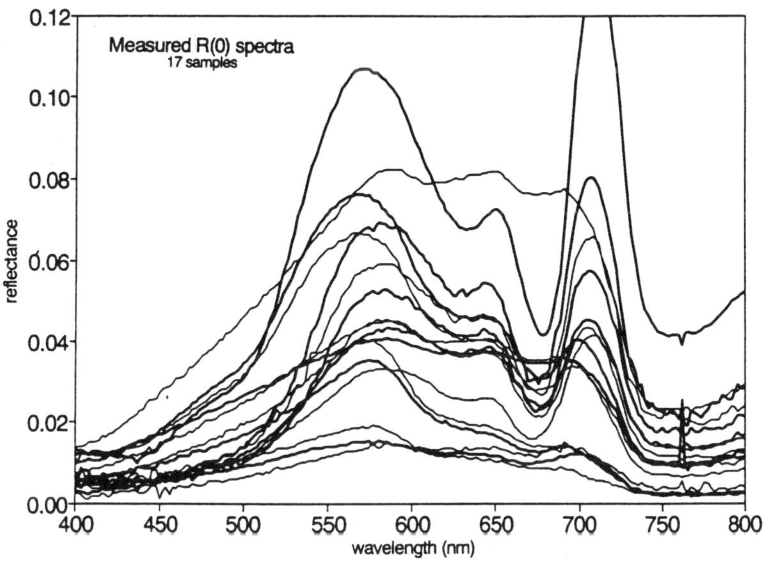

FIGURE 4f. Reflectance measurements at the same locations where the samples for the spectrophotometric measurements were taken. The measurements were made using a portable spectroradiometer (Spectron SE-590 with a 15° field-of-view aperture). Reflectance was calculated as subsurface upwelling radiance divided by above-surface downwelling radiance recalculated to subsurface downwelling irradiance using the method by Carder et al. (1991). The 31 measurements were measured under varying illumination conditions.

5.5 MODELED REFLECTANCE

Using the same model for calculating subsurface reflectance from absorption and backscattering for turbid waters as given in the section on the determination of the concentration of phycocyanin

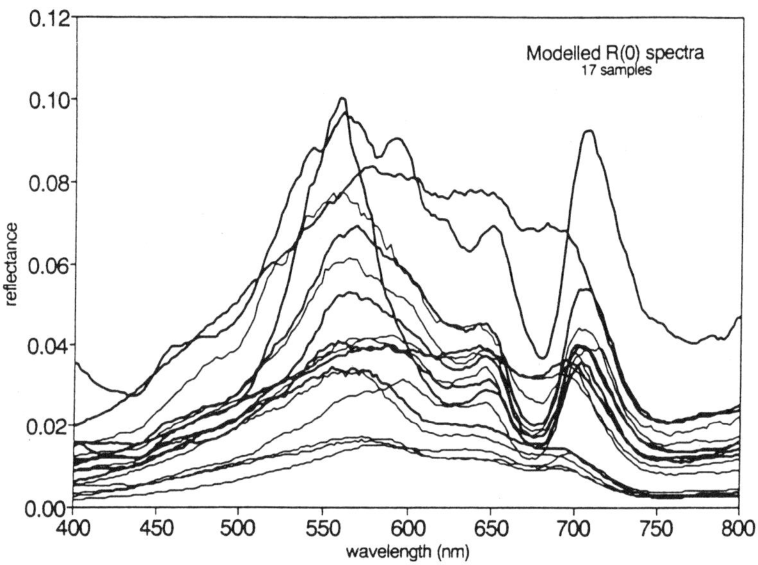

FIGURE 4g. Modeled reflectance measurements based on the spectrophotometric measurements (see text for details).

(i.c. $R(0) = \kappa\, b_b\, /\, (a + b_b)$ where $R(0)$ is subsurface irradiance reflectance, κ is a solar angle factor, calculated after Kirk (1991) and b_b is the backscattering coefficient and a is the total absorption coefficient) the modelled spectral reflectance can be calculated.

To calculate $R(0)$ the main problem is the determination of the ratio of backscattering to scattering. Scattering over an angle of 5-180 ° is known, but scattering over an angle of 0°-180° is needed in the $R(0)$ equation. The scattering over 0°-5° in the forward direction could be equivalent to the scattering over 5°-180° (Oishi, 1990). Because the volume scattering functions are unknown for these waters, the scattering over 5°-180° is assumed to be the total scattering. But it is possible $b_b{:}b$ is underestimated by half this way. b_b can be approximated by using the measured $R(0)$, the measured and calculated total a, the κ value for the solar angles and the scattering values multiplied by a factor to be determined by solving equation. Values found for $b_b{:}b$ ranged from 0.006 to 0.010 for shallow highly eutrophic lakes, from 0.0016 to 0.031 for the turbid detritus and silt rich Amsterdam Rhine Canal and river Vecht, 0.017 for a shallow mesotrophic clear lake and from 0.028 to 0.041 for clear oligotrophic to mesotrophic deep lakes. This indicates that the volume scattering functions for these water have significant differences. To assume a constant volume scattering function for turbid and/or algae-rich waters is probably erroneous as was predicted by Kirk (1983).

The similarity of the modelled $R(0)$ spectra to the measured $R(0)$ spectra is evident. It is expected that rigourous testing of the experimental procedures used to measure $R(0)$ will reveal a number of sources of error. This may lead to further improvement of the results.

Once the model parameters are known it is possible to do numerical experiments. Figure 4h shows the calculated reflectance if absorption by particulate matter would be absent. This type of reflectance curves would be found in waters with suspended mineral (white) sediment without organic particulate material. Figure 4h also shows that the water absorption shoulders at 610-620

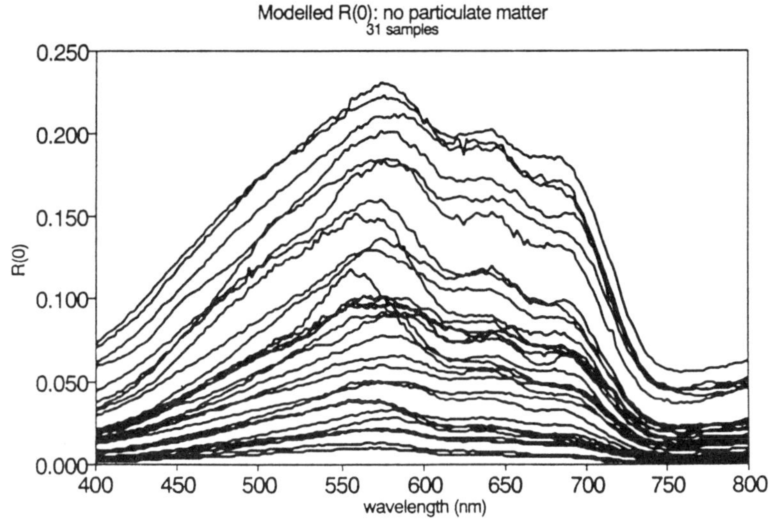

FIGURE 4h. Modeled reflectance measurements for the same waters, where the particulate matter absorption has been omitted. Note the decreases in reflectance centered at 625 and 675 nm, caused by the combination of the water and aquatic humus absorption and the spectral scattering.

and 660-670 nm combined with humus absorption and spectral scattering (which shows some effect of chlorophyll *a* absorption peak, see figure 4e) can cause decreases in reflectance at 625 and 675 nm. Such an effect might be misinterpreted as absorption by CPC and chlorophyll *a*.

5.6 IMAGING SPECTROMETRY REFLECTANCE MEASUREMENTS

The CASI is a pushbroom imaging spectrometer with three operating modes: spatial mode, spectral mode and a full frame mode. In the spatial mode up to 15 spectral bands may be defined and implemented immediately at the highest spatial resolution of 512 pixels across track. In this mission a spectral band selection was chosen that approximated the CAESAR Inland Water Mode (defined in Dekker et al., 1990) as close as possible. In spectral mode spectra at 1.8 nm intervals with a resolution of 3.6 nm are acquired over 39 lines in a "rake" across the scene. The 288 * 1.8 = 520 nm coverage may be positioned anywhere between 410 and 980 nm, although calibration is only guaranteed between 430 and 870 nm. A simultaneous "Scene Recovery Channel" is acquired in full spatial mode in a selected wavelength interval. This may be used in locating the position of the less easily visualised "raked" spectral lines. In full frame mode the entire CCD sensor area is digitized, acquiring 288 spectra samples for each of 512 spatial pixels. This is used in acquisition of calibration data and in applications where hyperspectral data sets are desired. This mode was not used here.

For the same water bodies as the measured and modelled $R(0)$ spectra upwelling radiance spectra were selected from a CASI flight on September 14th 1990 from an altitude of 500 m A.G.L. Using a groundbased PTFE panel reflectance measurement (calibrated to 100 % reflectance) as the downwelling radiance, the radiance reflectance was calculated. Figure 4i shows the CASI radiance reflectance spectra. No correction for atmospheric path radiance was made and the depicted spectra also contain a diffuse irradiance Fresnel reflectance component. The lowest reflectance spectrum, from the Water Reservoir, does, of course, give a maximum possible value for these two potential additional radiance components.

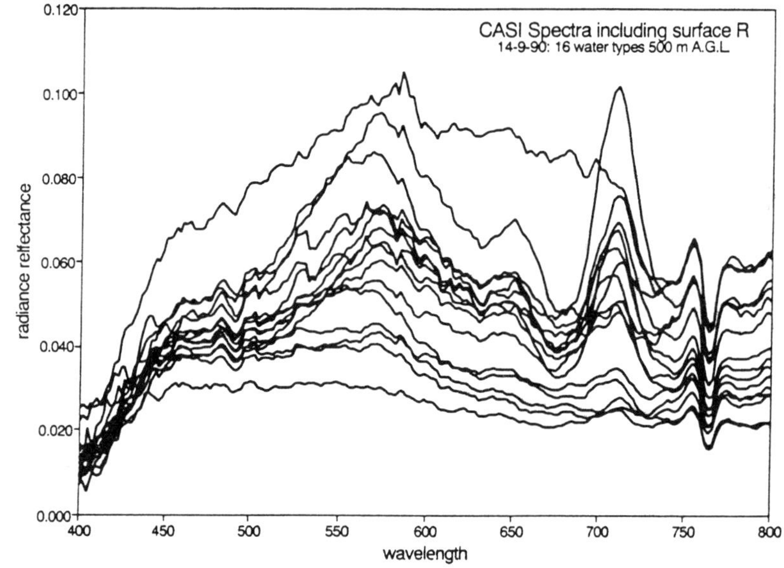

FIGURE 4i. For the same water bodies as the measured (figure 4f) and modeled $R(0)$ spectra (figure 4g) upwelling spectral radiance spectra were selected from a CASI flight on September 14th 1990 from an altitude of 500 A.G.L.

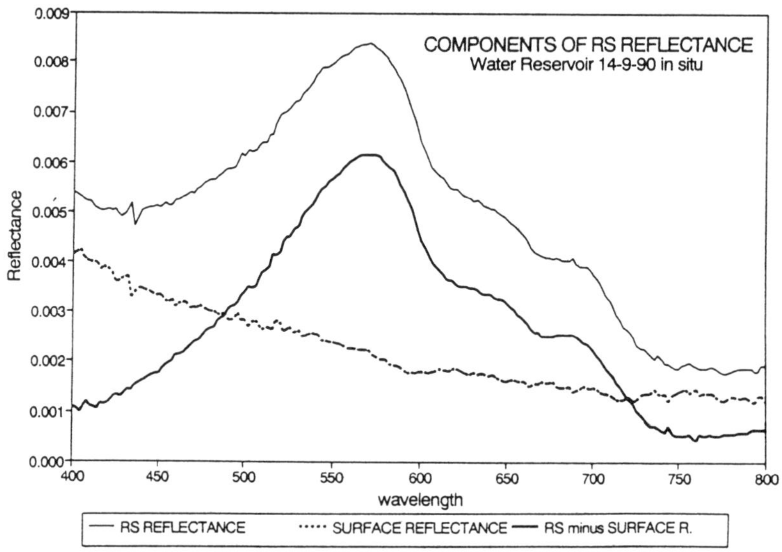

FIGURE 4j. The surface diffuse Fresnel reflectance component calculated for a ground-based measurement taken simultaneously with the CASI overflight.

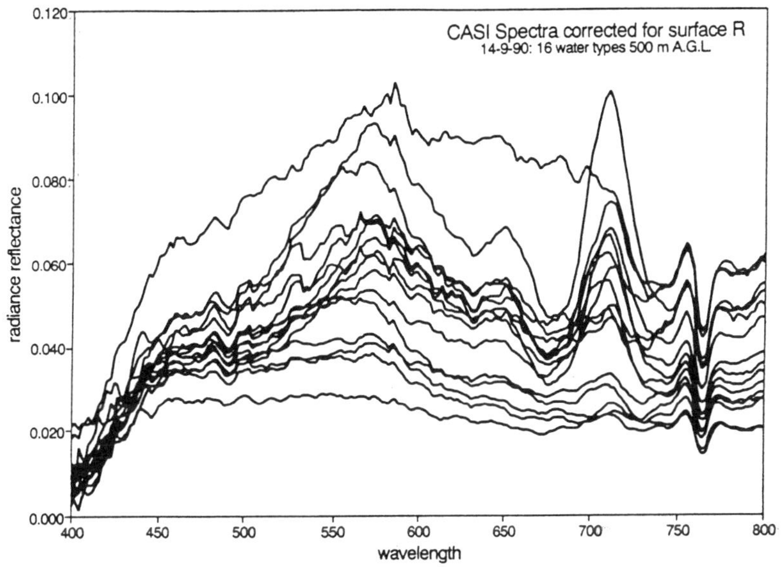

FIGURE 4k. The CASI reflectance spectra from figure 4i corrected with the diffuse reflectance component of figure 4j.

Using the fieldspectroradiometer upwelling radiance was measured both just below (+/- 1 cm depth) and above (+/- 100 cm) the water surface. From these measurements the diffuse reflectance can be estimated. This is given in figure 4j. Once this signal is subtracted from the CASI measured upwelling radiance the subsurface radiance reflectance can be calculated. This is presented in figure 4k. These spectra compare well to the measured and modelled $R(0)$ spectra. This agreement between airborne, surface and sample measurements is important for it forges the link between the results of research carried out on the underwater light climate by aquatic scientists and the remotely sensed signal.

6. Discussion

Future application of imaging spectrometry in practical water quality management will depend on the development of cheap and stable procedures to do so. This must include an absolute minimal gathering of field data to calibrate the images. It is argued that only complete development of the analytical approach can make this goal attainable. The final argument will be the cost of remote sensing data, as compared to the needs of water authorities. It should be kept in mind that this need for measurements is varied and also contains a host of observables that cannot be measured by remote sensing techniques.

Further development of imaging spectrometry requires the use of a multidisciplinary approach: at this stage the following measurements are essential: 1) intensive water quality data gathering and analysis, 2) field- and laboratory based underwater light climate measurements and 3) airborne imaging spectrometry. Additional measurements such as colour aerial photography and Landsat

TM or SPOT satellite imagery can be a convenient basis for interpreting imaging spectrometry measurements.

At present imaging spectroscopy is a research tool in development. As such it promises new results to and requires cooperation of (and co-funding by) organizations interested in the development of remote sensing, aquatic optics, optical theory and instrumentation, water research and by water authorities. On all aspects improvements are needed and possible. Only efficient execution of large multidisciplinary projects can assure progress at reasonable speed and cost.

7. Acknowledgements

The following persons and organisations contributed to the results presented: P.J.T. Verstaelen and R. Koeleman (amongst many others) at the Water Management and Pollution Control Authorities Amstel- en Gooiland, The Netherlands. L. van Liere and his staff at the Limnological Institute, Nieuwersluis (NL). T. Burger-Wiersma and A. van Egmond at the University of Amsterdam (NL). H.W. Balfoort and J. Snoek at the Laboratory for Aquatic Ecology at the University of Amsterdam (NL). J. Krijgsman and H. Hakvoort at the Faculty of Civil Engineering of the Technical University of Delft (NL). The Spectron SE-590 was provided and calibrated by the equipment pool of the National Environmental Research Council, UK. A grant from NERC made it possible for Dr T.J. Malthus to perform and analyze the measurements. The CASI was made available for research purposes by Aerospace Image Productions, Herrenberg, Germany. Part of this work was funded by the Dutch Remote Sensing Board (BCRS), the Water Management and Pollution Control Authorities Amstel- en Gooiland, The Netherlands and by the Institute for Inland Water Management and Wastewater Treatment.

8. References

Austin, R.W. (1974) 'The remote sensing of spectral radiance from below the ocean surface', In: Jerlov, N.G. and E. Steeman Nielsen (ed.), *Optical Aspects of Oceanography,* Academic Press, London and New York, 316-344.

Bennett, A. and L. Bogorad (1973) 'Complementary chromatic adaptation in a filamentous blue-green alga', *J. Cell. Biol.*, 58, 410-435.

Carder, K.L., S.K. Hawes, K.A. Baker,R.C. Smith, R.G. Steward, and B.G. Mitchell (1991) 'Reflectance model for quantifying chlorophyll a in the presence of productivity degradation products', *J. Geoph. Res.*, 96(C11), 20.599-20.611.

Davies-Colley, R.J., W.N. Vant, and R.J. Wilcock (1988) 'Lake water colour: comparison of direct observations with underwater spectral irradiance', *Water Resources Bulletin*, 24(1), 11-18.

Davies-Colley, R.J. and W.N. Vant (1987) 'Absorption of light by yellow substance in freshwater lakes', *Limnol. Oceanogr.*, 32(2), 416-425.

Davies-Colley, R.J., R.D. Pridmore, and J.E. Hewitt (1986) 'Optical properties of some freshwater phytoplanktonic algae', *Hydrobiologia*, 133, 165-178.

Dekker, A.G., 'Imaging spectrometry and multispectral remote sensing of surface water quality of eutrophic waters', thesis (in prep.)

Dekker, A.G., T.J. Malthus, and L.M. Goddijn (1992a) 'Monitoring cyanobacteria in eutrophic waters using airborne imaging spectroscopy and multispectral remote sensing systems', *Proceedings 6th Australasian Remote Sensing Conference*, Wellington, New Zealand, 2-6 November 1992.

Dekker, A.G., T.J. Malthus, and M.M. Wijnen (1992b) 'Spectral band location for remote sensing of turbid and/or eutrophic waters', *Proceedings First Thematic Conference on Remote Sensing for Marine and Coastal Environments*, New Orleans, Louisiana, USA, 15-17 June 1992.

Dekker, A.G., T.J. Malthus, M.M. Wijnen, and E. Seyhan (1992c) 'The effect of spectral band width and positioning on the spectral signature analysis of inland waters', *Remote Sensing of Environment*, 41(2/3), 211-226.

Dekker, A.G., T.J. Malthus, M.M. Wijnen, and E. Seyhan(1992d) 'Remote sensing as a tool for assessing water quality in Loosdrecht lakes', *Hydrobiologia*, 233, 137-159.

Dekker, A.G., T.J. Malthus, and E. Seyhan (1991) 'Quantitative modelling of inland water quality for high resolution MSS-systems', *IEEE Trans. on Geosc. and Rem. Sens.*, 29(1), 89-95.

Dekker, A.G. and S.W.M. Peters, 'The use of the Thematic Mapper for the analysis of eutrophic lakes: A case study in The Netherlands', *Int. J. Rem. Sens.*, in press.

Dekker, A.G., T.J. Malthus, and E. Seyhan (1990a) 'Improving quantitative analysis of inland water quality using high spectral resolution imaging and non-imaging data', *Proc. IGARSS' 90 Symposium*, Washington, May 20-24, 1990:p 117-120.

Dekker, A.G., T.J. Malthus, and E. Seyhan (1990b) 'An inland water quality bandset for the CAESAR system based on spectral signature analysis', *Proc. Int. Symp. Remote Sensing and Water*, Enschede, The Netherlands, August 1990, 597-606.

Doerffer, R. (1989) 'Imaging spectroscopy for detection of chlorophyll and suspended matter', In: F. Toselli and J. Bodechtel (eds.), *Imaging Spectroscopy: Fundamentals and Prospective Applications*, EuroCourses: Remote Sensing vol. 2, Kluwer Acad. Publ., Dordrecht, The Netherlands, 215-258.

Donze, M., Dubelaar, and Visser (1987) '*Anomalous behaviour of forward and perpendicular lightscattering of a cyanobacteria due to intracellular gas vacuoles*', BCRS Report 87-08 (Dutch Remote Sensing Board), NIWARS 42/28 (87-0).

Dubelaar, G.B.J. et al.(1989) 'Optical plankton analyzer: A flow cytometer for plankton analysis, II: specifications', *Cytometry*, 10, 529-539.

Gitelson, A.A. and K.Y. Kondratiev (1991) 'Optical models of water bodies', *Int. J. Rem. Sensing*, vol. 12, no. 3, 373-385.

Gitelson, A.A. and G.P. Keydan (1990) 'Remote sensing of inland surface water quality-measurements in the visible spectrum', *Acta Hydrophys.*, Berlin 34 (1.S), 5-27.

Gitelson, A.A., A.M. Nikanorov, G.Y. Szabo, and F. Szilagyi (1986) 'Étude de la qualité des eaux de surface par télédétection', *Proc. Budapest Symp. on Monitoring to Detect Changes in Water Quality Series*, July 1986, IAHS Publ.no. 157., 111-121.

Gordon, H.R. and A.Y. Morel (1983) *'Remote assessment of ocean color for interpretation of satellite visible imagery: a review'*, Lecture Notes on Coastal and Estuarine Studies, 4, Springer, New York.

Jerlov, N.G. (1976) *'Marine optics'*, Elsevier, Amsterdam, the Netherlands.

Kirk, J.T.O. (1980) 'Spectral absorption properties of natural waters: contribution of the soluble and particulate fractions to light absorption in some inland waters of South-eastern Australia', *Austr. J. Mar. Freshwater Res.*, 31, 287-296.

Kirk, J.T.O. (1981a) 'Monte Carlo Study of the nature of the underwater light field in, and the relationships between optical properties of, turbid yellow waters', *Austr. J. Mar. Freshwater Res.*, 32, 517-532.

Kirk, J.T.O. (1981b) 'Estimation of the scattering coefficient of natural waters using underwater irradiance measurements', *Austr. J. Mar. Freshwater Res.*, 32, 533-539.

Kirk, J.T.O. (1983) *'Light and photosynthesis in aquatic ecosystems'*, CSIRO, Canberra, Australia, Cambridge University Press.

Kirk, J.T.O. (1991) 'Volume scattering function, average cosines, and the underwater light field', *Limnol. Oceanogr.*, 36(3), 455-467.

Klepper, O., S.G. Vermij, and R. Lingeman (1984) 'The influence of light scattering on vertical extinction in Lake Maarsseveen', *Verh. Int. Ver. Limnol.*, 22, 82-86.

Kondratyev, K. Y. and D.V. Pozdniakov, (1990) 'Passive and active optical remote sensing of the inland water phytoplankton', *ISPRS J. of Photogramm. and Rem. Sensing*, 44, 257-294.

McGarrigle, M.L., E. O'Mongain, J.E. Walsh, T. Sommerville, and M. Bree (1990) *'National survey of lakes by remote sensing: Calibration of a low altitude water quality spectrometry'*, Environm. Res. Unit, St Martin's House, Waterloo Rd, Dublin, Rep. Ireland.

Melack, J.M. and S.H. Pilorz (1990) 'Reflectance spectra from eutrophic Mono Lake, California, measured with the Airborne Visible and Infrared Imaging Spectrometer (AVIRIS)', *SPIE,* Vol. 1298, Imaging Spectroscopy of the Terrestrial Environment, 202-212.

Miyazaki, T., H. Shimizu, and Y. Yasuoka (1987) 'High-speed spectroradiometer for remote sensing', *Applied Optics*, 26(22), 4761-4766.

Morel, A. and H.R. Gordon (1980) 'Report of the working group on water color', *Boundary Layer Meteorolgy*, 18, 343-355.

Oishi, T. (1990) 'Significant relationship between the backward scattering coefficient of sea water and the scatterance at 120°', *Applied Optics*,29(31), 4658-4665.

O'Neill, N.T., A.R. Kalinauskas, G.A.Borstad, H. Edel, G.A. Gower, and H. van der Piepen (1987) 'Imaging spectrometry for water applications', *Proc. 31st Ann. Int. Techn. Symp. Opt. & Optoelectr. Appl. Sc. & Engineering-Imag. Spectroscopy II*, San Diego, California. SPIE Proc. 834, August 1987.

Peacock, T.G., K.L. Carder, C.O Davis, and R.G. Steward (1990) 'Effects of fluorescence and water Raman scattering on models of remote sensing reflectance', *SPIE Proceedings*, Ocean Optics X, 1302, 303-319.

Petzold, T.J. (1972) *'Volume scattering functions for selected Ocean Waters'*, Rep. Visibility Lab. Scripps Oceanogr. Inst.,U. California, Refs. 72-78

Preisendorfer, R.W. (1976) *'Hydrologic Optics, Vol 1'*, Washington, Dep. of Commerce.

Prieur, L. and S. Sathyendranath (1981) 'An optical classification of coastal and oceanic waters based on the spectral absorption curves of phytoplankton pigments, dissolved organic matter and other particulate materials', *Limnol. & Oceanogr.*, 26(4), 671-689.

Seyhan, E., N.J.J. Bunnik, W. Verhoef, and J. van Kuilenburg (1974) 'Measurements of spectral signatures for water quality monitoring', *NIWARS Publication No. 24*, Delft, The Netherlands; Paper presented at the First General Conference of the Remote Sensing Society, Birmingham, Great Britain, Sept. 1974, 29.

Visser, S.A. (1984) 'Seasonal changes in the concentration and colour of humic substances in some aquatic environments', *Freshwater Biology*, 14, 79-87.

Vos, W.L., M. Donze, and H. Buiteveld (1986) 'On the reflectance spectrum of algae in water: the nature of the peak at 700 nm and its shift with varying concentration', *Comm. on San. Eng. and Water Managem.*, nr. 7, ISSN-0169-6246, TU Delft, 86-22.

Whitlock, C.H., L.R. Poole, J.W. Usry, W.M. Houghton, W.G. Witte, W.D. Morris, and E.A. Gurganus (1981) 'Comparison of reflectance with backscatter and absorption parameters for turbid waters', *Applied Optics*, 20 (3), 517-522.

FUTURE APPLICATIONS, SENSOR DEVELOPMENTS AND RESEARCH PROGRAMMES IN THE FIELD OF IMAGING SPECTROMETRY

JOHANN BODECHTEL and STEFAN SOMMER
AGF - Working Group Remote Sensing
Institute for General and Applied Geology
University of Munich, Luisenstr. 37
D-80469 München , Germany

ABSTRACT. The recent status of imaging spectrometry is characterised by the experimental to pre-operational deployment of airborne imaging spectrometers such as AVIRIS, CASI and GER-IS. In the next years also European airborne imaging spectrometers are expected to be provided in the commercial world to cover the requirements for high spatial and spectral resolution data acquisition for operational research and applications. In Europe national and international facilities such as the EC/ESA EARSEC programme are envisaged to satisfy the needs of the user community. In this context, full coverage of the VIS/NIR/SWIR (400 - 2500 nm) spectral region is increasingly requested by all disciplines related to land applications.
The next major step in the development of imaging spectrometry will be the launch of spaceborne instruments like MERIS (ESA) and MODIS (NASA) on the first ESA and NASA Polar Platforms scheduled for the end of the century. These sensors providing only moderate spatial resolutions (250 - 1000 m pixel size) essentially in the VIS/NIR spectral region (400 - 1050 nm) are mainly designed for the monitoring of global change in the framework of ESA's POEM and NASA's EOS programmes addressing mainly the requirement to establish long term global data bases. Besides these sensors, also high spatial resolution imaging spectrometers covering the full VIS/NIR/SWIR spectral range are considered indispensable for the monitoring, examination and assessment of sensitive dynamic processes and fluxes within the terrestrial ecosystem, such as soil erosion, sediment transport, nutrient flux and others. To cover these requirements, instruments like HIRIS (NASA) and HRIS (ESA) are under development and are discussed as candidates for the second generation of Polar Platforms.

1. Introduction

As demonstrated in the prior lectures, imaging spectrometry has been developed from experimental to pre-operational level in many application fields of earth remote sensing. Airborne campaigns such as EISAC 1989 and MAC-Europe 1991 considerably improved the capabilities of the European users in terms of data quality assessment, radiometric and atmospheric correction of imaging spectrometer data. Good progress has been made in the field of spectral signature modelling and in the definition and evaluation of relevant surface parameters retrievable with imaging spectrometry. Increasing efforts are focused on optimised methodologies for the exploitation of the data of future spaceborne imaging spectrometers with medium (MERIS/MODIS) and high spatial resolution (HRIS/HIRIS), which will be described in more details on the following pages.
Upon this basis, various options of instrument specifications of future, operational air- and spaceborne imaging spectrometers are being discussed in terms of optimising spectral coverage,

J. Hill and J. Mégier (eds.), Imaging Spectrometry - a Tool for Environmental Observations, 319–328.

spectral sampling, minimum band-sets and spatial resolution. For specific land applications of imaging spectrometry, especially the great variety of relevant phenomena and scales requires additional investigations to define instrument specifications suitable for the full range of disciplines involved. Preliminary minimum requirements to the specifications of imaging spectrometers dedicated to land applications are given below.

PARAMETER	REQUIREMENT
Swath Width:	≥100 km
Spatial Resolution:	<100 m
Spectral Resolution:	
a) VIS/NIR:	
Spectral Range	400 - 1050 nm
Bandwidth	5 - 10 nm
No. of Bands	≥10
b) SWIR I	
Spectral Range	1500 - 1800 nm
Bandwidth	<20 nm
No. of Bands	≥10
c) SWIR II	
Spectral Range	2000 - 2400 nm
Bandwidth	10 nm
No. of Bands	≥20
Sensor Sensitivity:	NER ≤0.001 [mW/cm2*sr*μ]
Radiometric Accuracy:	
Calibration	calibration accuracy better than 5% (in-band radiances)
Bandwidth	effective bandwidth must not exceed nominal bandwidth more than 20%
Band Position	deviation from nominal center wavelength ≤ ±3 nm
Signal-to-Noise ratio	>50 in all bands at scene minimum
Geometric Requirements:	
Band-toBand Registration	within one pixel between all bands
Spatial Resolution	real pixel size must not exceed nominal pixel size more than 10%

TABLE 1. Minimum user requirements to the specifications of spaceborne imaging spectrometers for land applications (after Bodechtel et al., 1992).

The requirements for atmospheric and oceanographic research are defined quite clearly and consequently led to the design of the spaceborne sensors Medium Resolution Imaging Spectrometer (MERIS/ESA) and Moderate Resolution Imaging Spectrometer (MODIS/NASA), which will be dedicated to large scale monitoring of global change. Due to the costs of spaceborne imaging spectrometer systems however, there is a tendency to build sensor systems capable of meeting the observational needs of the broad Earth science community rather than sensors dedicated to specific disciplines. As the subset of spectral bands/regions for each discipline is likely to be different, the need to build sensors covering the broader spectral region of interest, i.e. essentially VIS/NIR/SWIR (400 - 2500nm) and partly TIR, must be anticipated (Vane et al. 1991). The optimum performance of spaceborne instruments such as the High Resolution Imaging Spectrometers of ESA and NASA (HRIS/ESA; HIRIS/NASA), addressing the before-mentioned problems, is still under discussion.
Against this general background, the development of future imaging spectrometry shall be discussed with respect to new application and research programmes and the related sensor development.

2. Development of Application and Research Programmes

Recent plans for future applications and research programmes including imaging spectrometry cover a broad range of spatial and temporal scales. Global to continental observation with moderate spatial and high temporal resolution is required as well as regional observation with high spatial resolution, in order to better understand and monitor the complex dynamic processes, which are essential for global change.
At European regional level the operational use of imaging spectrometry is envisaged in the framework of agricultural monitoring and anti-fraud programmes as well as in the field of environmental monitoring and protection, e.g. monitoring of land degradation in the Mediterranean region (Hill et. al. 1991).
An important aspect of all planned application and research programmes is the synergistic use of imaging spectrometry data and remotely sensed data from other sources, particularly of SAR sensors. The complementarity of the different sensor types is currently being investigated in details under the aspect of necessary timing-conditions for optimum synergy between optical and microwave data for the modeling of natural processes.

2.1 GLOBAL SCALE PROGRAMMES

Undoubtedly the international Global Change programmes are the most advanced application for future spaceborne imaging spectrometry. The complementary ESA and NASA programmes POEM (Polar Orbit Earth Observation Mission/ESA) and EOS (Earth Observing System/NASA), both planned to incorporate several multisensor platforms including imaging spectrometers, are major parts of these international efforts including also numerous national contributions. These national and international research programmes are linked by a number of large scale umbrella programmes such as the World Climate Research Programme (WCRP) and the International Geosphere Biosphere Programme (IGBP).
Figure 1 illustrates the relationships of national and international programmes from the NASA/EOS point of view.

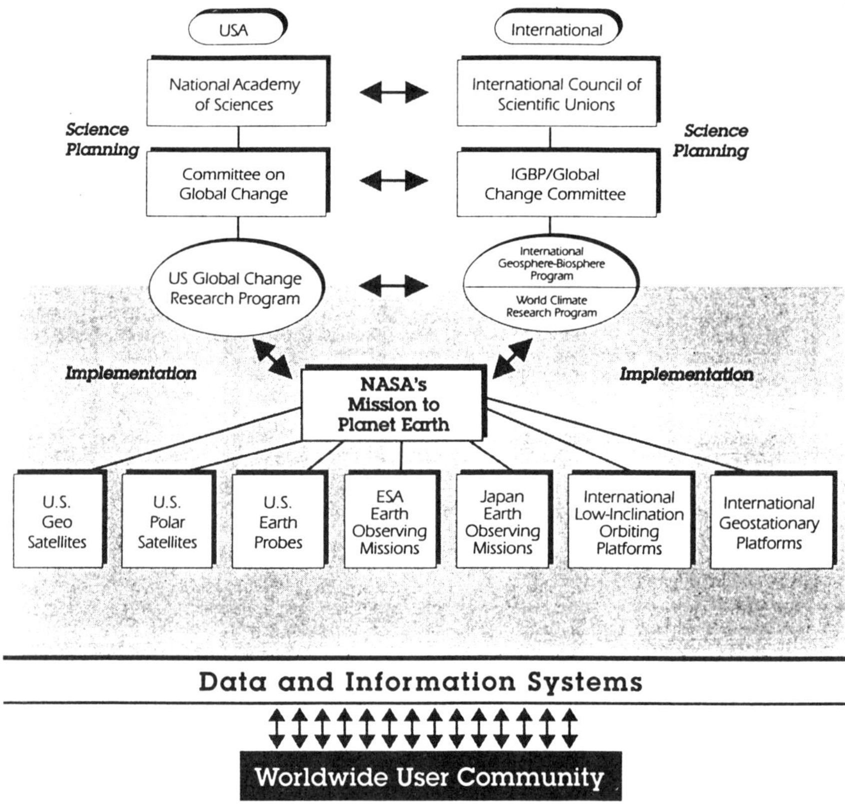

FIG.1: EOS AS AN ELEMENT OF THE GCRP AND IGBP

FIGURE 1. EOS as an element of the GCRP and IGBP (from EOS Reference Handbook 1991)

For ESA's POEM programme and similarly for the NASA EOS, the following principal objectives have been defined (Readings & Rast 1992):

- monitoring the Earth's environment on various scales, from regional to global.
- management and monitoring of the Earth's resources, both renewable and non-renewable.
- continuation and improvement of the service provided to the world-wide operational meteorological community.
- contribution to the understanding of the structure and dynamics of the Earth's crust and interior.

Detailed global change science priorities are given in Figure 2.

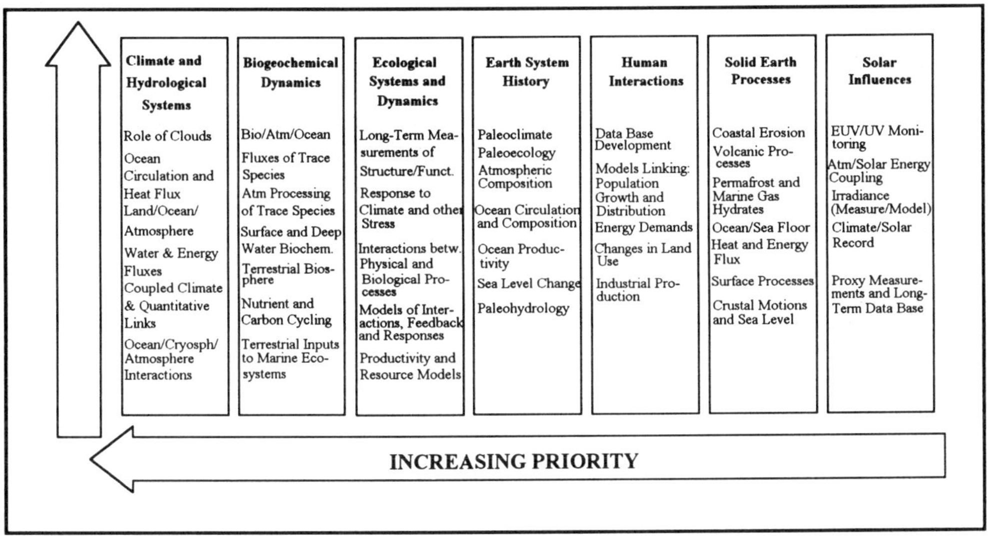

FIGURE 2. Global change science priorities (after EOS Reference Handbook 1991)

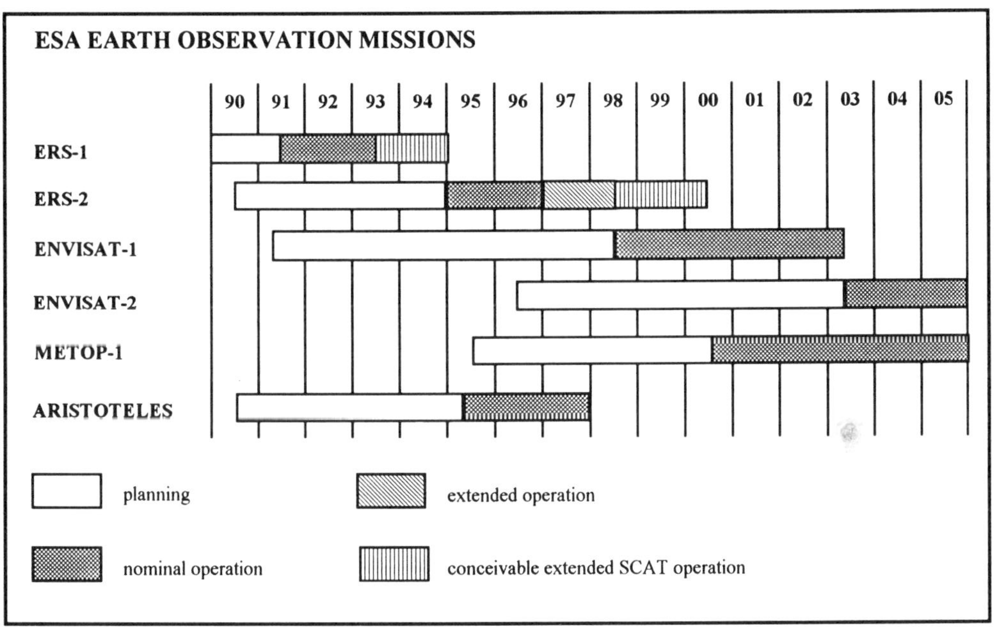

FIGURE 3. ESA earth observation missions (after ESF-EEOP 1992)

2.1.1 Imaging Spectrometry in the ESA POEM and in the NASA EOS Programme. In the ESA POEM programme a series of platforms is planned to be launched between 1995 and 2005 (see figure 3). According to the recommendations of the ESF-EEOP (1992) the polar platform missions shall be divided from the start into an environmental satellite series ENVISAT (first launch 1998) and an operational meteorological and climate monitoring series METOP (first launch 2000). In this framework, MERIS is an approved instrument on the first ENVISAT mission.

According to Rast (1992) the primary mission goals of MERIS include:

- Biological oceanography: Phytoplankton biomass and productivity via measurements of chlorophyll and yellow substance concentration.
- Land surface processes: Global scale vegetation monitoring of distribution, extent and condition.
- Atmospheric investigation on cloud and aerosol parameters (e.g. cloud top height, water vapour column content).

MERIS will provide the following main performance (Rast, 1992):

Spectral Range:	400 - 1050 nm
Total Field of View	$\pm$ 40.76 deg
Total Swath Width	1500 km
Spatial Resolution	250 and/or 1000 m
Spectral Sampling Interval	1.25 nm
Number of Transmitted Bands	15
Band Position and Width	programmable

The final bandsets are not yet fixed. Various proposals for bandsets adapted to land applications have been made, e.g. by Bodechtel et al. (1992).

Band Position	Bandwidth	Application/Feature
445 nm	10 - 20 nm	Vegetation/Water/Fe-Minerals
490 nm	10 - 20 nm	Vegetation/Water/Fe-Minerals
520 nm	10 - 20 nm	Vegetation/Water/Fe-Minerals
565 nm	10 - 20 nm	Vegetation/Water/Fe-Minerals
620 nm	10 - 20 nm	Vegetation/Water/Fe-Minerals
670 nm	10 - 20 nm	Vegetation/Water/Fe-Minerals
683 nm	5 - 10 nm	Vegetation/Water/Fe-Minerals
711 nm	5 - 10 nm	Vegetation
720 nm	5 - 10 nm	Vegetation
745 nm	5 - 10 nm	Vegetation/Atmospheric Correction
755 nm	5 - 10 nm	Vegetation/Atmospheric Correction
765 nm	5 - 10 nm	Vegetation/Atmospheric Correction
780 nm	5 - 10 nm	Vegetation/Water/Fe-Minerals
880 nm	10 - 20 nm	Vegetation/Water/Fe-Minerals
960 nm	10 - 20 nm	Vegetation/Water/Fe-Minerals
1035 nm	10 - 20 nm	Vegetation/Water/Fe-Minerals

TABLE 2. Proposal for a VIS/NIR band set optimised for land applications with high spectral resolution sensors (Bodechtel et al. 1992)

A candidate for a later series of POEM platforms is HRIS. HRIS providing both high spectral and spatial resolution in the spectral region between 400 and 2500 nm is specifically dedicated to land applications. The follwing performance requirements have been recognised ESA for Phase A studies:

HRIS will provide the following main performance (Rast, 1992):

Spectral Range:	400 - 2400 nm
Total Field of View	$\pm$ 30deg across track pointing
Total Swath Width	32 km
Spatial Resolution	32 m
Spectral Sampling	10 nm
Number of Transmitted Bands	30
Band Position and Width	programmable

The main disciplines to be covered by HRIS are (Rast 1992):

- Monitoring of dynamic land surface processes (soil erosion, land degradation, desertification)
- Vegetation characteristics (physiological and biochemical processes).
- Land use mapping.
- Geobotanical mapping to measure vegetation stress.
- Hydrology (inland water systems, snow and ice).
- Non renewable resource mapping by mineral identification.
- Atmospheric investigations (water vapour column abundance, cloud properties).
- Environmental processes monitoring (including atmospheric observations in connection with land and ocean observations).
- Oceanography (ocean dynamics, marine biology in coastal zones.

As already mentioned, the corresponding sensors to MERIS and HRIS on the NASA EOS platforms will be MODIS and HIRIS. The US sensors are quite similar to the European ones in their basic concept and with regards to their mission objectives. A basic difference between HIRIS and HRIS lies in the spectral resolution. The HIRIS concept foresees the simultaneous sampling over the entire spectral range with contiguous bands, whereas HRIS is planned to transmit simultaneously 30 to 40 selectable spectral channels.
MODIS is an approved instrument for the first EOS-A platform to be launched in 1998. HIRIS was tentatively scheduled for the second series of EOS platforms (EOS Reference Handbook 1991), however, the realisation of HIRIS becomes more and more doubtful due to financial and technical constraints.

2.2. LOCAL TO REGIONAL SCALE INVESTIGATIONS

The above mentioned global change programmes are and will continue to be underpinned by subprojects on regional to local scales, e.g. the national programmes KLEOPATRA (Germany) and TIGER (UK) or the EC sponsored EFEDA programme. In these cases airborne imaging spectrometers have been or will be deployed.
On European level the European Airborne Remote Sensing Capabilities (EARSEC) programme has been initiated by the EEC. In the framework of this programme a standing European airborne remote sensing facility will be established incorporating an imaging spectrometer named Digital Airborne Imaging Spectrometer (DAIS), being built by GER, covering the full spectral range from VIS/NIR/SWIR to TIR with 79 spectral bands.

In the course of the five years EARSEC programme, it is foreseen to support also the development of a European airborne imaging spectrometer covering the same spectral regions.
This facility shall mainly contribute to the following investigations:

- Test bed for future spaceborne sensors.
- Operational missions for EC forestry and agriculture programmes (e.g. anti-fraud control).
- Biochemistry of vegetation.
- Land degradation studies, environmental monitoring.

This EEC sponsored facility is envisaged to be supplemented by national facilities such as the planned Italian LARA (Laboratorio per la Ricercha Ambientale), which shall include a MIVIS imaging spectrometer with 102 spectral channels covering the VIS/NIR SWIR. The LARA programme will be dedicated mainly to environmental monitoring in the Mediterranean region.

3. Further Sensor Developments

Besides the described air- and spaceborne instruments, some additional imaging spectrometers, mainly airborne, are known to be under development for science and partly for the commercial market.
In the following table major characteristics of the most recent airborne sensors are given:

Instrument	Number of Bands	Spectral Coverage (nm)	Band Width (nm)	IFOV (mrad)	FOV (deg)	Sensor	Period of Operation
DAIS	32 8 32 1 6	498-1010 1000-1800 1970-2450 3000-5000 8700-12300	16 100 15 2000 600	3.3, 2.2,1.1 selectable	78.0	opto-mechanical scanner	first flights 1993
HYDICE	256	400-2500	4.7/11.7	0.5	70.0	CCD	under construction
IMS	64 64	800-1700 1500-3000	12.5 25.0	3.3/11.7	40.0 selectable		since 1991
MIVIS	20 8 64 10	433-833 1150-1550 2000-2500 8200-12700	20.0 50.0 8.0 400-500	2.0	70.0	opto-mechanical scanner	under construction
ROSIS	128	450-850	$\leq$ 5.0	0.55	32.4	CCD	since 1992
SFSI	122	1200-2400	10	0.4	11.7	CCD	first 1993

TABLE 3. Characteristics of new airborne imaging spectrometers

ROSIS (Reflective Optics System Imaging Spectrometer; DASA Germany) exists in one experimental prototype covering the VIS/NIR. Various air- and spaceborne versions have been designed upon this basis with a planned extention of the spectral range up to 2500 nm (Kunkel et al., 1992).
MIVIS (Multispectral Infrared and Visible Imaging Spectrometer) is built by DAEDALUS (USA) and will be available for the commercial market.

IMS (Imaging Spectrometer CNES, France),is an experimental instrument wich originally was developed for extraterrestrial research. It is rather profiling than imaging.
HYDICE is an airborne instrument with the characteristics and technical concept as planned for HIRIS. It is currently being constructed for the US Naval Research Laboratory for scientific military and civil deployment (Staenz, 1992).
The Canadian SWIR Full Spectrum Imager **(SFSI)** is under development at the Canada Centre for Remote Sensing and shall be optimised for application in vegetation studies and pedology/geology (Neville & Powell, 1992).

4. Conclusions

It is evident that in the next future the development of spaceborne imaging spectrometry will concentrate on the observation of global scale phenomena mainly in the VIS/NIR spectral range with moderate spatial resolution. For the most land applications of imaging spectrometry however, higher spatial resolution and spectral coverage up to the SWIR and even the TIR are desirable or even indispensable. Furthermore the necessity to improve the accuracy of atmospheric and radiometric correction of the data has been acknowledged by all participants of prior imaging spectrometry experiments. This means that until the advent of spaceborne sensors these gaps must be filled by the deployment of airborne sensors addressing, in preparation of future spaceborne missions, the following objectives:

- improvement of algorithms to extract meaningful geophysical/-chemical parameters from spectral signatures.
- modelling of dynamic processes observable at local to regional scales, such as land degradation, in their relationship to large scale changes.
- optimisation of sensor requirements, with respect to data reduction without loss of information.
- improvement of radiometric and atmospheric correction to come to robust, transportable correction algorithms that can be used on a routine basis to correct the data sets of each investigator.
- investigation of synergy effects between imaging spectrometry and other imaging remote sensing systems, particularly SAR.

Finally, it is noted that the further development of ancillary data-sets (e.g. spectral libraries, digital topography etc.) is essential to the further utilisation of imaging spectrometry.

5. References

Bodechtel, J., S. Sommer , H. Bach, and W. Mauser (1992) 'Use of Airborne Imaging Spectrometry to the Definition of Optimised Specifications for Land Applications with Future Spaceborne Imaging Spectrometers', *Proceedings of the European ISY Conference, Munich, March 30 - April 4 1992*, Volume II, 405-409.

ESF-EEOP (1992) 'A Strategy for Earth Observation from Space', *Strategy paper of the European Science Foundation (ESF) sub-panel European Earth Observation Panel (EEOP)*, Strasbourg September 1992.

Hill, J., and J. Mégier (1991) 'The Use of Imaging Spectrometry in Mediterranean Land Degradation and Soil Erosion Hazard Assessment', *Proceedings of the 5th International Colloquium - Physical Measurements and Signatures in Remote Sensing, Courchevel (F) 14-18 January*, ESA SP-319 Volume 1, 185-188.

Kunkel, B., F. Blechinger , and A. Schmitz-Peiffer (1992) 'Broadband Imaging Spectrometer Technology and its Use Potential for Environmental Change Missions', *Proceedings of the European ISY Conference, Munich, March 30 - April 4 1992*, Volume III, 1179 - 1185.

NASA-EOS (1991) *Earth Observing System Reference Handbook*, NASA Goddard Space Flight Center, May 1991, 147 pages.

Neville, R.A., and L. Powell (1992) 'Design of SFSI: An Imaging Spectrometer in the SWIR', *Canadian Journal of Remote Sensing*, 18, 210-222.

Rast, M. (1992) 'ESA's Activities in the Field of Imaging Spectroscopy', in , J. Bodechtel and F. Toselli (eds.) *'Imaging Spectroscopy: Fundamentals and Prospective Applications'*, 167-191, Kluwer Academic Publishers Dordrecht/Boston/London.

Readings, C.J., and M. Rast (1992) 'Observations of Atmospheric Chemistry on the First European Polar Platform', *Proceedings of the European ISY Conference, Munich, March 30 - April 4 1992*, Volume III, 1093-1097.

Staenz, K. (1992) 'A Decade of Imaging Spectrometry in Canada', *Canadian Journal of Remote Sensing*, 18, 187-198.

Vane, G., F. Baret, and M. Rast (1991) 'High Spectral Resolution from Visible to Thermal Infrared - Recent Progress and Future Trends', *Proceedings of the 5th International Colloquium - Physical Measurements and Signatures in Remote Sensing, Courchevel (F) 14-18 January*, ESA SP-319 Volume 2, 819-820.

INDEX

EURO

COURSES

REMOTE SENSING

1. A.S. Belward and C.R. Valenzuela (eds.): *Remote Sensing and Geographical Information Systems for Resource Management in Developing Countries.* 1991 ISBN 0-7923-1268-6
2. F. Toselli and J. Bodechtel (eds.): *Imaging Spectroscopy: Fundamentals and Prospective Applications.* 1992 ISBN 0-7923-1535-9
3. V. Barale and P.M. Schlittenhardt (eds.): *Ocean Colour: Theory and Applications in a Decade of CZCS Experience.* 1993 ISBN 0-7923-1586-3
4. J. Hill and J. Mégier (eds.): *Imaging Spectrometry – a Tool for Environmental Observations.* 1994 ISBN 0-7923-2965-1

KLUWER ACADEMIC PUBLISHERS – DORDRECHT / BOSTON / LONDON